T0339749

Jawaid A. Khan
Jeanne Dijkstra
Editors

Handbook of Plant Virology

Pre-publication
REVIEW. . .

"This handbook provides an excellent overview of selected topics in plant virology written by scientists who are experts in their own field of research. Thus, each chapter provides firsthand narratives of different aspects of plant viruses in such a way that is valuable for a broad range of readers, from students to scientists. The concepts and principles described provide a better understanding of the subject and a scientific rationale behind many protocols routinely used in plant virus research. The figures depicting genome strategies of a broad range of viruses is a useful tool for teaching at both undergraduate and graduate levels. Descriptions of plant viruses are up-to-date and provide essential information on the taxonomy of plant viruses. Indeed, a highly valuable resource for teaching and for graduate students and scientists."

Naidu A. Rayapati, PhD
Assistant Professor,
Washington State University

Handbook of Plant Virology

FOOD PRODUCTS PRESS®
Crop Science
Amarjit S. Basra, PhD
Editor in Chief

Plant-Derived Antimycotics: Current Trends and Future Prospects edited by Mahendra Rai and Donatella Mares

Concise Encyclopedia of Temperate Tree Fruit edited by Tara Auxt Baugher and Suman Singha

Landscape Agroecology by Paul A. Wojtkowski

Concise Encyclopedia of Plant Pathology by P. Vidhyasekaran

Molecular Genetics and Breeding of Forest Trees edited by Sandeep Kumar and Matthias Fladung

Testing of Genetically Modified Organisms in Foods edited by Farid E. Ahmed

Fungal Disease Resistance in Plants: Biochemistry, Molecular Biology, and Genetic Engineering edited by Zamir K. Punja

Plant Functional Genomics edited by Dario Leister

Immunology in Plant Health and Its Impact on Food Safety by P. Narayanasamy

Abiotic Stresses: Plant Resistance Through Breeding and Molecular Approaches edited by M. Ashraf and P. J. C. Harris

Teaching in the Sciences: Learner-Centered Approaches edited by Catherine McLoughlin and Acram Taji

Handbook of Industrial Crops edited by V. L. Chopra and K. V. Peter

Durum Wheat Breeding: Current Approaches and Future Strategies edited by Conxita Royo, Miloudi M. Nachit, Natale Di Fonzo, José Luis Araus, Wolfgang H. Pfeiffer, and Gustavo A. Slafer

Handbook of Statistics for Teaching and Research in Plant and Crop Science by Usha Rani Palaniswamy and Kodiveri Muniyappa Palaniswamy

Handbook of Microbial Fertilizers edited by M. K. Rai

Eating and Healing: Traditional Food As Medicine edited by Andrea Pieroni and Lisa Leimar Price

Handbook of Plant Virology edited by Jawaid A. Khan and Jeanne Dijkstra

Physiology of Crop Production by N. K. Fageria, V. C. Baligar, and R. B. Clark

Plant Conservation Genetics edited by Robert J. Henry

Introduction to Fruit Crops by Mark Rieger

Sourcebook for Intergenerational Therapeutic Horticulture: Bringing Elders and Children Together by Jean M. Larson and Mary Hockenberry Meyer

Agriculture Sustainability: Principles, Processes, and Prospects by Saroja Raman

Introduction to Agroecology: Principles and Practice by Paul A. Wojtkowski

Handbook of Molecular Technologies in Crop Disease Management by P. Vidhyasekaran

Handbook of Precision Agriculture: Principles and Applications edited by Ancha Srinivasan

Dictionary of Plant Tissue Culture by Alan C. Cassells and Peter B. Gahan

Handbook of Potato Production, Improvement, and Postharvest Management edited by Jai Gopal and S. M. Paul Khurana

Drought Adaptation in Cereals edited by Jean-Marcel Ribaut

Handbook of Plant Virology

Jawaid A. Khan
Jeanne Dijkstra
Editors

CRC Press
Taylor & Francis Group
Boca Raton London New York

CRC Press is an imprint of the
Taylor & Francis Group, an **informa** business

First published 2006 by The Haworth Press

Published 2019 by CRC Press
Taylor & Francis Group
6000 Broken Sound Parkway NW, Suite 300
Boca Raton, FL 33487-2742

ISBN-13: 978-1-56022-978-0 (hbk)
ISBN-13: 978-1-56022-979-7 (pbk)

Cover design by Kelly E. Fye.

Cover images taken from Color Plates 3.1 and 3.2 and Figure 7.3.

Library of Congress Cataloging-in-Publication Data

Handbook of plant virology / Jawaid A. Khan, Jeanne Dijkstra, editors.
        p. cm.
    Includes bibliographical references and index.
    ISBN-13: 978-1-56022-978-0 (hard : alk. paper)
    ISBN-10: 1-56022-978-0 (hard : alk. paper)
    ISBN-13: 978-1-56022-979-7 (soft : alk. paper)
    ISBN-10: 1-56022-979-9 (soft : alk. paper)
    1. Plant viruses—Handbooks, manuals, etc. I. Khan, Jawaid A. II. Dijkstra, Jeanne, 1930-

    QR351.H33 2006
    579.2'8—dc22

                                                                                2005018323

Visit the Taylor & Francis Web site at
http://www.taylorandfrancis.com

and the CRC Press Web site at
http://www.crcpress.com

CONTENTS

ABOUT THE EDITORS

Jawaid A. Khan, PhD, is a plant virologist at the National Botanical Research Institute in Lucknow, India, and co-editor of *Plant Viruses As Molecular Pathogens* (Haworth). His research interests include characterization of viruses/phytoplasmas infecting ornamental, horticultural, and other economically important crops, and the development of virus detection systems.

Jeanne Dijkstra, PhD, was a plant virologist from 1957 to 1995 at the Department of Virology, Wageningen Agricultural University in the Netherlands, where she investigated early events in the infection process, the interference between viruses in a plant, and the identification and taxonomy of plant viruses. During that time, she also taught undergraduate and postgraduate courses in virology, plant pathology, crop protection, plant breeding, and horticulture and crop science. Dr. Dijkstra is co-editor of *Plant Viruses As Molecular Pathogens* (Haworth).

CONTRIBUTORS

Alan A. Brunt, DSc, FIBiol, Professor, Associate of Horticulture Research International, Warwick, UK.

Tessa M. Burch-Smith, graduate student, Dinesh-Kumar lab, Department of Molecular, Cellular and Developmental Biology, Yale University, New Haven, Connecticut.

A.F.L.M. Derks, Ir, Senior Plant Virologist, Wageningen UR, Applied Plant Research, Flower Bulbs, Lisse, the Netherlands.

S.P. Dinesh-Kumar, PhD, Associate Professor, Department of Molecular, Cellular and Developmental Biology, Yale University, New Haven, Connecticut.

Ricardo Flores, Ph.D, Research Professor, Institute of Plant Molecular and Cellular Biology (Polythechnical University of Valencia-CSIC), Spain.

Fernando García-Arenal, PhD, Professor, Depto. Biotecnología, E.T.S.Ingenieros Agrónomos, Ciudad Universitaria, Madrid, Spain.

Crisanto Gutierrez, PhD, Research Professor, CSIC (Spanish Research Council) at the Centro de Biologia Molecular "Severo Ochoa" (CBMSO), and Director, Instituto de Biologia Molecular "Eladio Viñuela" (CBMSO), Madrid, Spain.

Sara Hughes, PhD, Postdoctoral Research Fellow, School of Biological Sciences, Plant Science Laboratories, University of Reading, Whiteknights, Reading, Berkshire, RG6 6AS, UK.

Michael J. Jeger, PhD, Campus Dean and Professor, Division of Biology, Faculty of Life Sciences, Wye campus, Imperial College London, UK.

Kook-H. Kim, PhD, Associate Professor, School of Agricultural Biotechnology, Seoul National University, Seoul 151-742, Korea.

Huub J.M. Linthorst, PhD, Associate Professor of Plant Molecular Biology, Section of Plant Cell Physiology, Institute of Biology, Leiden, the Netherlands.

José M. Malpica, PhD, Department of Biotecnology, INIA, Crta. La Coruna Km 7.5, Madrid, Spain.

Chikara Masuta, PhD, Professor, Plant Virology Lab., Graduate School of Agriculture, Hokkaido University, Sapporo, Japan.

Mike A. Mayo, PhD, Hon. Research Fellow, Scottish Crop Research Institute, Dundee, UK.

Jennifer L. Miller, graduate student, Dinesh-Kumar lab, Department of Molecular, Cellular and Developmental Biology, Yale University, New Haven, Connecticut.

Francisco J. Morales, PhD, Head, Virology Research Unit, Centro Internacional de Agricultura Tropical Apartado Aéreo 6713, Cali, Colombia.

Vicente Pallás, PhD, Research Professor, Institute of Plant Molecular and Cellular Biology (Polythechnical University of Valencia-CSIC), Spain.

Dick Peters, PhD, Senior Research Scientist, Laboratory of Virology, Wageningen Agricultural University, the Netherlands.

Ayala L.N. Rao, PhD, Professor, Department of Plant Pathology, University of California Riverside.

Vijay S.B. Reddy, PhD, Assistant Professor, Department of Molecular Biology, The Scripps Research Institute, La Jolla, California.

Nicola Spence, PhD, Head, International Development Team, Plant Health Group, Central Science Laboratory, York, UK.

Masashi Suzuki, PhD, Associate Professor, Laboratory of Bioresource Technology, Department of Integrated Biosciences, Graduate School of Frontier Sciences, The University of Tokyo, Japan.

Jan P.H. van der Want, PhD, Professor Emeritus, Laboratory of Virology, Agricultural University, Wageningen, the Netherlands.

Marc H.V. van Regenmortel, PhD, Emeritus Director at the Biotechnology School of the University of Strasbourg, France.

Preface

When we had nearly completed *Plant Viruses As Molecular Pathogens,* our publisher, The Haworth Press, Inc., asked us whether we would be willing to prepare a handbook or encyclopedia of plant virology.

At first we were hesitant to accept the invitation, as we were not sure whether such a publication was necessary. However, by consulting colleagues we became convinced that a handbook emphasizing the explanation of terms and expressions commonly used in the literature of plant virology would fulfill a need in this field. In contrast to an encyclopedia with topics in alphabetical order, we preferred a handbook in which the terms are explained embedded in a theme along with other related terms.

The book has been written for research workers, teachers, and advanced students in the field of plant virology, plant pathology, crop protection, (molecular) biology, and plant breeding.

As the whole field of plant virology had to be covered, we selected a number of representative topics, each with specific terms pertaining to that particular subject. For the respective chapters, we approached a number of authors whose specific know-how would add another dimension to the topics.

The present book is a handbook, not a textbook, hence we have tried to avoid too many details and to present only background information necessary for a good understanding of the various terms.

The 19 chapters of the book cover the following topics.

Chapter 1 focuses on the early concept of a virus and recognition of a virus as an infectious macromolecule. Chapter 2 highlights a major review of plant virus taxonomy and provides an outline of the current families and genera. Chapter 3 describes and illustrates both externally and internally visible symptoms using colored plates. Chapter 4 deals with isolation and purification of plant viruses, and some protocols are shown in case studies. Chapter 5 describes the architecture of plant viruses, placing emphasis on capsid morphology and virus assembly. Chapters 6 and 7 focus on the replication and gene expression of RNA and DNA viruses by clarifying the different expression strategies followed by these viruses. In Chapter 8, various aspects of viroids are described, ranging from structure, replication, and pathogenicity to identification and diagnosis. Chapter 9 deals with the transmission of plant viruses mediated by arthropod vectors such as aphids, beetles, mites, planthoppers, thrips, and whiteflies. Chapter 10 describes virus transmission by fungal spores with emphasis on the interaction between virus and fungus and the role of viral genes and virus transmission by nematodes; special attention is paid to the role of genes and gene products in the interaction between virus and vector. Chapter 10 also deals with transmission of virus through the seed: both the biological characteristics and molecular basis of seed transmission are considered. Chapter 11 describes mechanical transmission and highlights newly developed approaches to transmitting viruses mechanically. It also presents transmission of plant viruses by grafting, vegetative propagation, and by dodder, a parasitic plant. Chapter 12 describe serology and discusses antibody structure and serological methods. Detection, identification of viruses, and disease diagnosis based on biological, physical, and chemical properties are featured in Chapter 13. Ecological and epidemiological aspects are discussed in Chapter 14, describing mode and quantification of virus spread, cultural practices, and environmental conditions. Recombination in RNA and DNA viruses is the topic of Chapter 15, with special emphasis on RNA viruses. Chapter 16, on virus variability and evolution, deals with mutation, genetic exchange, selection, population diversity, and genetic drift. Chapter 17 is concerned with recombinant DNA technology in plant virology and provides information on reverse genetics, genetic transformation, plant-virus-based gene vectors, and expression of antibodies. Chapter 18 is devoted to plant resistance to viral infections. It details, among other things, pathogen-derived resistance and resistance-gene-dependent responses. The economic importance of plant viruses and control strategies are elaborated in Chapter 19. It discusses the direct and indirect losses incurred by viruses, sources of infection, and disease forecasting, and the various control strategies are outlined. At the end of each chapter a brief bibliography is listed.

The second part of the book, the appendixes, is more in the character of a concise encyclopedia of plant virology. It describes the 18 families and 81 genera of plant viruses currently recognized by the International Committee on Taxonomy of Viruses (ICTV). However, unlike encyclopedias, the families/genera are not arranged in alphabetical order, as that would strain the taxonomic relationships and consequently would be less beneficial to the reader. The order of families/genera we chose is based on the one presented in Exhibit 2.2, starting with the group of positive-sense single-stranded RNA viruses, followed by double-stranded RNA viruses, negative-sense single-stranded RNA viruses, single-stranded DNA viruses and reverse transcribing viruses. The descriptions in this part of the book further illustrate the meaning of terms and expressions used for characterization and identification of viruses.

We recall with pleasure the cooperation from the contributors, who kindly wrote most of the chapters and with whom we had fruitful communication throughout the preparation of the manuscripts.

Chapter 1

A Historical Outline of Plant Virology

J. P. H. van der Want

EARLY CONCEPTS OF VIRUS

Virus, a Latin word originally meaning venom, slimy excretion, pus, and even disgusting smell, was borrowed to refer to contagious entities that developed in individuals exposed to adverse conditions. This virus concept coincided with the then-favored theory of spontaneous generation. Unfavorable conditions were believed to upset an equilibrium existing in healthy individuals, leading to the development of infectious entities, "viruses."

Jenner, an English physician engaged in the study of smallpox at the end of the eighteenth century, referred to the exudate from pox pustules in cattle as "vaccine virus"; he inoculated persons with it to protect them against smallpox infection.

In the second half of the nineteenth century, a new era was ushered in by Pasteur and Koch, who found that certain autonomous microorganisms were actually the incitants of infectious diseases in humans and animals. They showed that each disease was incited by a definable cellular germ. According to Pasteur, a virus was a *contagium vivum fixum,* a living, corpuscular, contagious agent.

Fungi were shown to play a role in causing diseases in crop plants. In this case it was also originally believed that these fungi spontaneously arose under unfavorable conditions of plant growth. However, mycologists such as De Bary, Berkeley, Kühn, and the Tulasne brothers proved that these fungi were autonomously acting organisms without spontaneous generation. It is interesting to note that these views of the mycologists working in the plant field preceded those of the bacteriologists engaged in human and animal diseases. Around 1880, certain bacteria were also found to act as disease incitants in plants.

The author gratefully acknowledges the help of Mrs. T. van Bemmel, secretary to the Laboratory of Virology, Wageningen University, in processing the manuscript.

1

In the last quarter of the nineteenth century a filtration technique was developed by Chamberland, an associate of Pasteur, to free water and watery solutions from microorganisms without heating. Liquids were passed through a candle-shaped filter of unglazed earthenware. Microbes could not pass through the extremely fine pores of the filter candle, as shown by the filtrate being free from germs.

In 1892 Ivanovskij published his experiments in which the agent of an infectious tobacco disease was found to be present in the filtrate after passing sap from a diseased plant through a filter candle. Ten years earlier that very disease had been described by Adolph Mayer, who had named it *Mosaikkrankheit* (mosaic disease) because the leaves of an affected plant showed a characteristic mottle. Mayer had proved its infectivity by transferring sap from diseased to healthy plants using fine glass capillaries. However, he was unable to find an inciting fungus or bacterium. Nevertheless he was convinced that a bacterium provoked the disease. Like Mayer, Ivanovskij thought that nondescript, very small bacteria or a toxic substance excreted by bacteria had passed through the filter pores.

Contagium Vivum Fluidum

Beijerinck performed the same filtering of sap from a diseased tobacco plant and also showed the filterability of the infectious agent. This brought him in 1898 to postulate that the agent was completely different from pathogenic bacteria or fungi. He named it *contagium vivum fluidum* (i.e., infectious living fluid). The entity was infective because it could be transmitted from plant to plant, living because it increased when introduced into the plant, and fluid because it could pass through the pores of the filter candle. However, it could not be grown on artificial media. Later, Ivanovskij preferred the term *contagium solutum* as he did not consider the infectious entity to exist in a liquid form.

Dawn of Virology

To distinguish this infectious entity from any pathogenic cellular microorganism, the pathogen from infected tobacco was often referred to as a "filterable virus." The prefix "filterable" was eventually dropped and the term "virus" got its present, common meaning of a pathogen of minute dimensions unable to grow on artificial media. Beijerinck's 1898 publication can be considered to mark the beginning of a new branch of science, later called virology.

Discovery of Viruslike Organisms

It took several years before the true nature of viruses could be unraveled. In the meantime, more and more diseases were ascribed to the action of viruses because they proved to be transmissible although no microorganisms could be held responsible for them. However, in 1967 a team of Japanese scientists reported that not a virus but mycoplasmalike, minute organisms (now called phytoplasmas) occurring in the phloem of plants showing witches' broom were presumably the incitants of this type of disease. In 1973 American phytopathologists described rickettsialike organisms, another type of minute microorganism, as the causal agents of Pierce's disease of grapevine and phony disease of peach, both formerly attributed to virus infections.

In 1967, Diener characterized a pathogen smaller than a virus in potato plants affected by potato spindle tuber. He named it "viroid," as it resembled a virus but consisted of ribonucleic acid (RNA) only. Earlier, this potato disease was also ascribed to a virus.

VIRUS AS PATHOGEN

In the course of a century many virus diseases in plants have been described. Depending on the properties of the virus, as well as on the genetic makeup of the host, symptoms may range from mosaic, yellowing, or ringspot patterns on leaves to more or less severe necrosis of leaves, stems, roots, tubers, or bulbs. Sometimes aberrations of flowers (discoloration, distortion), retardation of plant growth, stunting, and dwarfing were found to occur. Often, external symptoms appeared to be accompanied by internal signs (e.g., phloem necrosis in potato affected by leafroll). Intracellular bodies of various shapes proved to be characteristic for certain virus infections.

Quantification of Virus

Tobacco mosaic virus (TMV), a tobamovirus, the very first virus studied, has proven to be very useful in studies aiming at elucidating the nature of viruses. It appeared to be stable and easily transmissible, attaining a high concentration in infected plants. Holmes's discovery in 1929 that *Nicotiana glutinosa* is hypersensitive to TMV has been of great help. He reported that leaves of this plant rubbed with a suspension of TMV developed necrotic spots (so-called local lesions) within a few days and that their number was proportional to the virus concentration in the inoculum. Hence, relative virus concentrations could be determined, and the first quantitative assay was born.

Virus Transmission

As soon as tobacco mosaic was recognized as an infectious disease, transmission and spread of the contagious entity became important to agricultural practice. In this case transmission by sap proved to be easy and this explained the spread in the field (see Chapter 11, Mechanical Transmission).

In 1936, Rawlins and Tompkins found that dusting leaves with a suitable abrasive, such as carborundum powder, before applying inoculum facilitated transmission of sap-transmissible viruses.

Transmission at will was exercised long before the existence of viruses was known. In the seventeenth century, tulip growers cut bulbs into halves and grafted the half of a tulip bulb producing an evenly colored flower with its wounded side to that of a bulb yielding a variegated flower. In this way, the former bulb produced a variegated flower, which was much more in demand than the evenly colored one.

Similar experience was acquired in 1869 when a grower attempted to create a variegated *Abutilon striatum* together with a green specimen of that species on one stem by grafting. He failed because the latter became variegated too. Many decades later, the virus etiology of both the variegation of *Abutilon* and the color breaking of the tulip was established.

Virus transmission by grafting (of woody and herbaceous plants) is still practiced in experimental virus transmission. However, successful grafting is possible only when scion and stock belong to closely related plant species. In case of distantly related or unrelated plant species, runners of dodder (*Cuscuta* spp.), a parasitic vine, can be used to transmit a virus from a diseased plant to a healthy one, as reported by Bennett in 1940.

In nature, most viruses are spread by specific vectors. The oldest known virus vectors are leafhoppers. At the end of the nineteenth century Japanese scientists discovered the role of leafhoppers in the spread of a dwarfing disease of rice, although the virus nature of its incitant was not yet known. Around 1936 Fukushi found this agent, now known as *Rice dwarf virus,* a phytoreovirus, to circulate in the insect after acquisition. The virus appeared to be transmitted through the egg in successive generations; no transmission occurred via males.

Soon thereafter, other insect species were identified as potent vectors. Among them, aphids play a predominant role in various parts of the world.

Two main mechanisms of transmission were discerned by Watson and Roberts in 1939. One mechanism implies circulation of virus inside the aphid's body after acquisition by feeding on an infected plant. In the other one the aphid acquires the virus by probing an infected plant and transmits it immediately.

Also, mites have been found to be vectors of certain plant viruses.

Before 1940 it was shown that a number of plant virus infections had their origin in the soil. Several years later it was established that certain species of nematodes act as vectors of some of these viruses, whereas some soil-inhabiting fungi are responsible for transmitting others.

A great variety of investigations has elucidated the relationships between viruses and their vectors and the mechanisms involved in acquisition and transmission of a virus by its vector. Further, viruses can be differentiated according to the mechanism of transmission involved.

Virus Epidemiology

The notion of the existence of specific virus vectors was an impetus for studies on the epidemiology of virus diseases. It included the life history of vectors and their behavior toward plants. Combined with data collected on sources of infection, an understanding of virus ecology was established. It formed a firm basis for controlling virus diseases by avoiding infections. Actually, preventioin is the only means of control in agricultural practice because no chemicals are known to cure plants after they have contracted a virus infection in the field. The application of chemicals is directed only toward decimating possible virus vectors in the field.

Control of Viruses

Freeing an infected plant by heat treatment has been successful only in a few cases. For instance, Kassanis in 1950 succeeded in curing potato tubers from *Potato leafroll virus,* a polerovirus, by keeping them at 37.5°C for a prolonged period. However, such a treatment has not obtained practical application, because heating has an adverse effect on the quality of the tubers. In 1952 Morel and Martin found that apical meristems of shoots of some systemically infected plants are virus-free, which has proved to be of utmost importance to the control of virus diseases. They succeeded in growing excised meristems of dahlia plants systemically infected with *Dahlia mosaic virus,* a caulimovirus, on an artificial medium onto virus-free plants. In this way, a number of systemically infected, vegetatively propagated crop plants have been freed from viruses. Plant breeders have been and are still looking for varieties that are more resistant or even immune to viruses. In recent years efforts to engineer resistance by recombinant DNA technology (transgenic plants) have been undertaken.

VIRUS AS INFECTIOUS MACROMOLECULE

Nucleoproteinaceous Nature of Viruses

It took a decade to discover the nature of TMV particles. Around 1930 it was found that TMV has antigenic properties, meaning that the virus, injected into rabbits, provoked the production of specific antibodies. It was already known that such antibodies give rise to a specific precipitate when mixed with their matching antigen in vitro. This phenomenon pointed to a proteinaceous nature of the virus, which was substantiated in purification studies culminating in the finding by Stanley in 1935 that TMV could be salted out from clarified sap by adding ammonium sulphate. Stanley stated that the purified virus was in the form of a crystalline protein. However, in 1936 Bawden and co-workers discovered that besides the protein, a small amount of nucleic acid of the ribose type was invariably present in purified TMV preparations. Hence, they characterized the virus as a nucleoprotein.

Virus Morphology

Behavior of TMV suspensions in polarized light had already led to the conclusion that the virus consisted of rod-shaped particles. In addition, X-ray diffraction studies had shown that each particle had a width of about 15 nm and a length of at least ten times the width. In 1939 TMV particles were seen for the first time by Kausche, Pfankuch, and Ruska in the newly constructed electron microscope developed by Ernst Ruska and Max Knole.

In the following years many viruses were characterized in their intrinsic properties thanks to various purification techniques. The development of ultracentrifugation, particularly that in density gradients, by Brakke in 1951 has been a great asset.

Infectivity of Viral Nucleic Acid

Another great step forward was the discovery in 1955-1956 that the nucleic acid part of TMV isolated from the virus particles possesses the genetic properties of the virus. Analysis of nucleic acids derived from different viruses showed differences in base composition, as could be expected of genetic material.

In the light of genetic differences between viruses, occurrence of virus strains has to be considered. Since about 1924 it has been known that variants of plant viruses exist in nature. When *Beet curly top virus,* a curtovirus that produces severe symptoms in sugar beet, was passed through certain other hosts it seemed to be attenuated when reintroduced into beet. Carsner observed that the passaged virus caused milder symptoms than the original isolate. In 1926 James Johnson found that TMV became attenuated in tobacco plants kept for at least ten days at 36°C. After this treatment it produced only mild symptoms in plants growing at a lower temperature. Such phenomena may be explained by assuming that a virus population consists of variants (strains). A change in conditions, host, or temperature, as in the previous examples, may have led to the elimination of a part of the population, namely, the strain-provoking severe symptoms. A true mutagenic treatment was demonstrated by Kausche and Stubbe in 1939. They subjected TMV-infected plants or inocula to chemicals (e.g., nitrogen mustard) or irradiation with X-rays, which led to mutations in the viral genome.

In 1931, Thung described an interaction between virus strains now called cross-protection. Tobacco plants infected with a "green" (mild) strain of TMV produced no further symptoms if subsequently inoculated with a "yellow" (severe) strain. Such an antagonism has been demonstrated in many other cases.

Types of Viral Genome

Many plant viruses proved to contain single-stranded RNA. Double-stranded RNA was described for *Wound tumor virus,* a phytoreovirus, in 1963, double-stranded DNA for *Cauliflower mosaic virus,* a caulimovirus, in 1968, and single-stranded DNA for the geminiviruses in 1977.

Most viruses with single-stranded RNA have their genomes in one strand. However, some RNA viruses have their genome divided in more than one RNA species (viruses with a divided [multipartite] genome). Such viruses are named multiparticle (multicomponent) viruses when each species is packaged in a different particle. The method of density gradient centrifugation has been of great help in elucidating the multiplication behavior of viral components. In this way, in 1966 Lister could distinguish between the two types of rod-shaped particles (long and short) present in preparations of *Tobacco rattle virus* (TRV), a tobravirus. In 1968 he showed that the genetic code for the coat protein is furnished by the short particles, which, isolated from the long ones, are not infective. The long particles, although infective, give rise only to the production of their RNA, leading to an unstable infection. Hence, only when both parts of the bipartite genome of TRV are present in the infected plant do stable infections ensue.

In the same period as Lister did his experiments, other multiparticle viruses proved to exist, namely, *Cowpea mosaic virus,* a comovirus (with a bipartite genome) and *Alfalfa mosaic virus,* an alfamovirus (with a tripartite genome).

Viruses with a divided genome also exist, in which the genome segments are contained in one particle, for example, *Tomato spotted wilt virus* (TSWV), a tospovirus. The spherical particle has its single-stranded RNA segmented into three parts, as was shown in 1977 by Van den Hurk, Tas, and Peters. In contrast to most other plant viruses, TSWV possesses a membrane that envelops the protein coat. Another example of a virus with enveloped particles is *Lettuce necrotic yellow virus,* a cytorhabdovirus with bacilliform particles. The viral genome consists of one single-stranded RNA molecule, which is not infectious when isolated, as it is a negative RNA strand. However, an RNA polymerase inside the virus particle directs the replication process in the infected cell, as was shown by Francki and Randles in 1972.

Satellitism

The phenomenon of satellitism deserves special attention. In 1962, in a culture of *Tobacco necrosis virus* (TNV), a necrovirus, Kassanis found a second virus, which he isolated from TNV by density gradient centrifugation. It proved to be unable to replicate on its own, but did so in the presence of TNV as a helper. Therefore, Kassanis called the second virus the satellite of TNV. Devoid of the genetic information for replication (for which it depends on TNV), the genome of the satellite carries only the code for its own coat protein. Since then other satellites and their helpers have been described.

Even satellite RNAs have been identified. The first was found to occur in association with *Cucumber mosaic virus* (CMV), a cucumovirus, by Kaper and Waterworth in 1977. It proved to increase the virulence of its helper virus in CMV-infected tomato plants.

Cryptic Viruses

The cryptoviruses (family *Partitiviridae*) form an interesting group. They were discovered in England in 1968 when healthy-looking beet plants were found to contain small, viruslike particles. In 1977 Kassanis and coworkers detected beet plants that were free of such particles. However, efforts to transmit the viruslike particles to the latter plants by conventional methods failed. Only transmission between plants by pollen was successful to a certain extent. Transmission from particle-carrying plants to their progeny through seed was common, though. The purified, small isometric particles contained double-stranded RNA in two segments, each enclosed in a separate particle. Since then, comparable particles have been found in various plant species in different countries, even in firs (*Abies* spp.). Plants with these particles are generally symptomless.

Virus Characterization

Methods of differentiating plant viruses, including diagnosis of the diseases they provoke, kept pace with the increase of knowledge of viruses and their nature. Until 1930 the incidence of external and internal symptoms was used to determine host ranges in order to discriminate between viruses. Indicator plants, reacting with specific symptoms, were of great help, especially those that react with local lesions. Investigations solely based on symptoms were soon supplemented by determinations of the so-called physical properties of a sap-transmissible virus in the crude sap of an infected plant, as advanced by James Johnson in 1927. These properties (dilution end-point, thermal inactivation point, longevity in vitro) are of only relative importance, as they depend a great deal on the species of the source and assay plants and the conditions under which they are grown.

A far more reliable basis for describing viruses was found in the properties of the virus particles themselves. Electron microscopy furnished a means to establish shape and dimensions as well as their structure, including the presence of a surrounding membrane. Techniques applying both electron microscopy and serology, such as immunosorbent electron microscopy (ISEM), developed by Derrick in 1973; decoration and immunolabeling (1975 to 1982) proved to be very successful for identification and detection of viruses. In the same period, sensitive serological tests were developed for virus detection. Among them, the enzyme-linked immunosorbent assay (ELISA), made suitable for plant viruses by Clark and Adams in 1979, has found worldwide application.

Sequencing of viral nucleic acid—a tribute from molecular biology— was successfully accomplished by IJsebaert, Van Emmelo, and Fiers in 1980 for tobacco necrosis satellite virus RNA. This technique allows for the establishment of viral gene functions. The polymerase chain reaction (PCR), a very sensitive test for detection of nucleic acid, was developed in 1982 by Karry Mullis. In 1985, recombinant DNA technology made it possible to insert a viral gene into a plant. In many cases, the transgenic plant thus obtained showed resistance to the virus whose gene had been inserted into its genome.

The impressive studies made in understanding the nature of viruses and their genomes are greatly promoted by methods and knowledge borrowed from molecular biology. The reverse is also true. We may now have some idea, for instance, about the multiplication of "simple" RNA viruses: replication of RNA, production of coat protein, assembly of new particles, but many processes still remain in darkness.

With the fast-growing number of known plant viruses, the need of a proper classification and nomenclature also increased. In 1959 Brandes and Wetter proposed a system based on virus morphology and serological affinity. In 1966 the International Committee on Nomenclature of Viruses, with a subcommittee on plant viruses, was established. Later renamed International Committee on Taxonomy of Viruses (ICTV), it aims at making groups of viruses with similar characteristics. Now viruses have been assigned to orders, families and subfamilies, and/or genera. A complete catalog of all properly described viruses was published by ICTV in 2000, describing the names of 1550 virus species belonging to three orders, 56 families, nine subfamilies and 233 genera. Out of these, 648 are plant virus and viroid species belonging to 14 families and 71 genera.

Since Beijerinck launched his theory of *contagium vivum fluidum* a little more than a century ago, advanced molecular biological techniques have made it possible to penetrate into the essence of a virus. However, in spite of all knowledge of the viral genes, very little is known about the molecular interactions between the genomes of virus and host that eventually may lead to disease of a plant.

BIBLIOGRAPHY

Van der Want, JPH, Dijkstra, J (2006). A history of plant virology. *Archives of Virology.*
Van Regenmortel, MHV, Fauquet CM, Bishop DHL, Carténse E, Estes M, Lemon S, Maniloff J, Mayo MA, McGeoch D, Pringle CR, Wickner RB (eds.) (2000). *Virus taxonomy: Seventh report of the International Committee on Taxonomy of viruses.* New York: Academic Press.
Zaitlin M, Palukaitis P (2000). Advances in understanding plant viruses and virus diseases. *Annual Review of Phytopathology* 38: 117-143.

Chapter 2

Plant Virus Taxonomy

Mike A. Mayo
Allan A. Brunt

INTRODUCTION

The classification of objects of study is an absolute requirement for rational thought about them, and an agreed nomenclature is essential for communicating the results of the study to others. Thus, from the time viruses were first investigated, they have been subjected to taxonomic study. Earlier attempts to develop systems for plant virus taxonomy were reviewed by Francki (1981), and since 1966, when the International Committee on Nomenclature of Viruses (ICNV) was formed, international efforts have been made to develop an agreed-upon universal taxonomy for viruses. The ICNV soon realized that it was dealing with taxonomy as well as nomenclature, and in 1973 its name was changed to the International Committee on Taxonomy of Viruses (ICTV). Seven reports have been released by the ICTV and these chart the great expansion of knowledge of viruses as well as the continuing discovery of novel virus forms. In 1971, two families of viruses were recognized, but this number has now risen to 64. Plant viruses were not placed in genera and families until 1995, but the plant viruses and viroids that can be classified have now been assigned either to one of the 71 genera that are classified into 20 families or to one of a further 17 genera that are not yet assigned to families.

The creation of virus taxa and their naming are done by the ICTV following rules laid out in the International Code of Virus Classification and Nomenclature (ICVCN). The ICVCN is subject to updates and appears in ICTV reports (e.g., Van Regenmortel et al., 2000). The updates are published in articles in the Virology Division News section of *Archives of Virology*.

In this chapter, we discuss the nomenclature of virus taxa; the issue of virus species and the criteria used for species demarcation; orthographic conventions for names of virus taxa, in particular species; and the current classification of plant viruses. Further, the structure of ICTV and how to find out about and influence the changes that are in prospect are also described.

NOMENCLATURE

Five hierarchical levels of taxa exist: order, family, subfamily, genus, and species. However, all levels are not used in every classification. Orders are used only where it is very clear that they are appropriate. At present there are four orders of viruses, only one of which (order *Mononegavirales*) contains plant viruses (the plant rhabdoviruses). The taxon subfamily is used only where a taxonomic problem in a family justifies the extra complication of having an extra taxonomic level. Sometimes it can be clear that a species belongs to a particular family, but it is uncertain into which existing or new genus the species should be classified. Such viruses are "unassigned in a family" and this category is, in effect, a temporary holding position for those viruses for which more data are awaited. However, more commonly, it is unclear what family is appropriate for a particular genus and the genus level is the highest level of classification for these viruses.

The hierarchical level occupied by a taxon is indicated by the suffix added to its name stem. These are, for orders *"-virales,"* for families *"-viridae,"* for subfamilies *"-virinae,"* and for genera *"-virus."* The corresponding name forms for viroid taxa are families *"-viroidae,"* subfamilies *"-viroinae,"* and genera *"-viroid."*

As new types of viruses continue to be discovered, new names must be created for taxa; several rules in the ICVCN govern the construction of these names. Sometimes, names that have been in use for many years conflict with these rules, but the ICVCN allows the retention of these in the interests of nomenclatural stability.

The main rules concerning the names of taxa are as follows:

3.9 Existing names of taxa and viruses shall be retained whenever feasible.
3.10 The rule of priority in naming taxa and viruses shall not be observed.
3.11 No person's name shall be used when devising names for new taxa.
3.12 Names for taxa shall be easy to use and easy to remember. Euphonious names are preferred. (In general, short names are desirable and the number of syllables should be kept to a minimum.)
3.13 Subscripts, superscripts, hyphens, oblique bars, and Greek letters may not be used in devising new names.

3.14 New names shall not duplicate approved names. New names shall be chosen such that they are not closely similar to names that are in use currently or have been in use in the recent past.

3.18 New names shall be selected such that they, in whole or in part, do not convey a meaning for the taxon which would either (1) seem to exclude viruses which lack the character described by the name but which are members of the taxon being named, (2) seem to exclude viruses which are as yet undescribed but which might belong to the taxon being named, or (3) appear to include within the taxon viruses which are members of different taxa.

3.19 New names shall be chosen with due regard to national and/or local sensitivities. When names are universally used by virologists in published work, these or derivatives shall be the preferred basis for creating names, irrespective of national origin.

3.23 A species name shall consist of as few words as practicable but be distinct from names of other taxa. Species names shall not consist only of a host name and the word "virus."

3.24 A species name must provide an appropriately unambiguous identification of the species.

3.25 Numbers, letters, or combinations thereof may be used as species epithets where such numbers and letters are already widely used. However, newly designated serial numbers, letters, or combinations thereof are not acceptable alone as species epithets. If a number or letter series is in existence it may be continued.

In recent years, two forms of name have been used for virus species. Thus *Cucumber mosaic virus* has also been written as cucumber mosaic cucumovirus. The latter is described as a non-Latin virus binomial (NLVB) because the genus name is included within the English name of the species. Much debate has occurred in recent years regarding whether or not the NLVB system for naming virus species should be formally adopted by ICTV (i.e., imposed on all virology). In an opinion poll organized by ICTV in 2002 at the International Congress of Virology in Paris, around 85 percent of respondents favored the formal adoption of this scheme. Currently, ICTV is consulting its constituent Study Groups in order to assess opinions in a more representative way. The current position is that either name form is permissible but, as with other taxa, any new name for a species must be submitted to ICTV in a taxonomic proposal and accepted by ICTV before it can be used as the species name.

In brief, the arguments in favor of converting to NLVB are that the species is a taxon and its name should reflect that a taxon and a virus are different entities. The taxon is an abstraction whereas the virus is a real (concrete) object. Also, the inclusion of the genus name in the species name would impart information in a text about the species. The arguments against adopting NLVB formally are that it would involve the formal change of many names, species unassigned to a genus would need a different name form until assigned, and when species change genus, as has been necessary for a number of species, the name must be changed.

VIRUS SPECIES

For many years, debates have continued about virus species—whether or not they exist and, if they do, how might they be defined. The outcome was the adoption by ICTV of the following definition, which was first proposed by Van Regenmortel (1990:249): "A virus species is a polythetic class of viruses that constitutes a replicating lineage and occupies a particular ecological niche."

The key feature of the definition is the recognition that the species is a polythetic class. That is a class whose members always have several properties in common, although no single property need be common to all members. No single criterion can be used to assign a virus to a species. Of course, a sufficient similarity in certain criteria might very well indicate membership (e.g., 99 percent identical genome nucleotide sequence), but when the virus has similarity closer to the criterion threshold, other criteria would also be used to assess its classification. Thus, for every genus a list of criteria exists that would indicate whether or not two viruses belong to the same or different species. Examples of these demarcation criteria are given in Exhibit 2.1. The criteria used are broadly similar for the two genera, but with some qualitative and quantitative differences. It is clear that viruses in different genera vary to different extents, presumably responding to constraints working against the appearance of sequence variants. A result of this important conclusion is that there cannot be key values for quantitative characters (such as percentage sequence similarity) that can be applied universally to all virus genera.

The discriminatory criteria are established by ICTV study groups, which comprise virologists with experience in working with viruses in a particular genus or family and are listed in ICTV reports. The composition of the cluster of viruses that constitute the membership of a species is decided in consultation with working virologists. It is essentially a pragmatic decision and can be revoked in the light of new information.

EXHIBIT 2.1.
Criteria Used for the Demarcation of Virus Species

Genus *Potyvirus*

1. Genome sequence relatedness
   - Amino acid sequences of coat proteins less than 80 percent identical
   - Nucleotide sequences of the genomes less than 85 percent identical
   - Cleavage sites in the polyprotein differ
2. Natural host range
   - In a few instances host ranges differ (mostly this discriminates strains)
3. Pathogenicity and cytopathology
   - Inclusion body morphologies differ
   - The viruses do not cross-protect one another
   - Abilities to be transmitted via seeds differ
   - The reactions of key hosts (e.g., *R* gene carriers) to infection differ
4. Antigenic properties
   - Reactions with particular discriminatory antibodies differ

Genus *Tombusvirus*

1. Genome sequence relatedness
   - Amino acid sequences of coat proteins less than 87 percent identical
   - Amino acid sequences of the polymerases less than 96 percent identical
2. Natural host range
   - Host ranges differ
3. Pathogenicity and cytopathology
   - Structures of multivesicular bodies differ
   - The reactions of key hosts to infection differ
4. Antigenic properties
   - Serological differentiation index of greater than 3
   - Distinctive reactions in ELISA tests

Source: Adapted from Berger et al., 2000 (genus *Potyvirus*) and Lommel et al., 2000 (genus *Tombusvirus*).

ORTHOGRAPHY

The names of virus orders, families, and genera have always been written in italic script and with an uppercase initial letter. In this respect, the practice in virology differs from those in disciplines regulated by the codes of nomenclature for botany, bacteriology, and zoology. However, the application of this style to species names, as in the revisions to ICVCN, has provoked considerable criticism in recent years; for example, most recently from Gibbs (2004), Bos (2003) and Van Regenmortel (2003). The rule is: "3.40 Species names are printed in italics and have the first letter of the first word capitalized. Other words are not capitalized unless they are proper nouns, or parts of proper nouns."

Taxa are abstractions and thus when their names are used formally, these are written distinctively using italicization and capitalization. In other senses, such as an adjectival form, italics and capital initial letters are not needed. Thus the phrase "the potyvirus, *Potato virus Y*" is correct. An equivalent phrase would be "*Potato virus Y* (genus *Potyvirus*)." When the name is converted into a derivative (e.g., potyviruses) it should never be in italics, as this is not the formal taxon name. Equally, when the name is that of an isolate (e.g., potato virus Y-NTN), the name is not that of the species and italics should not be used. However, ambiguity occurs in phrases such as "the tobacco mosaic virus 30K protein." This can be rewritten as "the 30K protein of tobacco mosaic virus," as it is not possible for an abstract taxon to have a protein. However, if the phrase is expanded it becomes "the 30K protein of viruses belonging to the species *Tobacco mosaic virus*," in which case the use of italic script is required.

The problem here is that the names of species are often the same as the vernacular names of viruses that are members of that species. However, in many instances, the problem is resolved by the use of abbreviations. These are not formal taxon names and should not be written in italics, even if referring to the species.

THE CURRENT GENERA
AND FAMILIES OF PLANT VIRUSES

Exhibit 2.2 shows the current genera and families in which plant viruses and viroids are classified. Currently, 88 genera of plant viruses and viroids have been identified, an increase of nine genera since the publication of the seventh ICTV report (Van Regenmortel et al., 2000). Most of these are classified into one of 20 families, an increase of three since the seventh ICTV report. Exhibit 2.2 also lists the type species of each genus.

EXHIBIT 2.2.
Families, Genera, and Type Species
of Plant Viruses and Viroids

(+) sense ss RNA viruses

Potyviridae	*Potyvirus—Potato virus Y*
	Rymovirus—Ryegrass mosaic virus
	Macluravirus—Maclura mosaic virus
	Tritimovirus—Wheat streak mosaic virus
	Ipomovirus—Sweet potato mild mottle virus
	Bymovirus—Barley yellow mosaic virus
Sequiviridae	*Sequivirus—Parsnip yellow fleck virus*
	Waikavirus—Rice tungro spherical virus
Comoviridae	*Comovirus—Cowpea mosaic virus*
	Fabavirus—Broad bean wilt virus 1
	Nepovirus—Tobacco ringspot virus
Luteoviridae	*Luteovirus—Barley yellow dwarf virus-PAV*
	Polerovirus—Potato leafroll virus
	Enamovirus—Pea enation mosaic virus-1
Tymoviridae	*Tymovirus—Turnip yellow mosaic virus*
	Marafivirus—Maize rayado fino virus
	Maculavirus—Grapevine fleck virus
Tombusviridae	*Tombusvirus—Tomato bushy stunt virus*
	Carmovirus—Carnation mottle virus
	Necrovirus—Tobacco necrosis virus A
	Machlomovirus—Maize chlorotic mottle virus
	Dianthovirus—Carnation ringspot virus
	Avenavirus—Oat chlorotic stunt virus
	Aureusvirus—Pothos latent virus
	Panicovirus—Panicum mosaic virus
Bromoviridae	*Bromovirus—Brome mosaic virus*
	Alfamovirus—Alfalfa mosaic virus
	Cucumovirus—Cucumber mosaic virus
	Ilarvirus—Tobacco streak virus
	Oleavirus—Olive latent virus 2
Closteroviridae	*Closterovirus—Beet yellows virus*
	Crinivirus—Lettuce infectious yellows virus
	Ampelovirus—Grapevine leafroll-associated virus 3
Flexiviridae	*Carlavirus—Carnation latent virus*
	Potexvirus—Potato virus X
	Capillovirus—Apple stem grooving virus

(continued)

(continued)

	Trichovirus—Apple chlorotic leaf spot *Foveavirus*—Apple stem pitting virus *Allexivirus*—Shallot virus X *Vitivirus*—Grapevine virus A *Mandarivirus*—Indian citrus ring spot virus
Unassigned genera	*Tobamovirus*—Tobacco mosaic virus *Tobravirus*—Tobacco rattle virus *Hordeivirus*—Barley stripe mosaic virus *Furovirus*—Soil-borne wheat mosaic virus *Pomovirus*—Potato mop-top virus *Pecluvirus*—Peanut clump virus *Benyvirus*—Beet necrotic yellow vein virus *Sobemovirus*—Southern bean mosaic virus *Idaeovirus*—Raspberry bushy dwarf virus *Ourmiavirus*—Ourmia melon virus *Umbravirus*—Carrot mottle virus *Sadwavirus*—Satsuma dwarf virus *Cheravirus*—Cherry rasp leaf virus

ds RNA viruses

Reoviridae	*Phytoreovirus*—Wound tumor virus *Fijivirus*—Fiji disease virus *Oryzavirus*—Rice ragged stunt virus
Partitiviridae	*Alphacryptovirus*—White clover cryptic virus 1 *Betacryptovirus*—White clover cryptic virus 2
Unassigned genus	*Endornavirus*—Vicia faba endornavirus

(−) sense ss RNA viruses

Rhabdoviridae	*Cytorhabdovirus*—Lettuce necrotic yellows virus *Nucleorhabdovirus*—Potato yellow dwarf virus
Bunyaviridae	*Tospovirus*—Tomato spotted wilt virus
Unassigned genera	*Ophiovirus*—Citrus psorosis virus *Tenuivirus*—Rice stripe virus

(continued)

(continued)

Varicosavirus—Lettuce big-vein associated virus

ss DNA viruses

Geminiviridae
Mastrevirus—Maize streak virus
Curtovirus—Beet curly top virus
Topocuvirus—Tomato pseudo-curly top virus
Begomovirus—Bean golden mosaic virus

Nanoviridae
Nanovirus—Subterranean clover stunt virus
Babuvirus—Banana bunchy top virus

Reverse transcribing viruses

Caulimoviridae
Caulimovirus—Cauliflower mosaic virus
Soymovirus—Soybean chlorotic mottle virus
Cavemovirus—Cassava vein mosaic virus
Petuvirus—Petunia vein clearing virus
Badnavirus—Commelina yellow mottle virus
Tungrovirus—Rice tungro bacilliform virus

Pseudoviridae
Pseudovirus—Saccharomyces cerevisiae Ty1 virus
Sirevirus—Glycine max SIRE1 virus

Metaviridae
Metavirus—Saccharomyces cerevisiae Ty3 virus

Viroids

Pospiviroidae
Pospiviroid—Potato spindle tuber viroid
Hostuviroid—Hop stunt viroid
Cocadviroid—Coconut cadang cadang viroid
Apscaviroid—Apple scar skin viroid
Coleviroid—Coleus blumei viroid 1

Avsunviroidae
Avsunviroid—Avocado sunblotch viroid
Pelamoviroid—Peach latent mosaic viroid

In order to create a higher taxon such as a family, it is necessary to be clear what lower taxa are to be included and also what taxa are not. For some genera it has been unclear where the boundaries should be drawn (for example, genera *Tobamovirus* and *Tobravirus*). As a result of this uncertainty, 17 genera are not assigned to any family. This cautious approach has the great advantage of not having to change taxonomic structures, and the names associated with the structures, when a higher classification does become possible. In general, families have been created only when virologists who have experience of the genera to be clustered come to a conclusion. The recent formation of the family *Flexiviridae* (currently awaiting ratification by the full ICTV membership) to contain genera *Carlavirus, Potexvirus, Capillovirus, Foveavirus, Trichovirus, Vitivirus,* and *Mandarivirus* (some of which have been in existence since the initial creation of plant virus groups in 1966) is an example of this caution.

THE MECHANICS OF VIRUS CLASSIFICATION

Virus taxonomy is administered by the ICTV, which is a committee of the Virology Division of the International Union of Microbiological Societies. Proposals for change to the existing taxonomy (usually based on newly acquired information) come mainly from ICTV Study Groups that comprise virologists with particular expertise with certain groups of viruses. However, any virologist can make a proposal to ICTV. Such a proposal would normally be put to the appropriate Study Group for an opinion. Study Groups report to a subcommittee (SC) of the Executive Committee (EC) of ICTV (the Plant Virus SC for proposals about plant-infecting agents) and proposals are put to the EC by chairs of the SC. The EC vote on accepting proposals, usually after some modification by the proposers, and accepted proposals are put to the membership of ICTV, comprising national members from the constituent societies of the Virology Division, life members of ICTV, and EC members. This voting can be done at plenary sessions held at the three yearly International Congresses of Virology or by postal voting.

The membership of ICTV at these different levels is shown, along with other information about ICTV, on the ICTV Web site at www.danforthcenter. org/iltab/ictv.net.

In recent years, ICTV has increased the possibility of the involvement of the virology community in making decisions on taxonomic proposals. Currently these proposals may be viewed on a public message board on the ICTV Web site for some time between when the EC first sees a proposal and when it makes a decision about whether or not to recommend accepting the proposals.

A further current initiative is the development of a database on ICTV (ICTVpd) (see Büchen-Osmond et al., 2000). In this project it is proposed that virologists will enter information into the database about particular virus isolates with which they are familiar by responding to an electronic questionnaire template. When the database contains sufficient data about many isolates, it will be a resource both for reference to compare new viruses with those existing and a means of developing better taxonomic structures.

BIBLIOGRAPHY

Berger PH, Barnett OW, Brunt AA, Colinat D, Edwardson JR, Hammond J, Hill JH, Jordan RL, Kashiwazaki S, Makkouk K, et al. (2000). Family *Potyviridae*. In Van Regenmortel MHV, Fauquet CM, Bishop DHL, Carstens E, Estes M, Lemon S, Maniloff J, Mayo MA, McGeoch D, Pringle CR, Wickner RB (eds.), *Virus taxonomy: Seventh report of the International Committee on Taxonomy of Viruses* (pp. 703-724). New York, San Diego: Academic Press.

Bos L (2003). Virus nomenclature; Continuing topicality. *Archives of Virology* 148: 1235-1246.

Büchen-Osmond C, Blaine L, Horzinek M (2000). The universal virus database of ICTV (ICTVdB). In Van Regenmortel MHV, Fauquet CM, Bishop DHL, Carstens E, Estes M, Lemon S, Maniloff J, Mayo MA, McGeoch D, Pringle CR, Wickner RB (eds.), *Virus taxonomy: Seventh report of the International Committee on Taxonomy of Viruses* (pp. 19-24). New York, San Diego: Academic Press.

Francki RIB (1981). Plant virus taxonomy. In Kurstak E (ed.), *Handbook of plant virus infections and comparative diagnosis* (pp. 3-16). Amsterdam, The Netherlands: Elsevier/North Holland Biomedical Press.

Fauquet CM, Mayo MA, Maniloff J, Desselberger U, Ball LA (2005). *Virus taxonomy: VIIIth report of the International Committee on Taxonomy of Viruses*. London: Elsevier/Academic Press.

Gibbs AJ (2003). Viral nomenclature, where next? *Archives of Virology* 148: 1645-1653.

Lommel SA, Martelli GP, Russo M (2000). Family *Tombusviridae*. In Van Regenmortel MHV, Fauquet CM, Bishop DHL, Carstens E, Estes M, Lemon S, Maniloff J, Mayo MA, McGeoch D, Pringle CR, Wickner RB (eds.), *Virus taxonomy: Seventh report of the International Committee on Taxonomy of Viruses* (pp. 791-825). New York, San Diego: Academic Press.

Mayo MA, Horzinek M (1998). A revised version of the International Code of Virus Classification and Nomenclature. *Archives of Virology* 143: 1645-1654.

Murphy FA, Fauquet CM, Bishop DHL, Ghabrial SA, Jarvis AW, Martelli GP, Mayo MA, Summers MD (eds.) (1995) *Virus taxonomy - the classification and nomenclature of viruses: Sixth report of the International Committee on Taxonomy of Viruses*. Vienna: Springer-Verlag.

Van Regenmortel MHV (1990). Virus species, a much overlooked but essential concept in virus classification. *Intervirology* 31: 241-254.

Van Regenmortel MHV (2003). Viruses are real, virus species are man-made taxonomic constructions. *Archives of Virology* 148: 2481-2488.

Van Regenmortel MHV, Fauquet CM, Bishop DHL, Carstens E, Estes M, Lemon S, Maniloff J, Mayo MA, McGeoch D, Pringle CR, Wickner RB (eds.) (2000). *Virus taxonomy: Seventh report of the International Committee on Taxonomy of Viruses.* New York, San Diego: Academic Press.

Chapter 3

Symptomatology

Jeanne Dijkstra
Jawaid A. Khan

INTRODUCTION

A plant is infected with a virus when the virus can replicate in its cells. When an infected plant possesses extreme resistance, the virus is unable to move from cell to cell and remains restricted to the cells into which it was introduced. An infection of this type is called subliminal and is hard to establish.

In plants with a lesser degree of resistance, the virus is often restricted to a number of cells near the site of entry as a result of host response. Such an infection usually leads to visible spots on the inoculated leaves. The spots (local lesions) may be of two types: chlorotic, as a result of loss of chlorophyll in the infected cells, or necrotic, due to death of those cells (see Color Plate 3.1, parts 1 and 2, at end of chapter). Necrotic local lesions are formed when the infected plant kills a number of its cells as a reaction to the presence of the virus (hypersensitive reaction). Such a plant is called field resistant. Plants that restrict virus movement and/or replication to some extent display suppressive (quantitative, partial) resistance.

A plant is immune when a particular virus cannot replicate in its cells or isolated protoplasts. In practice, however, it is difficult to distinguish between a subliminal infection and no infection at all. An immune plant is called a nonhost.

A plant is susceptible when the virus is able to move through the "system" (vascular bundles) of its host plant and to replicate in other plant parts. Systemically infected plants may show a gamut of different symptoms (sensitive plants) or practically no symptoms at all (tolerant plants). In the latter case, the infection is latent.

Induction of symptoms is a complex process wherein the genetic makeup of the virus as well as that of the host interact under the influence of the environment.

A certain virus strain may give rise to different symptoms in the same host plant under different environmental conditions, whereas different strains of the same virus may induce similar symptoms in some hosts under similar environmental conditions.

Mainly because of their relatively small size, a large number of virus genomes have been fully sequenced and functions of some of their genes have been defined. However, much remains to be investigated, for instance, how viruses interact with their hosts and how these interactions eventually result in the formation of symptoms in the host plants.

INTERNALLY VISIBLE SYMPTOMS

All symptoms are, of course, reflections of a disturbed physiology of the infected plant. Before the appearance of externally visible (macroscopic) symptoms, metabolic disturbances may lead to anatomical (i.e., cytological and histological) aberrations. The deviations may consist of abnormal enlargement of cells (hypertrophy) or reduction in their size (hypotrophy), an abnormal increase in the number of cells (hyperplasia or proliferation) or a decrease in their number (hypoplasia), degeneration of organelles (e.g., chloroplasts), abnormal accumulation of metabolites in phloem vessels, and necrosis of cells.

Besides changes in existing cells and tissues, new virus-induced structures, so-called inclusion bodies, may be present in infected cells. They are found either in the cytoplasm or in the nucleus. Inclusion bodies may be either (para)crystalline or noncrystalline with a round, oval, or irregularly shaped granular appearance, formerly referred to as X-bodies (see Color Plate 3.2, parts 11 and 12). The (para)crystals consist mainly of regularly arranged virus particles (e.g., the crystals induced by *Tobacco mosaic virus* (TMV), a tobamovirus in tobacco) and those of *Red clover mosaic virus,* a carlavirus, in pea (see Color Plate 3.2, part 12). The noncrystalline inclusions, however, contain either a mixture of virus particles, virus-coded proteins, and degraded cellular material (e.g., the irregularly shaped inclusions of *Bean yellow mosaic virus,* a potyvirus, in *Vicia faba*) (see Color Plate 3.2, part 11), or viral genome-coded proteins (e.g., the so-called pinwheel inclusions of potyviruses) or cytoplasmic structures where virus synthesis and assembly take place (e.g., the viroplasms of caulimoviruses).

Many of these inclusions can be seen by light microscopy and are of diagnostic value as they are characteristic of virus groups and independent of the host plant species.

EXTERNALLY VISIBLE SYMPTOMS

Most of the previously mentioned internal aberrations eventually lead to external (macroscopically visible) symptoms (see Color Plates 3.1 and 3.2). A virus-diseased plant seldom displays one type of symptom only. Usually, symptoms develop sequentially, sometimes starting with local chlorotic spots, followed by a variety of systemic symptoms that change with time. The sum total of all symptoms shown by a diseased plant is called the syndrome, which is, to some extent, characteristic of the virus and may be used for diagnostic purposes. However, as the type of symptoms greatly depends on environmental conditions, such as temperature and light, as well as the age of the host plant and its genotype, diagnosis should not lean too heavily on symptomatology. Moreover, many virus-induced symptoms are atypical and they may be the result of nonviral physiological disturbances caused by, for instance, herbicides, nutritional deficiencies (e.g., lack of magnesium or iron), toxins secreted by mites or insects, or genetic abnormalities. Very often these nonviral aberrations are characterized by abnormally colored leaves (sometimes even with mosaic patterns), necrosis, or malformations (e.g., leaf curling), making it difficult to distinguish viral symptoms from nonviral aberrations. In general, however, in virus-diseased plants discoloration of the leaves is not uniform; it may vary from leaf to leaf and is seldom prominent (no bright mosaics or vein banding, but diffuse mottling). Moreover, the type of symptoms may change with time.

In some cases, symptoms shown by infected plants may disappear. After an initial acute phase with severe symptoms (shock phase) the plant may recover and the newly formed leaves may show mild symptoms or even no symptoms at all, although the virus is still present in those leaves.

In spite of all the limitations of symptomatology for reliable diagnosis, symptoms are the only initial clue to a possible virus disease and they should, therefore, be described accurately.

As for an unambiguous description of symptoms standardized terms are indispensable, the terminology of Bos (1978) is used in this cluster.

Color Deviations

Color deviations may occur in all parts of a diseased plant, but they are commonly observed in leaves. Affected leaves may show variegation; that is, irregular patches of different colors (usually shades of green and yellow). Depending on the size and shape of the patches, the variegation is called a mosaic, flecking, or mottling (see Exhibit 3.1, end of chapter). Stems of diseased plants may display striping or mottling. In monocots, variegation of the parallel-veined leaves usually manifests itself as striking or streaking.

Molecular studies have demonstrated that a mutation in the movement protein gene of *Turnip yellow mosaic virus,* a tymovirus, results in increased severity of mosaic in Chinese cabbage. In millet plants infected with *Panicum mosaic virus,* a panicovirus, chlorosis of the leaves is much more severe when Panicum mosaic satellite virus (PMSV) is also present. It has been shown that the capsid protein (CP) of this satellite virus is responsible for the strong reaction.

Variegation of flower petals may lead to so-called breaking of the color, well-known from the famous feathered and flamed patterns in red and purple tulips infected with *Tulip breaking virus,* a potyvirus. Less spectacular variegation of flowers is also found in plants infected with other viruses; for example, gladiolus with broken pink color of the petals (see Color Plate 3.2, part 4).

In the ripe fruits of tomato infected with *Tomato spotted wilt virus,* a tospovirus, there may be pale red or yellow patches in the normal red skin.

Wilting and Withering

Many virus-infected plants, for example, cucumber infected with *Cucumber mosaic virus* (CMV), a cucumovirus, show loss of turgidity due to water deficiency. This happens when loss of water in transpiration exceeds water absorption. Reduced water supply may be due to abnormalities in the xylem vessels, such as necrosis or gum formation. When loss of water is near total, the affected plants desiccate and wither, as can be seen in, for instance, pea plants infected with *Pea early-browning virus,* a tobravirus.

Local desiccation of leaf tissue can be observed in the center of necrotic local lesions.

Irregular desiccation of superficial leaf tissue may give rise to fine line patterns called etching; for example, in tobacco plants infected with either *Tobacco etch virus,* a potyvirus, or *Tobacco rattle virus* (TRV), a tobravirus (Exhibit 3.1; Color Plate 3.2, part 6).

Necrosis

Death of cells and tissues characterized by black or (grayish) brown discoloration is called necrosis. Unlike in desiccation, in necrosis the plant has usually produced melaninlike substances responsible for the black or brown color. Pseudorecombinant studies have shown that the systemic necrosis in *Nicotiana edwardsonii* infected with CMV is determined by RNA2.

In some virus-plant combinations virus movement in a leaf is restricted by a hypersensitive response (HR) of the plant consisting of killing of its own cells around the site of entry of the virus, by which a necrotic local lesion is formed. Such a scorched-earth policy of the plant to prevent further spread of the virus by killing infected cells resembles to some extent programmed cell death (apoptosis) in animals. However, in plants not all of the infected cells are killed. The HR conferred by the *N* gene of *Nicotiana glutinosa* and directed against most tobamoviruses is one of the best-studied HR systems. Specific amino acid substitutions within the CP of *Tobacco mosaic virus,* a tobamovirus, are responsible for recognition of the host and evocation of HR. For the formation of (necrotic) local lesions, roles of replicase and CPs are suggested. For instance, the CP of PMSV (Panicum mosaic satellite virus) elicits necrotic local lesions on *N. benthamiana.* In the combination pea or cowpea with CMV, it is the replicase that is responsible for the formation of local lesions.

Pseudorecombinants generated between CMV and *Brome mosaic virus* (BMV), a bromovirus, show that the local lesions in *Chenopodium hybridum* are induced by the CP gene of BMV.

When plants are systemically infected, necrosis extends to the veins and spreads to other parts of the plant. As soon as the growing point is killed, the plant usually dies (Color Plate 3.2, part 7). In some cases, necrosis may be restricted to the sieve tubes and companion cells of stem and leaf; for example, in potato plants infected with *Potato leafroll virus* (PLRV), a polerovirus. Besides phloem degeneration, PLRV-infected plants may also show excessive accumulation of callose on the sieve plates.

In potato tubers infected with TRV, ring-shaped necrosis often develops.

Growth Reduction and Malformation

Virus-infected plants usually show a reduction in size, either in all their organs (proportional growth reduction), leading to dwarfing or stunting (Color Plate 3.1, part 3), or in parts of leaves or in the tip. In the last two cases, the growth reduction may cause malformation of organs.

When the growth of stem internodes is impeded, it will result in their shortening and, in extreme cases, to absence of internodes, so that the leaves are placed in a rosette (rosetting).

Leaf rolling and downward curling of leaves (leaf curling, epinasty) are the results of uneven growth of the leaf blade (Color Plate 3.2, part 7). Impeded development of the lamina tissue of a leaf results in leaf narrowing, which may even lead to leaves that have their main veins intact but are devoid of lamina (shoe stringing); for example, in tobacco infected with TMV or in papaya infected with *Papaya ringspot virus,* a potyvirus (Color Plate 3.2, part 8). When, however, the veinal growth is reduced, the lumpy leaves show inflated interveinal tissue (blisters) (Color Plate 3.1, part 7) and sunken veins (rugosity).

Degeneration of parts of the cambial layer of virus-infected trees may lead to the formation of elongated pits and furrows that are visible on the surface of the wood (stem pitting); for example, in citrus trees infected with *Citrus tristeza virus,* a closterovirus.

A special type of malformation are the so-called enations, that is, small outgrowths on leaves or stems. They are often the result of hyperplasia and hypertrophy of phloem and bordering parenchyma cells. Enations are usually found on the underside of leaves, either on the leaf veins, for example, in pea infected with both *Pea enation mosaic virus-1* and *Pea enation mosaic virus-2,* an enamovirus and umbravirus, respectively, or on the interveinal tissue, for example, in sunn-hemp *(Crotalaria juncea)* infected with *Sunn-hemp mosaic virus,* a tobamovirus (Color Plate 3.2, part 10).

BIBLIOGRAPHY

Bos L (1978). Symptoms of virus diseases in plants, with indexes of names of symptoms in English, Dutch, German, French, Italian and Spanish (3rd ed.). Wageningen, the Netherlands: Agricultural Publishing and Documentation PUDOC.

Christie RG, Edwardson JR (1982). *Light and electron microscopy of plant virus inclusions.* Florida Agricultural Experiment Stations Monograph Series No. 9 (2nd printing).

Edwardson JR, Christie RG, Purcifull DE, Petersen MA (1993). Inclusions in diagnosing plant virus diseases. In Matthews REF (ed.), *Diagnosis of plant virus diseases* (pp. 101-128). Boca Raton, FL: CRC Press.

Smith KM (1972). *A textbook of plant virus diseases* (3rd ed.). London: Longman.

Wood KR (1990). Pathophysiological alterations. In Mandahar CL (ed.), *Plant Viruses,* Volume II (pp. 23-63). Boca Raton, FL: CRC Press.

COLOR PLATE 3.1. 1: Chlorotic (yellowish) local lesions caused by *Cherry leaf roll virus,* a nepovirus, in *Chenopodium quinoa*; 2: Necrotic local lesions caused by *Tobacco mosaic virus,* a tobamovirus, in *Nicotiana glutinosa*; 3: Growth reduction (dwarfing, stunting) and yellowing caused by *Bean common mosaic virus,* a potyvirus, strain blackeye cowpea, in *Vigna unguiculata;* healthy plant below; 4: Interveinal chlorosis and yellowing caused by *Potato leafroll virus,* a polerovirus, in *Physalis floridana*; 5: Vein clearing symptom in *Nicotiana tabacum* "White Burley" infected with *Tobacco mosaic virus,* a tobamovirus; healthy leaf on left; 6: Vein chlorosis in *Dahlia variabilis* systemically infected with *Dahlia mosaic virus,* a caulimovirus; healthy leaf on left; 7: Mosaic and malformation of the leaf with dark green "blisters" caused by *Cowpea severe mosaic virus,* a comovirus, in *Vigna unguiculata;* healthy leaf below; 8: Stripe and line pattern in banana infected with *Cucumber mosaic virus,* a cucumovirus; 9: Vein mosaic caused by *Cowpea mottle virus,* a carmovirus, in *Vigna unguiculata.* (*Source:* From Jeanne Dijkstra and Cees P. de Jager's *Practical Plant Virology: Protocols and Exercises,* p. 39. © 1998, Springer-Verlag. Reproduced with permission.)

COLOR PLATE 3.2. 1: Vein banding caused by *Bean common mosaic virus,* a potyvirus, strain NY15, in *Phaseolus vulgaris*; 2: Rings and oak-leaf type of line pattern in *Sambucus racemosa* infected with *Cherry leaf roll virus,* a nepovirus; 3: Concentric rings and yellow blotch symptoms in *Capsicum annuum* infected with *Tomato spotted wilt virus,* a tospovirus; 4: Color breaking in flower of *Gladiolus* spec. caused by *Bean yellow mosaic virus,* a potyvirus; 5: Local necrotic ringspots caused by *Tobacco rattle virus,* a tobravirus, in *Nicotiana tabacum* 'White Burley'; 6: Etching in *Nicotiana tabacum* 'Samsun NN' systemically infected with *Tobacco rattle virus,* a tobravirus; 7: Top necrosis and epinasty and vein necrosis in a primary leaf caused by *Bean common mosaic necrosis virus,* a potyvirus, strain NL3, in *Phaseolus vulgaris*; 8: Leaf narrowing (shoe stringing) and mottling symptoms in *Carica papaya* infected with *Papaya ringspot virus,* a potyvirus; 9: Leaf rolling in potato infected with *Potato leafroll virus,* a polerovirus; healthy plant on left; 10: Enations on the underside of a leaf of *Crotalaria juncea* caused by *Sunn-hemp mosaic virus,* a tobamovirus; 11: Granular cytoplasmic inclusion bodies caused by *Bean yellow mosaic virus,* a potyvirus, in leaf epidermis of *Vicia faba* stained with methyl green pyronin; bar represents 10 µm; *N,* nucleus; Crystalline inclusion bodies caused by *Red clover vein mosaic virus,* a carlavirus, in stem epidermis of *Pisum sativum* stained with phloxine-methylene blue; bar represents 10 µm; *N,* nucleus. (*Source:* From Jeanne Dijkstra and Cees P. de Jager's *Practical Plant Virology: Protocols and Exercises,* p. 41. © 1998, Springer-Verlag. Reproduced with permission.)

EXHIBIT 3.1.
Color Deviations in the Leaves of Virus-Infected Plants

Even Color Distribution

Blanching: whitish due to absence of pigment
Chlorosis: yellowish-green (Color Plate 3.2, parts 1, 4)
Yellowing: yellow color predominates (Color Plate 3.1, parts 3, 4)
Reddening: color due to enhanced expression of anthocyanins
Browning: color due to death of (part of) the leaf (necrosis)
    (Color Plate 3.1, part 2; Color Plate 3.2, parts 5, 7)
Bronzing: color due to death of epidermis

Uneven color distribution-

Etching: whitish small curved lines and figures (Color Plate 3.2,
    part 6)
Mosaic: alternation of darker and lighter colored angular patches
    (Color Plate 3.1, part 7)
Striping, streaking: same as mosaic, but with elongated patches
    (Color Plate 3.1, part 8)
Flecking: same as mosaic but with rounded, sharply delimited
    patches
Mottling: same as mosaic but with diffusely delineated patches
    (Color Plate 3.2, part 8)
Blotching: diffuse mottling (Color Plate 3.2, part 3)
Line pattern: undulating or serrated yellow or necrotic bands along
    the primary and/or secondary veins (Color Plate 3.1, part 8;
    Color Plate 3.2, part 2)
Oak-leaf pattern: line pattern resembling an oak leaf (Color Plate 3.2,
    part 2)
Rings: usually yellow or necrotic rings with normally colored (green)
    center (Color Plate 3.2, part 2)
Concentric rings: yellow or necrotic rings alternating with normal
    (green) tissue (Color Plate 3.2, parts 3, 5)
Lesions: isolated chlorotic or brownish/grayish necrotic spots
    (Color Plate 3.1, parts 1, 2)
Vein clearing: veins become translucent (Color Plate 3.1, part 5)
Vein chlorosis: yellowish-green tissue adjacent to secondary and
    tertiary veins (Color Plate 3.1, part 6)
Vein mosaic: irregularly shaped, light-colored patches along the lat-
    eral veins (Color Plate 3.1, part 9)
Vein yellowing: bright yellow tissue adjacent to secondary and tertiary
    veins, sometimes forming a yellow network
Vein banding: broad bands of light or dark tissue along the primary
    and secondary veins (Color Plate 3.2, part 1)
Vein necrosis: brown veins due to death of vascular tissue
    (Color Plate 3.2, part 7)

Chapter 4

Isolation and Purification
of Plant Viruses

Sara Hughes
Nicola Spence

INTRODUCTION

To measure the physical and biochemical properties of a virus it is important that a virus can be separated from host constituents without it being physically or chemically damaged; this is also essential for the production of antisera. If a virus is stable and is present at a high concentration in the host, for example, *Tobacco mosaic virus* (TMV), a tobamovirus, various purification methods can be used. However, if a virus is less stable and/or present at low concentrations in the host it is more difficult to purify and more specific methods have to be used. In this chapter we explain virus purification procedures in general and give specific examples of procedures for some plant viruses in different groups.

PLANT MATERIAL

Propagation of Virus

In general, successful virus purification is easier when a high concentration of virus is present in the host. The choice of host in which to propagate

We thank Dr. Dez Barbara for his valued input and critical review of this manuscript.

the virus is therefore critical, and several factors should be taken into consideration when choosing. The host should preferably be

- free from inhibitors that will inactivate the virus or cause it to be lost during purification, for example, organic acids, mucilages, tannins, gums, latex, or phenolic compounds;
- free from other viruses; and
- easy and quick to grow from seed, for example, *Chenopodium quinoa.*

Many viruses are mechanically transmissible and the propagation host can get infected relatively simply, as outlined in the following. For other viruses more difficult methods of inoculation may be required, for example, aphids, whiteflies, or grafting. Phosphate buffer (1g di-potassium hydrogen orthophosphate anhydrous [K_2HPO_4], 0.1g sodium sulphite [$Na_2SO_3 \cdot 7H_2O$] dissolved in 100 mL distilled water) is frequently used as an aid in mechanical inoculation because it has been shown to enhance the infectivity of many viruses. In general, inoculum is prepared by grinding infected leaf material in this buffer using a mortar and pestle. Host plant leaves are dusted with carborundum powder (300 to 600 mesh), which acts as an abrasive to break the outer walls of host cells, and the inoculum is applied to the host leaves using inoculum-saturated muslin, other soft material, or gloved fingers.

It is preferable to determine the optimal period to maximize the amount of virus in the host at harvest. For mechanically transmissible viruses the amount of infective virus at different times can be determined using local-lesion hosts inoculated by the method described previously; the number of lesions obtained is normally related to the concentration of virus. Some typical local-lesion hosts are *Chenopodium* spp., *Nicotiana glutinosa*, *Vigna unguiculata.* If no local lesion host is available, then the inoculum can be serially diluted, inoculated to groups of plants, and the number of plants that become infected recorded. Where viruses are already partially characterized, then biophysical tests might be used to estimate viral titre (e.g., enzyme-linked immunosorbent assay [ELISA] and polymerase chain reaction [PCR] with broad-spectrum primers). However, in some cases, for example a completely new nonmechanically transmissible virus, it may be necessary to rely on nonspecific methods (e.g., electron microscopy) or unreliable indicators of titre (e.g., symptoms).

EXTRACTION OF INFECTED PLANT MATERIAL

Systemically infected leaf material is normally used for purification because virus concentration is often higher than in inoculated leaves. The first stage of virus purification is to extract the virus from the host cells, normally at 3-5°C, by homogenizing infected leaf material in an extraction buffer.

Several different methods can be used for homogenizing the sample, depending on the quantity and robustness of the virus being purified. A mortar and pestle can be used for small-scale preparations and delicate viruses; kitchen blenders, mincers, or liquidizers can be used for intermediate-scale preparations; and colloid mills or meat mincers can be used for large-scale preparations and robust viruses.

A suitable extraction buffer is dependent on the virus being purified and will be determined by experimentation or comparison with similar viruses. The molarity of the extraction buffer is important, solutions of between 0.1 M and 0.5 M are frequently used, as is the pH of the extraction buffer, because many viruses are stable only over a narrow pH range. For each virus the particles will have no net charge at a particular pH value, the isoelectric point (IP), and this must be avoided when choosing an extraction buffer because the particles may irreversibly precipitate. The IP is normally on the acid side of neutrality, so the extraction buffers are often neutral or slightly alkaline (pH 7 to 8.5), but either high or low pH values may cause the virus to dissociate or alter the physical structure. For many viruses or virus groups, specific problems will determine the precise nature of the extraction buffer; for example, carlaviruses are prone to aggregation (and subsequent loss of particles during purification) and borate-based buffers are often used to counter this.

Different additives can be added to the extraction buffer to improve extraction of particles or protect them from damage or loss:

Reducing agents, for example, thioglycollic acid, sodium sulphite, or mercaptoethanol, can be added at low concentrations (approximately 1 to 10 g·L^{-1}) to prevent virus inactivation by oxidation. They may also prevent binding of host constituents to virus particles.

Chelating agents, for example, EDTA (ethylene diamine tetra-acetic acid), assist in the removal of host ribosomes and prevent virus aggregation.

Mild detergents, for example, Triton X-100 (octyl phenol ethoxylate), and Tween-80 (polyoxyethylene sorbitan monooleate), may help in the release of virus particles from host constituents and decrease particle aggregation. However, strong surfactants, for example, sodium dodecyl sulphate (SDS), will often degrade particles.

Enzymes can also be added to improve purification of virus; for example, trypsin improves the purification of *Turnip mosaic virus,* a potyvirus, by digesting host constituents to aid in the release of virus.

CLARIFICATION OF PLANT EXTRACT

Once the virus has been extracted from the host cells it needs to be separated from host constituents by removing as much of the host material as possible. This process is called "clarification." At its simplest, filtering homogenates through muslin can remove large cell debris. Homogenates can be heated to 50-60°C for a few minutes or K_2HPO_4 can be added to coagulate host material. The sample is then subjected to a low-speed centrifugation (typically 1,000 to 10,000 × g for 5-15 min) to sediment the large plant debris. The pellet (host debris) is discarded and the supernatant is retained. However, particular viruses can sediment with the debris and can then be recovered by resuspending in the same or different buffer. The resuspended pellet is subjected to another low-speed centrifugation; the supernatant is retained and combined with the original supernatant. Organic solvents, for example, ethanol, butanol, or chloroform, can be added to the supernatant, hopefully causing host constituents to coagulate but leaving the virus in suspension. Although this must be adopted with care, as some viruses may be damaged or precipitated by this approach, vigorous stirring of the sap with organic solvents for 10-30 min often improves clarification.

After adding an organic solvent the preparation is again subjected to low-speed centrifugation (typically 5,000 to 10,000 × g for 10-20 min), after which the homogenate usually forms two phases with an interface of debris or precipitated material. The bottom phase comprises the organic solvent and plant materials, for example, chlorophyll and waxes, and the upper (aqueous) phase contains the virus. The aqueous phase is retained and is suitable for use in some diagnostic studies (e.g., electroblot immunoassay). However, the virus is only very partially purified and needs further purification for most purposes.

CONCENTRATION OF THE VIRUS AND FURTHER PURIFICATION

After clarification the virus needs to be concentrated and separated from soluble, low-molecular-weight host contaminants. Several procedures can be used to do this:

> *Differential centrifugation* uses cycles of high- (ultracentrifugation) and low-speed centrifugation. The aqueous phase is centrifuged, usually in an angle rotor at high centrifugal forces (typically

75,000 × *g* for 1-2 h, but this is very dependent on the morphology and size of the virus). The virus particles form a pellet, which is retained, and the low-molecular-weight plant material is left in suspension. The pellet is resuspended in a small volume of dilute buffer and is then subjected to a low-speed centrifugation (typically 5,000 × *g* for 10 min) and the aqueous phase is retained. Some viruses may be broken by the stresses caused by this procedure and some can become highly aggregated.

Polyethylene glycol (PEG, Mr 6000) precipitation is one of the most common methods for concentrating the virus. PEG (Mr 6000) is typically added at between 40 g and 80 g per liter of clarified aqueous solution with 0.1 M sodium chloride and is stirred for approximately 2 h at 3- 4°C. The solution is then centrifuged at low speed (typically 15,500 × *g* for 20 min). The PEG pellet is retained as this contains the virus, a buffer is added (to approximately 10 percent of the original volume), and the pellet is allowed to resuspend overnight. Then this solution is centrifuged again (typically 9,000 × *g* for 10 min) to remove insoluble material. PEG-concentrated virus is typically further purified by differential centrifugation.

Where very pure virus preparations are required for biochemical studies and antiserum production, the virus is often further purified by density gradient centrifugation. Different viral components may also be separated by these procedures. In general, two types of density gradients are used:

1. Equilibrium density gradients, which separate components according to their buoyant densities, are usually made with caesium chloride, but other salts and materials can be used. These gradients can require lengthy periods of centrifugation, unless a vertical rotor is available.

2. Other gradients can be used to separate the virus from debris on the basis of rates of sedimentation; most commonly such gradients are based on increasing concentrations of sucrose, the function of which is to provide stability as the materials are moved by centrifugal forces along the tube. The virus, sometimes visible as an opaque layer in the gradient, is collected with a hypodermic syringe, but more sophisticated apparatuses are available which both fractionate the gradients while simultaneously recording absorbance. After dilution the virus is usually separated from gradient material by high-speed centrifugation.

ASSAYING THE VIRUS
IN PURIFIED PREPARATIONS

The normal procedure for estimating the concentration of many purified viruses is by spectrophotometry. For well-characterized viruses the extinction coefficient is often known; for others the extinction coefficient can be estimated by comparison with viruses of similar morphology (e.g., a coefficient of 2 OD [optical density] per milligram will give a good estimate of concentration for many filamentous RNA-containing viruses). The shape of a spectrum taken from 240 to 320 nm can be used to judge purity and predict some properties such as percentage nucleic acid content.

Electron microscopy is also used to assess the number of particles present in a purified preparation. For unidentified viruses passive adsorption to grids in the presence of a wetting agent is used, but if a suitable antiserum is available, then immunotrapping can be used (although it is generally less quantitative). If it is necessary to assess the infectivity of the preparation, then mechanical inoculation (described previously) can be used, if available. For some viruses infectivity can be measured only with great difficulty, for example, by allowing vectors to feed on artificial virus-containing diets through a membrane (membrane feeding).

MAINTENANCE OF PURIFIED VIRUS

Once the virus is purified, it is best to use it as quickly as possible. Purified virus often loses infectivity if it is stored at 4°C because fungi and bacteria can grow and contaminate preparations but this can be averted with soluble additives, for example, dilute sodium azide. The purified preparation may be stored at a low temperature by adding an equal volume of glycerol. If the preparation is to be used for analytical studies it is best stored frozen.

CASE STUDIES

Purification of Cauliflower mosaic virus (CaMV), a Caulimovirus (S. Muthumeenakshi, personal communication)

Buffers

1. *Extraction buffer:* Dissolve 68 g KH_2PO_4 in 900 mL distilled water and adjust pH to 7.2 with 1 M KOH. Add 7.5 g sodium sulphite and make to 1 L with distilled water.

2. *Gradient buffer:* Dissolve 0.7 g KH_2PO_4 in 400 mL distilled water, adjust pH to 7.2 with 10 M KOH and make to 500 mL with distilled water.
3. *Sucrose gradients:* Make 10, 20, 30, and 40 percent (w/v) sucrose solutions in gradient buffer (i.e., dissolve 10 g sucrose in 100 mL gradient buffer to make 10 percent sucrose solution). Using a 36 mL plastic centrifuge tube (soft enough to be pierced with a needle) make gradients by putting 9 mL 10 percent sucrose into the tube, then underlay 9 mL 20 percent sucrose, followed by 9 mL 30 percent sucrose and 9 mL 40 percent sucrose. Make four sucrose gradients and store overnight at 4°C to allow the different concentrations of sucrose to diffuse, forming a continuous sucrose gradient.

Method

Harvest CaMV-infected leaves, weigh, and cut using scissors. Add the roughly cut leaves to a food processor with the extraction buffer in a ratio of 1:1.5 (leaf material:extraction buffer) and, at 4°C, blend at low speed to a smooth consistency. Filter the sample through miracloth for approximately 4 h at 4°C. Measure the volume of filtrate and add 2.5 mL Triton X-100 and 6 g urea per 92 mL filtrate. Stir the sample overnight at 4°C. Centrifuge the sample at approximately $1,700 \times g$ for 20 min. Retain the supernatant. Centrifuge the supernatant at $65,000 \times g$ for 90 min. Discard the supernatant and drain the pellet for approximately 5 min. Resuspend the pellet in 5 mL distilled water and stir the sample overnight at 4°C. Centrifuge the sample at $3,100 \times g$ for 10 min. Collect the supernatant and divide it into four aliquots (each aliquot should be approximately 700 µL). Overlay each aliquot onto a sucrose gradient and centrifuge the gradients at $60,000 \times g$ for 120 min. The virus should appear as an opaque band in the middle of the gradient. Pierce the tube with a syringe needle at the level of the opaque band, draw the band off and transfer it into a 10 mL centrifuge tube. Dilute the sample at least 1:1 with distilled water, making sure the centrifuge tube is filled to the top. (This prevents the tube from collapsing when the sample is subjected to high-speed centrifugation). Centrifuge at $93,000 \times g$ for 60 min. Discard the supernatant and drain the pellet for approximately 5 min. Resuspend the pellet in 400 µL distilled water.

Purification of Potato virus Y (PVY), a potyvirus (A. Baker, personal communication)

Buffers

1. *Extraction buffer:* Make 0.5 M K_2HPO_4 by dissolving 87 g in 1 L of distilled water and 0.5 M KH_2PO_4 by dissolving 68 g in 1 L of distilled

water. Add 84 mL 0.5 M K_2HPO_4 per 16 mL KH_2PO_4 (final pH will be 7.5) and add 2.5 g Na_2SO_3 per liter.

2. *Clarification buffer:* Make buffer as in (1) except add 0.37 g disodium EDTA per 100 mL of buffer in place of Na_2SO_3.

3. *Dilution buffer:* Make 0.25 M K_2HPO_4 by dissolving 43.5 g in 1 L of distilled water and 0.25 M KH_2PO_4 by dissolving 34 g in 1 L of distilled water. Add 84 mL 0.25 M K_2HPO_4 to 16 mL 0.25 M KH_2PO_4 (final pH will be 7.5) and add 0.37 g disodium EDTA per 100 mL buffer.

Method

Homogenize PVY-infected potato leaf material in the ratio 1:2:1 (leaf material:extraction buffer:chloroform). Centrifuge at $1,200 \times g$ for 10 min. Filter the sample and centrifuge the filtrate at $5,000 \times g$ for 10 min. Filter the sample and measure the volume of filtrate. Add PEG (Mr 6000) at 60 $g \cdot l^{-1}$ and sodium chloride at 40 $g \cdot l^{-1}$ and stir for 90 min at 4°C. Centrifuge sample at $13,000 \times g$ for 10 min. Discard supernatant and resuspend pellets in 2 mL of extraction buffer, pool the samples and stir overnight at 4°C. Divide preparation into two centrifuge tubes and dilute to the top using clarification buffer. Centrifuge at $9,000 \times g$ for 10 min. Recover supernatant and then centrifuge at $75,000 \times g$ for 120 min. Resuspend pellets in 2 mL of clarification buffer, pool, and stir overnight. Dilute sample to a total volume of 9 mL and add 4 g of caesium chloride. The volume and weight can be scaled up or down depending on the size of the tubes, but should be kept in proportion. It is important to ensure that the tube is full to prevent its collapsing under high-speed centrifugation. Centrifuge at $202,000 \times g$ for 18 h. Draw off virus bands and dilute using dilution buffer. Centrifuge at $75,000 \times g$ for 120 min. Resuspend pellet in 150 μL distilled water.

Purification of Watermelon chlorotic stunt virus *(WmCSV)*, a begomovirus *(A. Baker, personal communication)*

Buffers

1. *Extraction buffer:* Make a 0.5 M stock solution (in relation to NaOH) of sodium citrate buffer by dissolving 52 g citric acid and 20 g NaOH in 500 mL distilled water; adjust pH to 6.5 using 0.5 M NaOH (almost an equal volume of 0.5 M NaOH will be required to reach the desired pH). Dilute to 0.1 M, add disodium EDTA to 25 mM, sodium sulphite to 0.01 M and 1mL β-mercaptoethanol.

2. *Dilution buffer:* Dilute the 0.5 M stock solution of sodium citrate buffer to 0.1 M.

Method

Grind WmCSV-infected watermelon leaves to a fine powder in liquid nitrogen and then homogenize 100 g leaf material in 200 mL of extraction buffer using a food processor at low speed for 3 min at 4°C. Add 40 mL chloroform per 200 mL of extraction buffer, homogenize the suspension for a further 3 min at low speed and centrifuge at $500 \times g$ for 10 min. Filter the supernatant through two layers of muslin and measure the volume of filtrate. Add PEG (Mr 6000) at 70 g·l^{-1} and sodium chloride at 11.7 g·l^{-1}. Stir the preparation overnight on a magnetic stirrer at 4°C. Centrifuge the preparation at $10,000 \times g$ for 25 min and resuspend the precipitate in 2 mL of extraction buffer. Pool the resuspended precipitates and make the volume up to the top of the centrifuge tubes with extraction buffer. Centrifuge the sample at $40,000 \times g$ for 120 min. Resuspend the resultant pellets in 1 mL of extraction buffer, pool them and stir overnight at 4°C. Put 1 mL of the preparation onto each of two 10 to 40 percent sucrose gradients (made as described previously) and centrifuge at $100,000 \times g$ for 180 min. Draw off the virus bands with a syringe and dilute with dilution buffer to fill two centrifuge tubes then centrifuge at $100,000 \times g$ for 120 min. Resuspend the resultant pellets in 250 μL of distilled water.

BIBLIOGRAPHY

CMI/AAB *Descriptions of plant viruses* 1-354. Warwick, UK: Association of Applied Biologists, Wellesbourne.

Dijkstra J, De Jager CP (1998). *Practical plant virology: Protocols and exercises.* Heidelberg: Springer-Verlag.

Hull R (2002). *Matthews' plant virology* (4th ed.). London: Academic Press.

Walkey DGA (1991). *Applied plant virology* (2nd ed.). London: Chapman & Hall.

Walkey DGA (1992). *Plant virus diseases of Yemen and associated areas.* London: Overseas Development Administration.

Chapter 5

Architecture of Plant Viruses

Ayala L. N. Rao
Vijay Reddy

INTRODUCTION

Virus particles are assemblies of diverse degrees of complexity, composed mainly of highly organized protein subunits and nucleic acid molecules and enriched with a number of functional properties. At present, recognized plant viruses are classified into 81 genera. Of these, viruses belonging to 17 genera exhibit helical symmetry (rigid rod or flexuous rods) and the remaining 64 genera have icosahedral symmetry (spherical). Whatever the ultimate architecture of the virus, assembly is a complex process involving protein subunits interacting with one another and with viral nucleic acids. Some steps in the assembly process are common to all viruses having icosahedral symmetry, but the mechanisms might be widely different. Assembly of infectious virions is governed by two important principles: (1) the interaction of protein and nucleic acid has to be highly specific in order to avoid packaging of unrelated nucleic acid molecules; (2) the size of the virion is tightly regulated, at least in viruses with icosahedral symmetry, to make it insecurely stable in harsh environments. Once assembled, the resulting architecture has two primary roles: (1) it should be flexible enough to disassemble upon entering uninfected host cells for the release of the genome; (2) it provides an optimal interaction with the host machinery in promoting successful infection. The purpose of this chapter is not to belabor with details of viral structures and assembly but to update readers regarding how plant virologists exploited the information gained from structural studies.

Unlike viruses infecting humans or animals, most plant viruses are non-enveloped (absence of a membrane surrounding the capsid) and are therefore simple in their composition. The infectious virion is composed of a genome (RNA or DNA) and structural protein subunits often referred to as capsid (coat) protein (CP) subunits. The majority of plant viruses contain single-stranded, positive-sense (+ss) RNA as the genome, whereas only a limited number of viruses have been reported to have double-stranded RNA or DNA genomes.

CAPSID MORPHOLOGY

Capsid morphology of plant viruses, and viruses in general, belongs to one of the four types: isometric (icosahedral), rod-shaped (helical), filamentous, and bacilliform.

Isometric Viruses

The simple spherical viruses that display icosahedral architecture have their 60 CP subunits in identical (equivalent) environments related by the fivefold, threefold, and twofold axes of symmetries (e.g., Tobacco mosaic satellite virus). These types of capsids are commonly referred to as T = 1 capsids. The triangulation (T) number usually refers to the number of unique structural environments in a capsid. According to the quasi-equivalent theory, $T=h^2+hk+k^2$, where h and k are any integers. Caspar and Klug's quasi-equivalence theory explains the ways of arranging the subunits in these kinds of capsids with minimal changes in the subunit environment. The majority of spherical viruses have multiples of 60 subunits (e.g., 180), and still obey icosahedral symmetry (e.g., bromoviruses, tombusviruses). Since only the capsids with 60 subunits can occupy identical environments, the subunits of the capsids with more than 60 copies should occupy more than single environments and still display icosahedral symmetry. In the case of T = 3 capsids, the 180 subunits occupy three distinct (A, B, and C) but quasi-equivalent environments (see Figure 5.1). Table 5.1 shows the list of plant viruses, whose structures have been determined at high resolution by X-ray crystallography, compiled from the VIPER Web site (viperdb.scripps. edu/). All the spherical plant viruses that have been structurally characterized to date belong to either T = 1 or T = 3 capsid architectures with diameters ranging from 18 to 34 nm. The protein subunits of all the capsids have the so-called jelly roll-like β-barrel fold also seen in many of the animal and insect viruses.

The bromo- and cucumoviruses of the family *Bromoviridae* form the T = 3 capsids. However, the virions of *Alfalfa mosaic virus* (AMV, an alfamovirus) usually form bacilliform particles with variable sizes. Interestingly, the N-terminal (1 to 26) deletion mutant of the capsid protein exclusively forms T = 1 particles, due to elimination of positively charged residues. The structure of the T = 1 particle of AMV was determined at 4Å resolution.

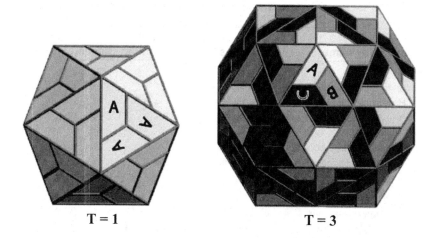

FIGURE 5.1. Schematic representation of T = 1 and T = 3 icosahedral lattices. Each trapezoid corresponds to a subunit. The environments of the subunits are all identical, identified by the letter A, in T = 1 capsids. However, the chemically identical subunits (same gene products) occupy three structurally distinct environments (A, B, and C) in T = 3 capsids.

Members of the *Comoviridae* form true T = 1 capsids, but because of their capsid proteins contain three β-barrel domains; their capsid architecture is also referred to as pseudo T = 3 or P = 3 capsids. The structures of these capsids and their subunits resemble the animal picorna viruses. The polyprotein of comoviruses is cleaved into a single domain (small) protein and two domain (large) proteins, whereas it remains as an intact 3-domain protein in nepoviruses. While the C-terminal β-barrel domain (small protein) is located at the fivefold vertices, the N-terminal and the middle β-barrel domains are located near the twofold and threefold axes, respectively. The capsids of sobemo-, tombus-, and tymoviruses all form the typical T = 3 capsids, while the satellite viruses form T = 1 capsids.

Rod-Shaped Viruses

The most studied rod-shaped plant virus is the *Tobacco mosaic virus* (TMV), a tobamovirus. The nucleoprotein capsids of TMV are rod shaped and display helical symmetry and are 70-300 nm in length with a diameter of 18 nm. The subunits of TMV are arranged into a right-handed helix surrounding the viral genome (+ss) RNA with a helical pitch of 2.3 nm.

TABLE 5.1. List of plant viruses structurally characterized at high resolution.

Virus/family name	Genus	Capsid architecture	Average diameter (in nm)
Bromoviridae			
Alfalfa mosaic virus	*Alfamovirus*	T = 1	21.0
Brome mosaic virus	*Bromovirus*	T = 3	27.4
Cowpea chlorotic mottle virus	*Bromovirus*	T = 3	27.8
Cucumber mosaic virus	*Cucumovirus*	T = 3	29.4
Tomato aspermy virus	*Cucumovirus*	T = 3	29.4
Comoviridae			
Bean pod mottle virus	*Comovirus*	T = 1 (P = 3)	28.0
Cowpea mosaic virus	*Comovirus*	T = 1 (P = 3)	27.8
Red clover mottle virus	*Comovirus*	T = 1 (P = 3)	28.0
Tobacco ringspot virus	*Nepovirus*	T = 1 (P = 3)	28.2
Sobemovirus			
Cocksfoot mottle virus	*Sobemovirus*	T = 3	29.4
Rice yellow mottle virus	*Sobemovirus*	T = 3	29.0
Southern bean mosaic virus	*Sobemovirus*	T = 3	29.4
Sesbania mosaic virus	*Sobemovirus*	T = 3	29.2
Satellites			
Panicum mosaic satellite virus	–	T = 1	16.0
Tobacco mosaic satellite virus	–	T = 1	16.8
Tobacco necrosis satellite virus	–	T = 1	18.4
Tombusviridae			
Carnation mottle virus	*Carmovirus*	T = 3	33.6
Tomato bushy stunt virus	*Tombusvirus*	T = 3	33.6
Tobacco necrosis virus	*Necrovirus*	T = 3	28.8
Tobamovirus			
Tobacco mosaic virus	*Tobamovirus*	rod-shaped	70-300
Tymovirus			
Desmodium yellow mottle virus	*Tymovirus*	T = 3	29.6
Physalis mottle virus	*Tymovirus*	T = 3	29.8
Turnip yellow mosaic virus	*Tymovirus*	T = 3	29.2

Bacilliform Viruses

Among the viruses that form bacilliform particles, AMV is structurally well studied. Based on electron microscopic analysis, a model for the organization of subunits in a bacilliform particle has been proposed. This model assumes a cylindrical hexagonal (P6) lattice on the tubular part of the particle and ½ icosahedral caps with the threefold axis are coincident with the cylindrical axis closing both the ends of the tube. Figure 5.2 shows such a model generated using the AMV CP subunit structure.

INTERACTIONS PROMOTING VIRUS ASSEMBLY

Prior to the recombinant DNA era, the assembly of virions by CP had been envisioned to mainly protect its infectious genome from adverse conditions of extracellular milieu. However, the advent of recombinant DNA technology provided many tools to manipulate and analyze viral genomes

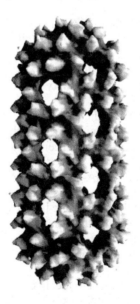

FIGURE 5.2. Surface representation of the bacilliform particle of *Alfalfa mosaic virus* (AMV) based on the structure of the AMV capsid protein subunit. The shaft of the bacilliform particle obeys cylindrical hexagonal (P6) symmetry, whereas the ends are capped with ½ icosahedral T = 1 particle. The pitch of the tube is 10 nm with 3.45 hexameric (six) units per turn.

(RNA or DNA) to determine the functions of their gene products in the virus infection cycle. Several advances made toward this end clearly established that viral CP is multifunctional, and most important, the assembly and the resulting architecture of mature virions is critical for the survival of the virus and its ability to cause disease. Thus, knowledge of the detailed mechanism by which viruses assemble into structurally stable virions is an important prerequisite for understanding the overall biology of plant viruses.

Assembly of mature virions that involves RNA-protein interaction is an important phase in the virus infection cycle. In fact, assembly has been shown to be obligatory for several plant viruses to move from cell to cell and/or for long-distance transport as well as acquisition by insect vectors for dissemination to new hosts. Thus, the study of virion assembly provides basic information concerning how RNA and CP interactions lead to the assembly process. Many types of interactions involving protein and RNA dictate the assembly process:

1. Protein-protein interactions play a major role in virion assembly, but their contribution relative to RNA-protein interactions varies among different plant virus groups. For example, members of tymo- and comoviruses are predominantly stabilized by protein-protein interactions and can therefore form empty capsid shells in the absence of RNA.
2. RNA-protein interactions dictate assembly in some viruses, such as tobamo-, alfamo-, and bromoviruses. In these viruses capsid formation requires RNA, and therefore empty virions are never found in vivo.
3. Sequence-independent RNA-protein interactions are envisioned to stabilize encapsidated RNAs. Basic N-terminal arms found in the coat proteins of several RNA viruses are implicated in the interaction with RNA phosphates and allow the assembly of infectious virions.
4. Finally, sequence-dependent RNA-protein interactions are critical for initiating the viral assembly process.

Since viral RNA and CP subunits are localized in the same compartment of the cell, cellular tRNA (transferRNA), mRNAs (messenger RNA), or rRNA (ribosomal RNA) species could potentially be copackaged with viral genomic RNAs. Specific sequence and/or structural-dependent interactions between RNA and CP are envisioned to ensure that the majority of assembled virions exclusively contain viral RNA. It is thought that specific recognition between viral RNA and CP leads to the formation of a complex that nucleates the binding of additional CP molecules, resulting in the assembly of a complete particle. Despite the fact that several plant viruses have been thoroughly characterized with respect to translation, replication, movement, and the biology of symptom expression, the mechanism of virus assembly has been studied in only a few viruses.

PATHWAYS

Among plant viruses, the best-defined system of virus assembly is that of TMV, a helical plant virus particle in which the assembly is initiated by specific interaction between the CP disk and an internal sequence in the RNA genome. The study of virus assembly in TMV was pioneered by Frankel-Conrat by in vitro reconstitution of TMV from dissociated protein and RNA. It was observed that native TMV CP alone was able to aggregate at neutrality into discs and rod-shaped virions. Additional experiments provided first evidence that the length of the rod-shaped virions is determined by the length of the RNA molecules, which also limits and stabilizes the CP aggregation.

By contrast to helical viruses, the assembly of icosahedral viruses has been studied in more detail. The first experimental evidence that purified RNA and CP of an icosahedral virus (*Cowpea chlorotic mottle virus* [CCMV], a bromovirus) can reassemble in vitro to produce infectious particles was provided as early as 1967. Since then bromoviruses have been used as model systems for understanding the assembly of small spherical viruses. Virions of bromoviruses are predominantly stabilized by RNA-protein interactions, and therefore RNA is required for formation of icosahedral capsids in vivo. RNA-protein interactions are thought to neutralize the negative charge of the RNA and allow tight packaging of nucleic acid within the virus particle. The assembly process is independent of homologous replication machinery and no indication has been found of any other virus-encoded proteins being involved in this process. Bromoviruses assemble into icosahedral particles with T = 3 quasi-symmetry. The structure of CCMV has been determined to 3.3Å resolution by X-ray crystallography. It was hypothesized that the assembly of CCMV is initiated by formation of a hexamer (six subunit proteins surround each vertex) of dimers (two subunit proteins). However, polymerization kinetics (the rate at which many protein subunits assemble) observed by light-scattering and gel-filtration assays suggested that capsid assembly in CCMV is nucleated by the formation of a pentamer (five subunit proteins surround each vertex) of dimers. Assembly studies with *Brome mosaic virus* (BMV), a bromovirus, indicated that assembly of RNA-containing virions requires a highly conserved tRNA-like structure (TLS). The transient, yet critical involvement of the TLS in BMV assembly suggests a role in nucleating a higher-order arrangement of CP dimers that serves as an intermediate on the encapsidation pathway. Unlike capsids such as those of the tymoviruses, which are stabilized by strong protein-protein interactions, the assembly of BMV and other bromovirus virions is characterized at neutral pH by a strong involvement of polyanionic nucleating agents. RNA is presumably the physiological nucleating agent, but this role may be served even by the polymers polyvinylsulfate or dextran sulfate in vitro.

GENOME PACKAGING

Genome packaging is considered to be a highly specific process. As mentioned earlier, during packaging viral nucleic acids must be distinguished from other cellular RNA molecules present in the compartment in which assembly takes place. Viral nucleic acids are exclusively packaged into a great majority of the particles. Such discrimination is the result of specific recognition of sequences or structures unique to viral nucleic acids, often termed as origin of assembly sequence (OAS) or packaging signals. Although specific sequences might promote packaging of viral genomic RNAs, their presence does not guarantee packaging since several other parameters also govern the encapsidation process. The fixed dimensions of icosahedral capsids impose an upper limit on the size of viral nucleic acid that can be accommodated. Consequently, nucleic acids larger than a wild-type genome cannot be packaged, even when they contain appropriate packaging signals. The OAS or the packaging signal of TMV has been characterized to be a stem-loop structure consisting of 69 nucleotides (nt) localized within the 3' half of the movement protein open reading frame (ORF) sequence. In the icosahedral *Turnip crinkle virus* (TCV), a carmovirus, a 186 nt fragment at the 3' end of the TCV CP-coding region has been shown to be responsible for specific packaging of viral RNA. Likewise, in *Southern bean mosaic virus* (SBMV), a sobemovirus, a 24 nt region within an ORF2 has been found to specifically interact with the CP and is predicted to serve as a nucleating site during virus assembly. Unlike viruses having one genomic segment (e.g., TMV, TCV, SBMV), those with a genome divided among multiple nucleic acid species have evolved mechanisms to balance distribution of the genome segments into either a single virion or among multiple virions. The genomes of plant viruses belonging to the genera *Bromovirus, Cucumovirus, Hordeivirus,* and *Alfamovirus* are divided among three RNA segments. In each of these genera, the genomic and the subgenomic RNAs are distributed into separate particles and their size and number varies with genera. For example, in bromo- and cucumoviruses the largest two genomic RNAs are packaged individually into two virions and the third genomic RNA and its subgenomic RNA are predicted to copackage into a third virion. However, all three virions do not display any physical heterogeneity either in size or appearance. By contrast, in members of the genus *Alfamovirus* the three genomic and the subgenomic RNA are packaged individually into four distinctly sized virions. However, the mechanism(s) involved in maintaining a high degree of precision in distributing the four RNAs into three or four individual capsids is currently obscure.

BIBLIOGRAPHY

Baker TS, Nelson NH, Fuller SD (1999). Adding the third dimension to virus life cycles: Three-dimensional reconstruction of icosahedral viruses from cryo-electron microscopy. *Microbiology and Molecular Biology Reviews* 63: 862-922.

Caspar DLD, Klug A (1962). Physical principles in the construction of regular viruses. *Cold Spring Harbor Symposium on Quantitative Biology* 27: 1-24.

Choi YG, Dreher TW, Rao ALN (2002). tRNA elements mediate the assembly of an icosahedral RNA virus. *Proceedings of the National Academy of Sciences USA* 99: 655-600.

Fox JM, Johnson JE, Young MJ (1994). RNA/protein interactions in icosahedral virus assembly. *Seminars in Virology* 5: 51-60.

Johnson JE, Speir JA (1997). Quasi-equivalent viruses: A paradigm for protein assemblies. *Journal of Molecular Biology* 269: 665-675.

Lin T, Clark AJ, Chen Z, Shanks M, Dai JB, Li Y, Schmidt T, Oxelfelt P, Lomonossoff GP, Johnson JE (2000). Structural fingerprinting: Subgrouping of comoviruses by structural studies of red clover mottle virus to 2.4-A resolution and comparisions with other comoviruses. *Journal of Virology* 74: 493-504.

Lucas RW, Larson SB, Canady MA, McPherson A (2002). The structure of tomato aspermy virus by X-ray crystallography. *Journal of Structural Biology* 139: 90-102.

Reddy VS, Natarajan P, Okerberg B, Li K, Damodaran KV, Morton RT, Brooks CL III, Johnson JE (2001). Virus Particle Explorer (VIPER), a Web site for virus capsid structures and their computational analyses. *Journal of Virology* 75: 11943-11947.

Chapter 6

Replication and Gene Expression of Plant RNA Viruses

Kook-Hyung Kim

INTRODUCTION

Over the past decade, considerable progress has been made in understanding both the replication and gene expression strategies of plant RNA viruses. Sixty-six genera of plant viruses with RNA genomes, of which 46 were classified into 13 families, have been distinguished since the publication of the seventh International Committee on Taxonomy of Viruses (ICTV) report. Genomes of plant RNA viruses show considerable variations and thus have to employ a wide variety of strategies for their gene expression. They also show significant variations in capsid morphology, ranging from the icosahedral viruses to the filamentous viruses, the rod-shaped viruses. These RNA viruses share at least two common steps in their infection cycle (i.e., replication of viral RNA and synthesis of mRNA [messenger RNA]). Plant RNA viruses accomplish these processes by employing virus-encoded RNA-dependent RNA polymerase (RdRp) which is not usually found in healthy plants. It is known that the expression of all viruses has to pass through an mRNA stage. The positive-sense single-stranded (ss) RNA of plant RNA viruses can act as mRNAs directly upon entry into their host plants. This chapter, emphasizes explanation of specific terms used in each section and attempts to present a summary of the recent information on the various strategies involved in replication and gene expression of the major families of RNA viruses that infect plants.

GENOME STRUCTURE

Unlike the genomes of all cells, which are composed of DNA, the majority of known plant viruses contain their genetic information encoded in RNA. Recent expansion on virus genome sequencing led us to better understand the structure of the genome. To date, the genomes of at least 300 species have been fully sequenced, including type species of most plant RNA virus genera. Information obtained from genome sequences shed light on genome organization, phylogenetic relationship with other viruses, and the specific regulatory or recognition signals important for successful virus replication. Their RNA genomes show a wide variation in their genome structure. The composition and structure of virus genomes is more varied than any of those seen in the DNA viruses. They may have different terminal structures such as cap structure or genome-linked viral protein (VPg) at the 5' end, and a polyadenylate (poly A) tail, transfer RNA-like structure or no specific structure at the 3' end of their RNA genomes. The viral RNA genomes may be ss or double-stranded (ds), and in a linear, circular, or segmented configuration. Viruses with ss genomes may be positive (+)-sense, negative (–)-sense, or ambisense. Virus genomes vary greatly in size, encoding a wide range of proteins. All produce RdRp, capsid protein (CP) or CPs, and most probably encode one or more proteins potentiating virus movements (cell to cell and long distance) in the plant. Each of these variations has consequences on the replication of genomes and expression of virus genes.

Positive-Sense ssRNA Viruses

The ultimate size of ssRNA genomes is limited by the fragility of RNA and the tendency of long strands to break. In addition, RNA genomes tend to have higher mutation rates than those composed of DNA because they are copied less accurately, which also tends to drive RNA viruses toward smaller genomes. Single-stranded RNA genomes range in size from those of closteroviruses (up to approximately 20,000 nucleotides [nt] long) to carmoviruses, such as *Carnation mottle virus* (approximately 4,000 nt). Most of the (+)-sense ssRNA viruses encode four to seven proteins, but some closteroviruses encode 12 proteins. Such genomes from different virus families share a number of common features:

1. Purified (+)-sense viral RNA is directly infectious when applied to susceptible host cells in the absence of any virus proteins.
2. Nontranslated regions (NTRs) at the 5' and 3' ends of the genome do not encode any proteins, and have conserved regions. These regions may be involved in virus replication, including ribosome and replicase recognition sites, and CP recognition sites as an assembly signal. Sequences in intergenic (IG) regions are also involved in both RNA synthesis and the translation.

3. Both ends of (+)-sense ssRNA virus genomes contain specialized structures.

The 5' end of some plant viral RNAs have a methylated blocked terminal group, known as a cap, while members of several other ssRNA virus groups have a protein (VPg) of relatively small size (35-24 kDa) covalently attached to the 5' end of RNA genomes. PolyA sequences or tRNA-like structures have been reported at the 3' ends. These signals allow viral RNA to be recognized by host cells and to function as mRNA.

All nondefective, wild-type, (+)-sense ssRNA viruses encode RdRp. Information on the genome organization and sequence similarities of the nonstructural proteins show that the overall sequence similarity of viral proteins among the (+)-sense ssRNA viruses is quite poor. Despite wide variation in genomic configuration and very different polypeptide sequences, common motifs have been identified, in particular of their RdRps and helicases, suggesting that most plant RNA viruses are phylogenetically related and appear to have possible evolutionary links with some animal RNA viruses. Some of these identified motifs were also present in DNA-dependent DNA polymerases and RNA polymerases. The crystal structure of RdRps revealed that they have the same overall shape as other polymerases, although some domains, including fingers and thumb, are distinctive. Based on multiple alignments of RdRps and helicases, these viruses were divided into three large supergroups, and it was suggested that members in each supergroup employed similar RNA replication strategies. Although grouping of (+)-sense ssRNA viruses using other nonstructural proteins is also possible, RdRp is the only domain allowing an all-inclusive phylogenetic analysis. Common features shared by these supergroups are listed in Table 6.1.

Negative-Sense ssRNA Viruses

Negative-sense RNA viruses are a large and diverse group of enveloped and nonenveloped viruses. They include both segmented and nonsegmented viruses found in hosts from the plant and animal kingdoms, and they have a wide range of morphologies, biological properties, and genome organizations. These viruses seem to be an evolutionarily recent development, as they infect only higher eukaryotes, such as arthropods, vertebrates, and higher plants. They all have flexible nucleocapsids with helical symmetry surrounding the ssRNA genome. The nucleocapsid is surrounded by an envelope and their genomes range from 10,000 to 15,000 nt in size. These common structural features suggest they may have evolved from the same

TABLE 6.1. Characteristics of plant RNA virus superfamilies using RNA-dependent RNA polymerases (RdRps).

Group	Virus genus	Common features
Super1 (Picornavirus supergroup)	*Comovirus, Nepovirus Potyvirus, Bymovirus Sobemovirus, Polerovirus*	VPg at 5' end, 3'-poly(A) No sgRNAs, no overlapping ORFs Polyprotein processing
Super2 (Carmovirus supergroup)	*Dianthovirus, Carmovirus Tombusvirus, Necrovirus Luteovirus, Avenavirus Machlomovirus*	No 3'-poly(A), often capped at 5'-one to several genome segments
Super3 (Alphavirus supergroup)	*Tymovirus, Carlavirus Potexvirus, Capillovirus Tobamovirus, Hordeivirus Tobravirus, Closterovirus Alfamovirus, Bromovirus Cucumovirus*	Cap structure at 5' end sgRNAs, no overlapping ORFs readthrough expression strategy (most of them)
(–)-sense ssRNA	*Cytorhabdovirus Nucleorhabdovirus Tospovirus, Tenuivirus Ophiovirus*	Genomic RNA with nucleocapsid protein Membrane-bound particles Virion RdRp Ambisense arrangement (some)
dsRNA	*Fijivirus, Oryzavirus Phytoreovirus, Alphacryptovirus Betacryptovirus, Varicosavirus*	Segmented genome Monocistronic (most of them)

ORFs: open reading frames; sgRNAs: subgenomic RNAs; VPg: viral protein genome linked.

ancestral virus. The viruses infecting plants show close association with insects and host plants. Two families *(Rhabdoviridae* and *Bunyaviridae)* and three unassigned genera *(Tenuivirus, Ophiovirus,* and *Varicosavirus)* of plant viruses are reported to have (–)-sense ssRNA genomes. Viruses with (–)-sense RNA genomes are slightly more diverse than (+)-sense RNA viruses. Possibly because of the difficulties of expression, they tend to have larger genomes encoding more genetic information. Because of this, segmentation is a common though not universal feature of such viruses.

Double-Stranded RNA Viruses

Although the double-stranded nature of the genomes greatly increases stability of nucleic acids, only a few members (two families, *Reoviridae* and

Partitiviridae, and one unassigned genus, *Endornavirus*) of viruses with dsRNA genomes are reported. Although it is tempting to speculate that these viruses are the monophyletic survivors of a pre-DNA dsRNA genome era, at least some of them may descend from (+)-sense ssRNA viruses. The relative scarcity of dsRNA viruses may be due to (1) the difficulty in translation of dsRNA genomes resulting from the coexistence of equimolar amounts of both senses or (2) the induction of gene silencing caused by dsRNA genomes. Viruses with dsRNA contain multiple linear dsRNA segments (i.e., 10 to 12 dsRNA segments for reoviruses) that are monocistronic in most cases. These viruses are certainly polyphyletic in origin, and almost certainly a wide variety of mechanisms is used for expression and replication. The complete nt sequences of 9 out of 12 *Wound tumor virus* (WTV) genome dsRNAs have been sequenced. Each genome RNA apparently encodes a single open reading frame (ORF) and is separately encapsidated. The mechanism, by which only one copy of each genomic RNA segment is packaged, is not clear. However, it appears that the information necessary for their replication and packaging resides in the terminal domains of these RNAs.

EXPRESSION STRATEGIES

The replication of RNA viruses is intimately linked with the expression of their genomes. In this context, viruses must produce all or most of the components of an RdRp, and often other proteins as well, in order to transcribe full-length complementary RNA molecules from RNA templates (see Figure 6.1). One of the distinctive features of eukaryotic cells is that they are generally of monocistronic nature (i.e., one mRNA, one polypeptide chain). Similarly, viral RdRps have limited ability in initiating RNA synthesis on internal sites of template RNAs. Many plant RNA viruses, however, encode more than one gene on a single genomic RNA and thus downstream genes on viral genomes could not be expressed unless they deploy specific mechanism(s) to overcome the monocistronic nature of the host cell. A paucity of data is available on the expression of plant RNA virus genomes, but these data indicate that many plant RNA viruses deploy several different strategies to express genes located at the downstream region of the genome. Genomes of RNA viruses are fully or partially translated upon entry into the cell to produce the RdRp, followed by synthesis of full-length complementary RNA and then of full-length genomic (g) RNA, and often also of subgenomic (sg) RNA(s), and thus are usually infectious as naked RNA. To overcome the limitation that the eukaryotic translation machinery is heavily biased to expressing only the 5'-proximal ORF, these

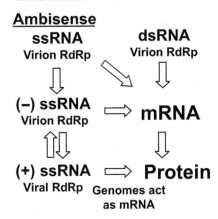

FIGURE 6.1. Pathways of information flow for plant RNA viruses. All plant RNA viruses must synthesize mRNA that can be transcribed by the plant's cellular ribosomes.

viruses have evolved several strategies for expressing their whole genomes. Three main strategies have developed in replication of RNA viruses, as discussed earlier.

1. With some virions (potyviruses), the virion RNA is messenger and is translated monocistronically into a large peptide, which is then cleaved into distinct proteins.
2. Virion transcribed to yield monocistronic mRNAs by initiating transcription at different places.
3. The genome consists of separate RNA fragments, which give rise to monocistronic RNAs.

Transcriptional Regulation of Gene Expression

The use of sgRNA(s) is widespread in plant viruses as a strategy to follow the monocistronic nature of their host cells. During virus replication from a gRNA, sgRNAs that contain more than one ORF, giving 5' truncated, 3'-coterminal versions of the genome, are synthesized. In any case, all sgRNAs will be 3'-coterminal, as there appear to be no mechanisms for transcription termination. Several mechanisms have been proposed for sgRNA synthesis in different RNA virus systems (see Figure 6.2). The synthesis of sgRNA of *Brome mosaic virus* (BMV), a bromovirus, occurred by internal initiation on (–)-sense RNA based on studies utilizing an in vitro

1. Internal initiation

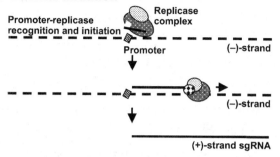

2. Premature termination

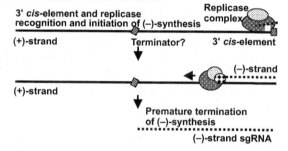

3. Discontinuous transcription

A. Leader priming

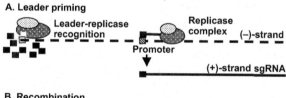

B. Recombination during minus-strand synthesis

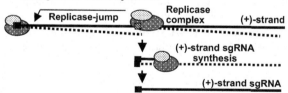

FIGURE 6.2. Mechanisms for sgRNA transcription. Solid and dotted lines indicate plus- and minus-strand RNA, respectively. Hatched squares represent potential promoters. Shaded circles with different sizes represent replicase complex. (*Source:* Partially adapted and modified from Miller and Koev, 2000.)

BMV replicase extract, but this process may require additional elements. Studies on *Alfalfa mosaic virus* (AMV), an alfamovirus, and *Turnip crinkle virus* (TCV), a carmovirus, have also provided experimental support for this model. Internal initiation on (–)-sense template clearly does not apply to all other RNA viruses. A discontinuous RNA transcription process involving long-distance interactions between the terminal leader and the IG sequences has been proposed for coronaviruses. Alternatively, sgRNA synthesis in this system may involve generation of truncated (–)-sense RNAs that serve as templates for sgRNA synthesis. A mechanism involving disruption of (–)-sense RNA synthesis may also apply to the plant RNA virus, *Red clover necrotic mosaic virus* (RCNMV), a dianthovirus, and *Tomato bushy stunt virus* (TBSV), a tombusvirus. In this model, a *trans*-acting sequence element or distal element directly interact with a complementary region in the sgRNA promoter region to terminate (–)-sense synthesis and generate templates for sgRNA synthesis. Base-pairing between the two RNA sequences creates termination structure, which makes the RdRp complex pause and dissociate from the template. Other examples of RNA-RNA long-distance interactions have been shown for transcription of sgRNA of *Potato virus* X (PVX), a potexvirus.

The segmented (–)-sense RNA viruses share the transcription initiation mechanism called "cap snatching," where virion-associated protein or replicase protein cleaves cellular mRNAs 12-18 nt from their 5' ends. The resulting capped leaders are used to prime transcription of nonpolyadenylated mRNA on the viral genome. These mRNAs are shorter than the genome segments, as transcription is apparently terminated by hairpin loops. It is not certain whether there is (1) a specific sequence requirement and (2) length and structure of suitable donor RNA. Recently, requirements for capped leader sequence for use during transcription initiation has been studied using *Tomato spotted wilt virus* (TSWV), a tospovirus. The length of cleaved leader could vary between 13 and 18 nt. Cleavage occurred preferentially at adenine residue. Two genera of the plant RNA virus genomes encode two proteins having one ORF in the virion sense and the other in the complementary sense (ambisense RNA strategy): tospoviruses and tenuiviruses transcribe an mRNA from the 3' end of the S segment (+)-sense RNA.

Translational Regulation of Gene Expression

The typical translational regulation strategy is "proteolytic cleavage" strategy followed by members of the family *Potyviridae*. These viruses make use of co- and posttranslational cleavages by virus-coded endoproteinases, as well as of sequential and sometimes alternative cleavages, in order to make a large number of proteins and to regulate their own replication. These viruses use an RdRp that makes use of a protein (VPg) as a primer for both (+)- and (−)-sense RNA production and tend to have 3'-poly(A) genome segments. The polyprotein strategy is shared by many plant viruses (see Table 6.2). The strategy results in near-equimolar amounts of the different proteins being made. Because the replicase complex for these viruses is

TABLE 6.2. Summary of genome expression strategies by (+)-sense plant single-strand RNA viruses.

Group	Number of strategies	Strategy	Virus genus
1	One	Proteolytic cleavage	*Potyvirus, Ipomovirus, Macluravirus, Rymovirus, Tritimovirus, Sequivirus, Waikavirus*
2	Two	sgRNA Proteolytic cleavage	*Sobemovirus, Marafivirus, Capillovirus*
3	Two	sgRNA, readthrough	*Tobamovirus, Avenavirus, Aureusvirus, Carmovirus, Tombusvirus*
4	Two	sgRNA, frameshifting	*Crinivirus, Umbravirus*
5	Two	sgRNA, segmentation	*Bromovirus, Alfamovirus, Cucumovirus, Ilarvirus, Oleavirus, Hordeivirus*
6	Two	Segmentation proteolytic cleavage	*Bymovirus, Nepovirus*
7	Two	sgRNA, internal initiation	*Potexvirus*
8	Three	sgRNA, readthrough segmentation	*Tobravirus, Furovirus, Pecluvirus, Pomovirus*
9	Three	sgRNA, frameshifting segmentation	*Dianthovirus, Crinivirus*
10	Three	sgRNA, frameshifting proteolytic cleavage	*Closterovirus, Sobemovirus*
11	Three	Proteolytic cleavage segmentation, two-start	*Comovirus*
12	Three	sgRNA, two-start proteolytic cleavage	*Tymovirus*
13	Four	sgRNA, internal initiation frameshifting, readthrough	*Luteovirus, Polerovirus, Enamovirus*
14	Five	sgRNA, segmentation readthrough, two-start proteolytic cleavage	*Benyvirus*

sg: subgenomic RNA.

"single-use," a freshly synthesized RdRp polyprotein is needed to initiate replication on each new template. As for encapsidation of the potyvirus genome, about 2,000 CP molecules are required and one molecule of each of ten other gene products has to be made to produce one CP subunit; this appears to be a very inefficient procedure. These viruses developed means to remove excess components by exporting the latter as "nuclear inclusion body" (NIa and NIb) or replicase protein subunits to the nucleus to be sequestered as insoluble aggregates. Many other viruses using polyprotein strategy have additional devices that can avoid this problem.

Some viruses utilize translational readthrough or frameshift to avoid terminator and to produce more than one protein. The termination codon of the 5' ORF may be leaky, and if there is an inframe ORF immediately downstream to the terminator, this event would produce a larger protein composed of the 5' ORF product as the N-terminal and the 3' ORF as the C-terminal half. This is termed "terminator readthrough" (shortened as readthrough). In the normal protein synthesis, a ribosome stops the protein chain elongation when it encounters one of the three terminators: UAG (amber), UGA (opal), and UAA (ochre). When the stop codon is read by suppressor tRNA instead of release factors, this will permit a terminator to be read as a code for an amino acid and therefore overcome the translation termination. This event occurs at a very low frequency as a means to regulate the gene expression at the translational level. The readthrough strategy is found in many RNA viruses (see Table 6.2). Viruses also regulate synthesis of their proteins by changing reading frames. Normal translation efficiently maintains a single reading frame. Sometimes, a ribosome may shift either −1 or +1 nt, or more radically hops to a distant location to change the reading frame. This phenomenon is termed "ribosomal frameshifting" (frameshifting for short). This translational frameshift occurs near the end of the first ORF. It allows a ribosome to bypass the stop codon and produces two proteins, a frame and a transframe fusion protein, translated from two or infrequently three overlapping ORFs. The frameshift strategy is also found in many virus groups (see Table 6.2).

Many regulatory mechanisms function at initiation since this is a rate-limiting step in the translation. These mechanisms include leaky scanning, ribosome shunting, internal ribosome entry, and non-AUG start. For many years it was suspected that the expression of a downstream ORF in a dicistronic mRNA was translated by an unusual mechanism termed "internal ribosome entry." Although the vast majority of plant mRNAs are monocistronic, many plant (+)-sense RNA viruses contain more than one ORF in overlapping reading frames. In this case, the 5' NTR was shown to contain a sequence that forms a complex secondary/tertiary structure, to which ribosomes and *trans*-acting factors bind, and promotes translation initiation. These 5' NTRs are called an internal ribosome entry site (IRES). Leaky scanning is the mechanism used by RNA viruses in translation of many polycistronic RNAs. The 40 S ribosomal subunits start scanning from the 5' end of the mRNA but do not start translation at the first AUG. This mechanism is most certainly facilitated if the first AUG is in a less favorable context than the second AUG. Some viral ORFs appear to start with non-AUG codon, although the translation initiation at these sites is inefficient compared to the conventional AUG start codon. Several examples of non-AUG translation initiation have been reported (see Table 6.2). In certain cases, ribosomes directly shunt or bypass to the appropriate initiation site if scanning ribosomes interfered with too many short ORFs in the leader sequence or if the AUG initiation codon is buried in the stem of an RNA stem-loop structure. Examples of ribosome shunting are found on plant pararetroviruses.

Regulation of Gene Expression by Genome Segmentation

Another way RNA viruses accommodate the monocistronic requirement of the eukaryotic translation system is to break the viral genome into several pieces. In the case of segmented genomes, each piece of the genome contains one or more ORFs and effectively becomes a monocistronic mRNA. At least 28 virus genera have multipartite genomes. Viruses with segmented genomes may produce single proteins from single segments or, in the case of bromoviruses or tobraviruses, for example, may have both monocistronic and multicistronic segments.

Most viruses use more than one of these strategies to express their genetic content. For example,

1. RCNMV utilizes segmented genome, sgRNA, and ribosomal frame-shifting.
2. *Cowpea mosaic virus* (CPMV), a comovirus, utilizes segmented genome and polyprotein processing.
3. *Tobacco mosaic virus* (TMV), a tobamovirus, utilizes sgRNA and terminator readthrough.
4. *Tobacco rattle virus,* a tobravirus, employs segmented genome, sgRNA, and terminator readthrough.
5. *Beet necrotic yellow vein virus* (BNYVV), a benyvirus, which is by far one of the most sophisticated viruses in terms of using various gene expression strategies.

The genome organization of BNYVV suggests the involvement of the five strategies discussed previously: genome segmentation, terminator read-through, sgRNA, two-start, and proteolytic cleavage.

VIRUSES WITH DIVIDED GENOMES

For some plant RNA viruses the genome is divided into two or more segments. Segmented virus genomes are those that are divided into two or more physically separate molecules of nucleic acid, all of which are then packaged into a single particle (multipartite viruses). Multipartite viruses with their genome segments packaged in separate particles are called multicomponent (multiparticulate) viruses. These discrete particles are structurally similar and may contain the same component proteins, but often differ in size depending on the length of the genome segment packaged.

Subdividing the genome into separate particles is the unique feature found only in plant (+)-sense RNA viruses, since animal and bacterial viruses rely on chance encounters between appropriate host cells and virions released into the environment. The multicomponent nature would reduce the probability of the same host cell being infected at the same time by all required viral particles for animal and bacterial viruses. Many of the (+)-sense plant RNA viruses, however, are transmitted from cell to cell, either by sap-sucking insects or by direct physical movement through plasmodesmata. This results in a large input of infectious virus particles, providing the opportunity for infection of an initial cell by more than one particle. Separating the genome segments into different particles gives some advantages, including (1) the ease and efficiency of translation of several small RNAs compared to one large RNA molecule, (2) the relative easiness of encapsidation, and (3) physical separation of gene expression during virus replication. However, multicomponent viruses also introduce a new problem in that all of the discrete virus particles must be taken up by a single host cell to establish a productive infection. This is perhaps the reason why multiparticulate viruses are found only in plants.

Many examples of segmented virus genomes have been found, including many human, animal, and plant pathogens, such as orthomyxoviruses, reoviruses, and bunyaviruses. Rather fewer examples of multiparticulate viruses have been found, all of which infect plants. These include bipartite viruses (which have two genome segments/virus particles) and tripartite viruses (three genome segments/virus particles). Viruses with segmented genomes include alfamoviruses, bromoviruses, bymoviruses, comoviruses, cryptoviruses, cucumoviruses, dianthoviruses, benyviruses, hordeiviruses, ilarviruses, nepoviruses, tenuiviruses, tobraviruses, reoviruses, tospoviruses, and enamoviruses.

REGULATION OF REPLICATION

Although the RNA replication is error prone (usually 10,000 times higher than that of DNA polymerases) and thus any individual virus particle may contain an average of one or more mutations during virus replication, replication and packaging of viral RNAs display striking specificity. This is generally due to the presence of the *cis*-acting signals present in viral genomes. It is likely that many of the control signals are located at both ends of the genome(s), especially at the 3' end. Three basic elements believed to be involved in virus replication are located at the 3' end, including tRNA-like structures, pseudoknots, and poly(A) tail. The RNA genomes of certain (+)-sense plant RNA viruses have tRNA-related properties that have been shown to be involved in (−)-sense RNA synthesis. A few viruses have pseudoknots upstream of pseudoknot present in the tRNA-like structures which appear to participate in RNA replication and/or translation. Poly(A) structures also have been shown to affect RNA replication dramatically in many RNA viruses. There might be other signals elsewhere in the genome (i.e., internal control region (ICR)-like sequences found in bromoviruses, cucumoviruses, tobamoviruses, tobraviruses, and tymoviruses). These signals selectively control replication and assembly of viral RNAs out of thousands of abundant cellular RNAs. Although advances in technology are leading to elucidate the high-resolution structure of some RdRps, molecular basis of specificity caused by *cis*-acting signals is still less well understood. Replication complexes are usually closely associated with membrane complexes derived from the endoplasmic reticulum (ER) or perhaps nuclear membranes. In some cases it has been shown that other viral proteins besides RdRp are associated with replication complexes, although they may or may not affect virus replication. For example, the movement protein (MP) of TMV is closely associated with replication complexes, but an MP-defective mutant replicates as well as wild-type virus, indicating that this gene is not absolutely required for replication. In contrast, regulation of the expression of RNA by virus-encoded CP has been reported for AMV. Association between MP and CP with replication complexes indicates the closely coordinated nature of RNA replication and virus movement and encapsidation.

The virus-encoded proteins required for RNA replication have been deduced from the composition of purified polymerases capable of copying genomic RNA to produce a (−)-sense RNA, from the use of mutants; for divided genome viruses from the minimum number of RNA segments needed to infect protoplasts; and from the presence of conserved sequence motifs found in polymerases in other systems. The observation that many viruses have a limited host range suggests that host-specific factors are involved in viral replication. Several studies have shown that host-encoded proteins copurify with RdRp complexes, including BMV, *Cucumber mosaic virus* (CMV), a cucumovirus, CPMV, PVX, TMV, and *Turnip yellow mosaic virus* (TYMV), a tymovirus. The requirement for host factors goes some way in explaining the inability of some extracted viral RdRps to fully complete a replication cycle. In many cases, identified host proteins are ribosome-associated or translation/transcription factors. Positive-sense ssRNA can act as a template for protein synthesis, as well as a template for (–)-sense RNA synthesis. This may lead to the ribosomes, translating 5' to 3' direction, and replication complexes, copying the RNA template in the 3' to 5' direction, and collisions unless simultaneous translation and replication are blocked. How the switch from translation to replication is achieved for plant (+)-sense RNA viruses is not clear. One of major functions of these host factors may be the modulation of translation and replication of (+)-sense RNAs. Recently, the modulation of BMV RNA translation and recruitment from translation to replication by the host's Lsm1p-7p and associated factors has been reported. BMV replication requires a p41 subunit of eIF-3, mab1-1, mab2-1, mab3-1, and Lsm1p. Involvement of Ydj1p factor in forming BMV replication complexes and its role in (–)-sense BMV RNA synthesis has also been reported. TMV RNA synthesis also requires host cell proteins. A highly purified replication complex of TMV contained three host proteins (56, 54, and 50 kDa [kilodaltons]) as well as 126 and 183 kDa viral proteins (p126 and p183, respectively). The 56 kDa protein was found to be closely related with yeast eIF-3. Mutations in the *Arabidopsis* *TOM1* and *TOM2* genes reduced TMV replication, but replication of CMV and TYMV was unaffected.

SYNTHESIS OF RNA

Eukaryotic cells synthesize RNA using DNA as a template and they usually do not contain enzymes that synthesize RNA using an RNA template. Most RNA viruses, therefore, encode protein(s) that are involved in virus replication. Regardless of the replication strategy of the RNA viruses, they must express their genome as functional mRNA upon host infection to make their gene products, as mentioned earlier (see Figure 6.1). The discovery of the RdRps marked a major breakthrough in understanding the replication of progeny RNA from genomic viral RNA. Replication of (+)-sense

RNA viruses involves an initial transcription of full-length (–)-sense RNA from an infecting (+)-sense RNA template, followed by transcription of (+)-sense RNA, and perhaps also sgRNA(s). Each of these steps is differentially regulated. For example, (–)-sense RNA accumulation of PVX plateaus by 6 hours postinoculation (hpi), while (+)-sense gRNA and sgRNA continue to accumulate until or beyond 24 hpi. The synthesis of (–)- and (+)-sense RNAs is likely accomplished by different RNA polymerase holoenzymes. Double-stranded RNA forms of viral genomes, designated replicative form (RF) RNA, and of sgRNA(s) and a heterogeneous family of RNAs, which are partly ds and partly ss, designated replicative intermediate (RI) RNA, can be isolated from infected cells for most plant ssRNA viruses. RF and RI RNAs are considered to be intermediates of RNA replication. However, it is not clear whether RF is formed in vivo or is an artifact resulting from the annealing of both (+)- and (–)-sense RNAs during extraction.

Enzyme Activities Are Required for RNA Synthesis

RNA of most (+)-sense RNA viruses bind directly to the ribosome to initiate translation. For all other RNA viruses mRNA must be transcribed in order to begin the process of translation. Many RNAs contain cap structure and poly (A) sequences. Cap consists of 7-methylguanosine, added to the 5'-terminus of primary transcript. To do these functions, three or more enzyme activities are required for the synthesis of RNA. RNA viruses encode protein(s) with RdRp, helicase, and methyltransferase activities. In general, these enzymes are collectively termed as "replicase." Assembly of active RdRp requires not only viral proteins but also viral RNA, either to direct some nontemplate function or to recruit essential host factors in the RdRp complex. The two proteins, p126 and p183, are required for efficient TMV replication. Although p126 is not required for RNA synthesis, excess levels of p126 significantly enhance replication. The p126 contains two domains; an N-terminal domain with methyltransferase and guanosyltransferase activities and a C-terminal domain with helicase activity. The helicase domain of p126 interacts with itself to produce hexamerlike oligomers, similar to the oligomeric structures assumed by many DNA and RNA helicases. The excess amount of p126 may lead to the formation of more efficient enzyme complex. The p183 contains amino acid motifs characteristic of RdRp. Potential RdRps have been described for many plant RNA viruses, including AMV, BMV, *Cowpea chlorotic mottle virus,* a bromovirus, CMV, *Cucumber necrosis virus,* a tombusvirus, PVX, RCNMV, TBSV, TMV, *Tomato spotted wilt virus,* a tospovirus, TCV, and TYMV.

Positive-Sense ssRNA Viruses

Replication of plant (+)-sense RNA viruses takes place in the cytoplasm of infected cells. As indicated earlier, RNA synthesis can be divided into three distinct steps: negative-strand synthesis, positive-strand synthesis, and sgRNA transcription for many plant RNA viruses catalyzed by an RdRp. They replicate by utilizing input gRNA as a template for synthesis of a complementary (–)-sense RNA, which subsequently serves as a template for synthesis of genomic (+)-sense RNA. Many virus RNA genomes contain more than one gene on a single RNA, but only the first gene is translated. Thus, the expression of the downstream genes is achieved by employing various translational events or by generating sgRNAs. For some (+)-sense RNA virus groups, (–)-sense RNA also serves as a template for synthesis of sgRNAs that contain genes frequently encoding structural proteins. Subgenomic RNAs of the (+)-sense RNA viruses have identical sequences at the 3' ends while containing truncations at their 5' ends. Thus, 5' ends of each sgRNA are in close proximity to the initiation codon of each downstream gene. Initiation of the synthesis of a (–)-sense RNA on a (+)-sense RNA template requires binding of the RdRp to a recognition site at the 3' end of the template. Many control signals are required for the synthesis of RNAs. BMV RNA synthesis depends on cis-acting signals in the 5', 3', and IG sequences within each BMV RNA. Negative- and (+)-sense RNA syntheses are promoted by a conserved 3' terminal tRNA-like structure and 5' terminal sequences with nontemplate guanylate, respectively. It also has been shown that the 3' tRNA-like domain and upstream pseudoknot domain play key roles in increasing RNA stability and in the regulation of translation, as well as acting as recognition sites for the TMV RdRp. Deletions introduced at the 5' NTR also reduced RNA synthesis to an undetectable level. Relatively little is known about cis-acting elements required for the synthesis of TMV sgRNAs. Multiple sequence and structural elements in the 5' NTR of the PVX RNA affect both gRNA and sgRNA accumulation. Levels of sgRNAs were reduced in protoplasts inoculated with mutants containing modifications in the 5' NTR elements, despite no reduction in (–)-sense RNA accumulation. The conserved octanucleotide sequence elements located upstream of the two major PVX sgRNAs have shown to be important for sgRNA accumulation in protoplasts and potentially some other aspect of the infection process in plants.

Negative-Sense ssRNA Viruses

Characteristically, the genetic information of (–)-sense RNA viruses is exclusively found in the form of a ribonucleoprotein complex (RNP) in which the genomic or antigenomic ssRNA is tightly encapsidated in a nucleoprotein (N). The (–)-sense genome has two functions: as template for mRNA transcription and as the template for replication. RNA from (–)-sense RNA viruses is not infectious RNA. All viruses with (–)-sense ssRNA genomes carry the RdRp in their virus particles, and thus one of the early events on entry into plant cells is the transcription by virion-associated RdRp of (+)-sense RNAs from genomic RNA(s) in the newly uncoated nucleoprotein complexes. Common features in expression and replication of (–)-sense RNA viruses appear to include distinct transcription and replication functions for the RdRp, probably triggered by binding of the virion N subunits. It still remains unclear how RdRp can synthesize RNAs on protein-associated RNA templates and how RdRps switch templates and perform both functions. Thus, both (–)-sense RNA and (+)-sense RNA may be found complexed with nucleoprotien N in replication complexes. For the segmented genomes of bunyaviruses, this usually means a single mRNA per segment; for the nonsegmented rhabdoviruses, this means multiple transcription initiation and termination events on a full-length (–)-sense RNA, at intergenic repeated sequences. Some RNA viruses are not strictly "(–)-sense" but ambisense, since they are partly (–)-sense and partly (+)-sense, with their RNA of opposite polarities covalently joined.

RNA Viruses

Reoviruses are the best-studied dsRNA viruses. However, transcription and replication of plant reoviruses are not well understood. Representatives of the family infect plants, animals, and insects, and many infect an insect vector as well as an animal or plant alternate host. The viroplasms are the sites of virus synthesis in cytoplasm, in both plant and insect cells. The mode of replication of plant reoviruses is unknown. However, by analogy with the animal reoviruses, the dsRNA is probably totally conserved inside the inner shell during replication. The first transcriptional event is the synthesis of viral mRNAs using the (–)-sense RNA as a template, like normal cellular transcription. Naked core particles in the cytoplasm are able to transcribe capped and nonpolyadenylated genome-segment-length monocistronic mRNAs via an RdRp activity. The transcription and replication of these viruses are closely coupled. The newly synthesized (+)-sense RNA is extruded from the core into cytoplasm, where it is translated by host ribosomes. When a sufficient amount of viral proteins has accumulated, the same (+)-sense RNA transcripts are packaged within an opened subviral particle (inner core). Once the core is formed with the complete set of viral RNA segments, the (–)-sense viral strands are synthesized to complete the formation of dsRNA. In members of the family *Reoviridae* the (–)-sense RNAs are synthesized by the viral replicase on a (+)-sense template.

VIRUSES WITH INCOMPLETE RNA GENOMES

Various small RNAs associated with plant-virus interactions exist, and they fall into two groups: those that can replicate independently and those that cannot. Self-replicating agents are plant viroids. The non-self-replicating agents are usually parasites of viruses. They depend on a host (helper) virus for replication and sometimes encapsidation. This type of subviral agent includes satellite RNA, satellite virus, virusoid, and defective (D) RNA and defective interfering (DI) RNA.

Viroids are independently replicating circular RNAs capable of causing diseases in infected plants. They consist of naked RNA, which does not code for any protein, nor is protein associated with it, and replicate independent of any associated plant viruses. Essentially, each viroid particle is a circular ssRNA molecule containing 246 to 375 nt. The nt sequences of about 30 members of the viroid group and those of numerous variants are known (see also Chapter 8 in this book, and for more sequence information visit http://nt.ars-grin.gov/subviral/viroids). They are highly resistant to enzymatic degradation because they have no free ends and because they have a very tight secondary structure. All RNA viruses have their own RdRp, which is made in the host, but viroids do not code any polypeptides of their own and thus they use the host's preexisting DNA-dependent RNA polymerases. The viroids are replicated in infected cells by host enzymes through double-stranded intermediates by a rolling circle mechanism. Two pathways of the rolling circle model exist. In the symmetrical pathway, the linear, multimeric, (–)-sense RNAs are processed and ligated to monomeric, circular, (–)-sense RNA, which serves as the template for synthesis of the linear, multimeric, (+)-sense RNAs. In the asymmetric model, the circular, (+)-sense RNA is transcribed into linear, multimeric, (–)-sense RNAs, which then act as the template for the generation of the linear, multimeric (+)-sense RNAs. Three enzyme activities are required to accomplish these pathways: RNA polymerase, an RNase to process multimeric RNAs into monomer RNAs, and RNA ligase to produce circular monomer. The two families of viroids are the *Pospiviroidae* and the *Avsunviroidae*, with five and two genera, respectively. The genera of the *Pospiviroidae* can be further divided into two subgroups, B1 and B2, based on the sequence of the conserved regions. The genera of the *Avsunviroidae* are capable of autocatalytic cleavage in vivo. Members of the genera of the *Avsunviroidae* probably replicate via a symmetric rolling circle pathway, whereas those of the *Pospiviroidae* probably use an asymmetric pathway.

Unlike viroids, satellite RNA, satellite virus, virusoid, and D RNAs or DI RNAs require a helper virus for replication and show a specific association between agent and helper virus. In all members of this group the helper virus provides RdRp. Some isolates of certain plant viruses contain satellite agents. Some satellite genomes encode proteins (CPs) used to encapsidate genomes and are called satellite viruses, while satellite RNAs do not encode any protein. Satellite agents have the following common properties. They are still dependent upon a helper virus for replication and may modify the symptoms of infection by their helper virus. They affect disease symptoms in some hosts and interfere with the replication of their helper virus to some degree. In many cases the coreplication of satellites suppresses the replication of the helper virus genome. This is usually paralleled by a reduction in the disease induced by the helper virus; however, notable exceptions exist in which the satellite exacerbates the pathogenicity of the helper virus, albeit on only a limited number of hosts. The genomes are not part of their helper virus genome and are therefore differentiated from DI particles that came from the helper virus genome. Genomes of satellite agents vary from 194 to approximately 1,500 nt in length. The larger satellites (900 to 1,500 nt) contain ORFs and express proteins in vitro and in vivo, whereas the smaller satellites (194-700 nt) do not appear to produce functional proteins. Some of these smaller satellites encode ribozymes and are able to undergo autocatalytic cleavage. The replication of satellite agents is poorly understood, but it is widely assumed that satellite replication must be carried by an RdRp coded for, at least in part, by a helper virus. Several satellite RNAs associated with some viruses have been shown to have viroidlike structural properties and have been termed virusoids. Virusoid genomes are 220-338 nt long, ss circular RNAs and possess a ribozyme activity. They can replicate in the cytoplasm using an RdRp. They depend on a helper virus for replication.

Interference may occur during replication by the generation of defective particles. In early serial passages, defective particles rapidly increase in titer, then decrease the yield of the infectious virus, and finally the total particle yield is progressively reduced. With some viruses, subviral RNA molecules are derived from parent genomic RNA. They usually contain the normal virion proteins but have a shorter genome. They are replication defective and require the helper functions of a normal virus coinfecting the same cell. In many cases, interference of a helper virus may occur during replication by the generation of these subviral RNAs; these are called DI RNAs and thus ameliorate viral symptoms. They deprive the regular virus of its replicase by binding to it more effectively. They do not make a replicase of their own because they are always defective in their replicase gene. If they do not interfere with the parent virus they are termed D RNAs. However, in the case of TCV generation and replication of D RNAs they are known to cause severe symptoms. The genomes of DI particles are internally deleted but retain both ends, which are essential for the replication of RNA viruses. The formation of defective genomes of RNA viruses is the consequence of the high variability of these genomes, and many types of DI genomes are continuously made and are very heterogeneous during viral multiplication. The DI and D RNAs are formed by

1. deletions—the polymerase jumps to a site beyond on the same template, skipping a fragment;
2. snapbacks—this occurs when the replicase, having transcribed part of the (+)-strand, switches to the newly synthesized (–)-strand as a template;
3. panhandle—this is formed by a similar mechanism, when the polymerase carrying a partially made (–)-strand switches back to transcribing the extreme 5' end of it, so that on annealing, the strand forms a panhandle; and
4. compounds—these genomes are made by a combination of deletions and snapbacks.

CONCLUDING REMARKS

RNA viruses are a taxonomically diverse group that includes major pathogens of plants, fungi, animals, and bacteria. The genome sequences of many viruses have made it clear that the taxonomic classification of these RNA viruses is a complicated task. However, these viruses employ RdRp to transcribe and replicate in infected host cells with some host-encoded factors. It has turned out that plant RNA viruses precisely recognize *cis-* and *trans-*acting regulatory elements and regulate viral RNA synthesis. We still do not know this precise regulation of mRNA transcription and viral replication. It seems that more than one host protein is involved for each step of these processes, and that each virus has evolved to use possible tools required and thus has a variety of signals and mechanisms for transcription and replication. The synthesis of RNA from RNA has not been considered to be a major mechanism in healthy plants. The RdRps of RNA viruses contain one or more viral proteins that contain polymerase, helicase, and, in many cases, methyltransferase motifs. Why these taxonomically diverse groups of plant RNA viruses use a relatively common mechanism during transcription and replication processes is still unknown. The better understanding of mRNA transcription and virus replication, especially from RNA virus-host interaction, genomics, and proteomics projects, will widen our understanding of the cellular processes occurring in RNA virus-infected plants.

BIBLIOGRAPHY

Ahlquist P, Noueiry AO, Lee WM, Kushner DB, Dye BT (2003). Host factors in positive-strand RNA virus genome replication. *Journal of Virology* 77: 8181-8186.

Buck KW (1999). Replication of tobacco mosaic virus RNA. *Philosophical Transactions of the Royal Society of London Series B* 354: 613-627.

Commandeur U, Rohde W, Fischer R, Prüfer D (2002). Gene expression strategies of RNA viruses. In Khan JA, Dijkstra J (eds.), *Plant viruses as molecular pathogens* (pp. 175-201). Binghamton, NY: The Haworth Press, Inc.

Diener TO (1999). Viroids and the nature of viroid diseases. *Archives of Virology Supplement* 15: 203-220.

Miller WA, Koev G (2000). Synthesis of subgenomic RNAs by positive-strand RNA viruses. *Virology* 273: 1-8.

Rubio T, Borja M, Scholthof HB, Feldstein PA, Morris TJ, Jackson AO (1999). Broad-spectrum protection against tombusviruses elicited by defective interfering RNAs in transgenic plants. *Journal of Virology* 73: 5070-5078.

Sivakumaran K, Sun J-H, Kao CC (2002). Mechanisms of RNA synthesis by a viral-dependent RNA polymerase. In Khan JA, Dijkstra J (eds.), *Plant viruses as molecular pathogens* (pp. 147-174). Binghamton, NY: The Haworth Press, Inc.

Chapter 7

Replication and Gene Expression of DNA Viruses

Crisanto Gutierrez

INTRODUCTION

Most plant viruses have genomes composed of one or more RNA molecules. However, a few families have been identified whose members have DNA genomes. Their multiplication cycle frequently involves interactions with either cytoplasmic or nuclear factors and cellular structures. This has led to the acquisition of very unique strategies to exploit, and in some cases subvert, the infected cell physiology to the benefit of the virus. Cellular pathways as diverse as gene expression, translation, replication, proliferation and differentiation control, and intra- and extracellular macromolecular trafficking, among others, can be interfered with by one or more viral proteins. Quite interesting from an evolutionary point of view is the striking resemblances of plant DNA virus-host interactions to those occurring in animal DNA viruses. This chapter describes several topics related to genome expression and replication of plant DNA viruses.

GENOME STRUCTURE

The nature of the genetic material and some of its biological properties are the main criteria used for virus taxonomy allowing the distinction of different families and genera. A summary of the main characteristics of the three families of plant DNA viruses, namely *Geminiviridae, Nanoviridae,* and *Caulimoviridae* is presented in Table 7.1.

Family Genimiviridae

The *Geminiviridae* family includes an increasing list of plant viruses that can produce significant reductions in economically important crops of both

TABLE 7.1. General characteristics of plant DNA virus groups.

DNA[a]	Molecules per genome	Size[b]	Replication intermediates	Virus family
ss circular	1 or 2	2.5-3	DNA	*Geminiviridae*
ss circular	>6	<2	DNA	*Nanoviridae*
ds circular	1[c]	6-8	DNA and RNA	*Caulimoviridae*

[a]ss and ds indicate single- and double-stranded, respectively.
[b]The genome size is given in kilobases.
[c]Some members contain linear dsDNA or two circular molecules.

monocotyledonous and dicotyledonous plants all over the world. In all cases, the morphology of the geminivirus particle has a geminate structure consisting of two quasi-icosahedral halves. The genome consists of one or two circular, single-stranded (ss) DNA molecules, 2,500 to 3,000 nucleotides (nt) in length. The geminivirus DNA replication cycle relies entirely on DNA replication intermediates and occurs within the nucleus of the infected cell, where transcription of viral genes is controlled by two main bidirectional promoter regions. The intricate and fine network of virus-host cell interactions are contributing to the use of geminiviruses as extremely useful molecular tools to delineate basic processes in plant biology, such as transcriptional regulation, cell cycle and differentiation control, gene silencing, plasmodesmata function and macromolecular trafficking. This family comprises four genera, namely *Mastrevirus, Curtovirus, Topocuvirus,* and *Begomovirus,* distinguished by their genome organization, host range, and insect vector. Examples of the genetic organization of each of them are depicted in Figure 7.1.

Within the genus *Mastrevirus* (type species: *Maize streak virus,* MSV, which causes one of the most important plant diseases in Africa), most of its 10 to 12 species are transmitted by leafhoppers and infect monocotyledonous plants in the Poaceae family, for example, *Wheat dwarf virus* (WDV). Some of them are adapted to infect dicotyledonous plants, such as *Tobacco yellow dwarf virus* (TYDV) and *Bean yellow dwarf virus* (BeYDV). The mastrevirus genome consists of one molecule, ~2,600 nt in length with two nontranscribed regions, the large intergenic region (LIR), and the small intergenic region (SIR), at opposite locations in the circular genomic DNA molecule, and an encapsidated genome contains an ~80 nt-long DNA sequence annealed to part of the SIR, which may play a role at the early stages of DNA replication. The genome encodes four proteins: RepA (replication-associated protein) and Rep (replication proteins), on the complementary-sense

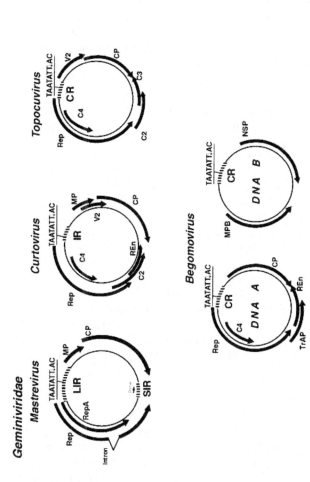

FIGURE 7.1. Genome organization of the four genera of the family *Geminiviridae*. The maps depicted correspond to those of each genus type species: *Maize streak virus* (MSV, genus *Mastrevirus*), *Beet curly top virus* (BCTV, genus *Curtovirus*), *Tomato pseudo-curly top virus* (TPCTV, genus *Topocuvirus*) and *Bean golden mosaic virus* (BGMV, genus *Begomovirus*). Description of protein names are given in the text. The invariant TAATATT↓AC sequence is also indicated. The downward arrow (↓) indicates the initiation site for rolling-circle DNA replication. The large (LIR) and the small (SIR) intergenic regions in mastreviruses, the intergenic region in curtoviruses (IR), and the common region (CR) for topocu- and begomoviruses are indicated by discontinuous lines.

(c-sense) transcript and MP (movement protein) and CP (capsid protein), on the viral-sense (v-sense) transcript.

Curtoviruses (type species: *Beet curly top virus,* BCTV) are also transmitted by leafhoppers and infect dicotyledonous plants. They have a genome consisting of a DNA molecule, ~3,000 nt in length with an intergenic region (IR), but its genetic organization differs from that of mastreviruses. The c-sense strand encodes four proteins, Rep, C2 (a possible transcriptional regulator), REn (replication enhancer) and C4, and the v-sense strand, a V2 protein in addition to MP and CP.

Tomato pseudo-curly top virus (TPCTV), is the sole species assigned to the genus *Topocuvirus,* the least-well-characterized genus. It is transmitted by a treehopper and infects plants in southern United States. The genome is monopartite and has an organization similar to that of curtoviruses, but contains only two genes in the v-sense strand.

Begomovirus (type species: *Bean golden mosaic virus,* BGMV) is the genus with more species (>70 to date). Members of this genus (*Tomato golden mosaic virus,* TGMV; *African cassava mosaic virus,* ACMV) are transmitted by whiteflies and infect dicotyledonous plants, causing several of the economically most serious diseases, including cassava mosaic and cotton and tomato leaf curl, among others. The genome is, in most cases, bipartite (DNA circles A and B, 2,500 to 2,800 nt in length), although some members contain a monopartite genome. The two circles share 200-400 nt, the so-called common region (CR) located between the two divergent promoters. The DNA A encodes at least four proteins: CP on the v-sense strand and Rep, TrAP (a transactivator protein), and REn (replication enhancer protein), on the c-sense strand. DNA B encodes proteins involved in genome movement: MPB (movement protein in DNA B) and NSP (nuclear shuttle protein).

Family Nanoviridae

The family comprises two genera: *Nanovirus* and *Babuvirus.* The genus *Nanovirus* is being described here (see descriptions of plant viruses in appendixes). The genus *Nanovirus* (type species: *Faba bean necrotic yellows virus,* FBNYV) comprises two more species, namely *Milk vetch dwarf virus* (MDV) and *Subterranean clover stunt virus* (SCSV), whereas *Banana bunchy top virus* (BBTV) is the sole species in the genus *Babuvirus.* The genus *Nanovirus,* and hence the family *Nanoviridae,* has been named after the genome of its members, which is multipartite with more than eight circular ssDNA molecules but very small (<1.8 kb each) and encoding a single protein in each circle. This is a unique situation among viruses and an example of their genomic organization appears in Figure 7.2. It has been proposed

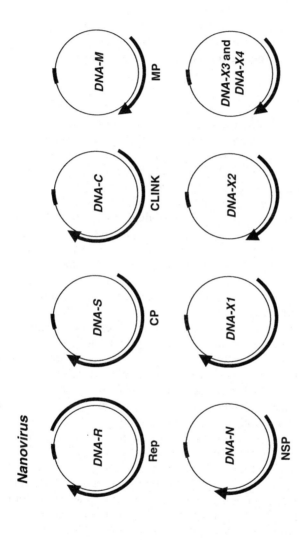

FIGURE 7.2. Genome organization of the genus *Nanovirus*. The circles correspond to the different genomic components of *Faba bean yellow necrotic virus*. Descriptions of protein names are given in the text.

that the name of each circle reflects the protein function encoded. Thus, in the case of FBNYV, DNA-R encodes the Rep protein, functionally equivalent to geminivirus Rep, DNA-S, a structural protein, DNA-C, the Clink (cell cycle link) protein, the DNA-M, the MP, and the DNA-N, the NSP. The remaining circles are named DNA-X1, DNA-X2, and so on, since they encode proteins of yet unknown functions. Nanoviruses are currently being exploited as efficient gene expression vectors in plants.

Family Caulimoviridae

Members of the family *Caulimoviridae* are the only plant viruses whose genome is double-stranded DNA (dsDNA). They are also known as pararetroviruses because RNA intermediates are formed during genome replication and are converted into dsDNA through a reverse transcription step. The family comprises six genera, namely *Caulimovirus, Soymovirus, Cavemovirus, Petuvirus, Badnavirus,* and *Tungrovirus.* Members of the first four genera are transmitted by aphids, have isodiametric (~50 nm) particles and form cytoplasmic inclusion bodies. The badna- and tungroviruses, on the other hand, are transmitted by mealybugs, have bacilliform particles (~130x30 nm) and do not produce inclusion bodies. In members of all the six genera, the genome is a circular dsDNA molecule, ~7,000 to 8,000 nt in length and, typically, the viral DNA contains one ssDNA strand discontinuity in one strand (the one with coding capacity) and one or more in the complementary strand, all of them at specific sites. These sites are important for genome replication during the (–)- and (+)-strand synthesis step.

The gene products of the three better-characterized genera, *Caulimovirus, Badnavirus,* and *Tungrovirus,* are described here (see Figure 7.3). The genome of *Cauliflower mosaic virus* (CaMV), type species of the genus *Caulimovirus,* encodes the following seven open reading frames (ORFs) (I to VII): I, the MP or cell-to-cell (CTC) protein, which associates with the cell wall; II, the insect-transmission factor (helper component, HC); III, the virion-associated protein (Vap), which most likely plays a role in virus assembly; IV, the CP; V, the virus replicase with several enzymatic activities, such as reverse transcriptase (RT), ribonuclease (RH), and aspartate protease (AP); VI, a matrix protein, the transactivator/viroplasmin protein (TAV) of the inclusion bodies; VII, product without a defined function.

The genome of *Commelina yellow mottle virus* (CoYMV), type species of the genus *Badnavirus,* has three ORFs, of which I and II encode products of unknown function while ORF III encodes in its polycistronic RNA the CTC, CP, RT, RH, and AP activities. The genome of *Rice tungro bacilliform virus* (RTBV), type species of the genus *Tungrovirus,* has four ORFs. ORF

Caulimoviridae

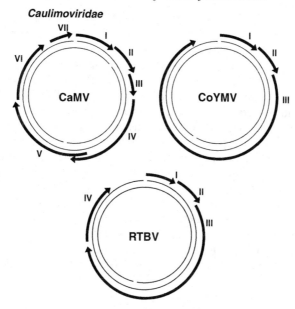

FIGURE 7.3. Genetic organization of family *Caulimoviridae.* The maps depicted correspond to members of each of the three genera of the family *Caulimoviridae,* namely *Cauliflower mosaic virus* (CaMV), a caulimovirus; *Commelina yellow mottle virus* (CoYMV), a badnavirus; and *Rice tungro bacilliform virus* (RTBV), a tungrovirus. Description of protein names are given in the text.

III, comparable to that of badnaviruses, encodes the polyprotein that is processed to give rise to CP, CTC, and the polymerase with RT and RH activities. ORF IV is expressed from the 35S RNA and may be involved in the control of RTBV genome expression. The functions of the expressed products of ORFs I and II are not known.

TRANSCRIPTION STRATEGIES

Geminiviridae

The transcriptionally active template is the dsDNA intermediate formed early after infection. Geminivirus transcription is complex and leads to multiple RNA species. The expression strategies of the different geminiviruses have in common that transcription is always bidirectional from two promoters (c-sense and v-sense promoters) located close to the ends of the intergenic regions. It depends on cellular RNA polymerase II, but differences exist in the form in which mastre- and begomoviruses express their genomes and regulate the process.

Most of the information available for mastreviruses is derived from studies with MSV, WDV, and *Digitaria streak virus* (DSV). One characteristic feature of the mastrevirus gene expression strategy is the presence of both unspliced and spliced transcripts. The c-sense promoter directs the synthesis of two 5'-coterminal mRNAs that encode the RepA (unspliced RNA) and Rep (spliced RNA) proteins. In some cases, heterogeneity at the 3' end also appears. In any case, this splicing event is rather inefficient and the processed mRNA represents a small proportion of the total c-sense transcript population. At least for MSV and DSV, the v-sense promoter, which lacks a good TATA box, directs the synthesis of two 3'-coterminal transcripts: a more abundant and smaller mRNA that encodes the CP, and a larger mRNA from which the MP is translated. However, in the case of WDV a single bicistronic mRNA is produced. In MSV, an intron located within the *MP* gene is also spliced out in some transcripts. The organization of the v-sense transcripts poses the question of how *CP* is translated and it has been proposed that some kind of frameshifting may occur.

The presence of large WDV Rep-DNA complexes detected in vitro around the c-sense and v-sense promoters (see next section) is consistent with Rep's role in modulating gene expression: repression of the c-sense promoter and activation of the v-sense promoter. However, this is still a matter that needs to be fully demonstrated. One possibility is that Rep, RepA, or both participate in regulating mastrevirus gene expression. In this context, it may be relevant to mention the property of RepA and Rep to form both homo- and hetero-oligomers.

Begomoviruses, TGMV for instance, produce a complex mixture of six and four mRNAs derived from the DNA A and B components, respectively. A single v-sense transcript is produced from each component. In contrast, c-sense transcription is more complex, with all mRNAs being 3'-coterminal. In the case of DNA A, the largest mRNA is the only one capable of translating full-length Rep, while two other RNAs contain the AL4 ORF. The two smallest RNAs are bicistronic and encode TrAP and REn. The TGMV promoter that produces the longest transcript (named AL61) is a strong promoter. A region of ~60 base pairs upstream from the transcription start site, where the TATA box and a G box are located, is required for full promoter function. Interestingly, this region overlaps the origin of DNA replication active during the rolling-circle step (see also Genome Replication). TGMV Rep autoregulates its own promoter in a virus-specific manner through binding to a region located between the transcription start site and the TATA box. It has been suggested that this Rep-mediated repression occurs through interference with the transcription machinery. The strategy delineated for TGMV may also be followed by other begomoviruses, as the latter have similarly organized c-sense and v-sense promoter regions. V-sense promoters in both genomic DNA components are transactivated by TrAP in a replication-independent manner.

Detailed studies of the transciption in nanoviruses are not available yet.

Caulimoviridae

The CaMV genome is expressed from two promoters, the 35S and the 19S. The 35S promoter, widely used for plant transformation, produces an RNA species which is ~180 nt longer that the viral genome length (the so-called pregenomic DNA). It encodes all the ORFs except ORF IV, encoded in the 19S transcript (the subgenomic DNA). In addition, other RNA species are formed whose functional significance is not clear yet. The CaMV 35S promoter has a modular organization. One domain constitutes the core minimal promoter while others, namely domains A and B, confer root- and leaf-specific expression, respectively. Domain B is further divided into several subdomains (B1 through B5). The tissue-specific expression seems to depend on the interaction with specific cell factors. Among these, transcription factor ASF1 (or TGA1) binds the TGACG sequence in domain A, ASF2 interacts with GATA sequences of subdomain B1 and the CA-rich binding factor to subdomain B3. The CaMV 19S promoter is rather weak compared to the 35S promoter and lacks the complex enhancer domain structure.

The RTBV genome is different from CaMV in that it is expressed from a single, polycistronic, longer-than genome 35S RNA, which is the template for both translation and reverse transcription. The promoter contains a core region and, at least, two elements that stimulate transcription: one is position-dependent and is located close to the transcription start site and the other, a CT-rich sequence, is position- and orientation-independent. Cellular factors interact with this region to regulate transcription. For instance, rice nuclear factors (RNFG1 and 2) seem to confer a tissue-specific expression that is stronger in rice shoots than in roots. ORFs I, II, and III are translated from this unspliced, long RNA. Typically, ORF I has an AUU start codon, instead of the standard AUG codon, present both in ORF II and III. However, translation of ORFs I and II is rather inefficient (~10 percent relative to ORF III). An ORF IV is translated after splicing the 35S RNA, which releases a large ~6 kb piece. This splicing event brings ORF IV close to the 5'-end of the spliced RNA. In any case, this spliced transcript represents a minor proportion of all RTBV transcripts.

Another characteristic feature of the pararetrovirus 35S RNA is the presence of several small ORFs in its 5'-leader sequence. These small ORFs modulate translation of the main ORF, which occurs by a "shunt" mechanism instead of the standard scanning process from the 5' end of the mRNA.

GENOME REPLICATION

Geminiviridae

Geminiviruses (and most likely members of the family *Nanoviridae*) are the only plant viruses whose genome replicative cycle relies entirely on DNA intermediates and occurs in the nucleus of the infected cell (Figure 7.4). The geminivirus replication cycle consists of two distinct stages. During the first one, the genomic ssDNA is converted into a covalently closed, supercoiled dsDNA molecule and depends exclusively on cellular factors. Not only the transport of genomic ssDNA to the nucleus but also the mechanistic details of this replication stage are very poorly understood in molecular terms. It requires a priming step at the so-called (–)-strand (or c-sense) origin. In mastreviruses this is located at the SIR in a region where a ~80 nt-long DNA molecule is already annealed to the encapsidated ssDNA. The presence of ribonucleoside monophosphate at its 5' end strongly suggests that it is synthesized by a primase activity of cellular origin before encapsidation. In other geminiviruses the situation is different, since genomic ssDNA does not contain any annealed piece of DNA and priming at this stage seems to occur in the common region. The minimal *cis*-acting signals that define a functional origin at this stage are completely unknown.

The second stage uses the dsDNA intermediate as a template for genome amplification through a rolling-circle replication (RCR) mechanism. The initial priming event depends on the interaction of the viral initiator protein (Rep for both gemini- and nanoviruses) with *cis*-acting signals of the genetically defined origin. An integral part of it is the 9 nt invariant sequence (TAATATT↓AC) present in all geminiviruses sequenced to date (the arrow indicates the initiation site). This sequence is flanked by inverted repeats of various lengths which, by unknown mechanisms, contribute to form a stem-loop structure to which Rep can get access and carry out the initiation reaction, a single-strand, site-specific, endonucleolytic cleavage that provides a free 3'-OH primer-terminus for further elongation during the RCR stage. Geminiviruses differ in the mechanism used for initiation of DNA replication during RCR as they have different sequence requirements, form different DNA-protein complexes, and require the participation of different viral and, perhaps, cellular proteins.   •

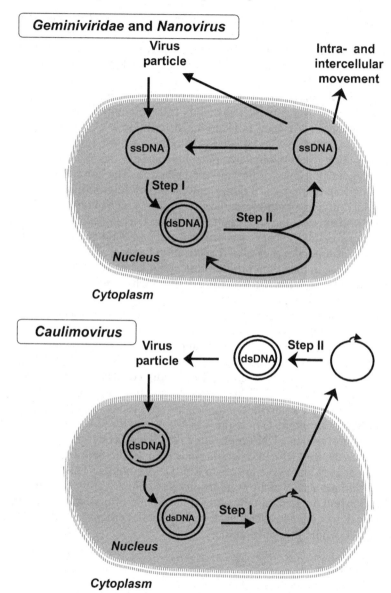

FIGURE 7.4. Replication strategies of plant DNA viruses. Simplified schemes of the DNA replication cycles of gemini- and nanoviruses (upper panel) and that of members of the family *Caulimoviridae* (lower panel).

Mastreviruses require a relatively large region of the LIR for origin activity consisting of three main modules: the ~200 bp (base pair) core, absolutely required and containing the stem-loop at its 3' end, flanked by two auxiliary sequences that stimulate DNA replication. Rep forms two kinds of complexes with the origin of DNA replication. Two of them are the high-affinity C and V complexes formed around the c-sense and v-sense promoters, respectively. RepA, which shares its DNA binding domain with Rep, also forms these complexes at locations differing in a few base pairs; it is not known whether they modulate Rep binding. A third, low-affinity Rep-DNA complex (the O-complex) assembles around the stem-loop and is most likely responsible for the initiation reaction. The current model proposes that Rep (perhaps as a dimer) interacts with high-affinity sequences and rapidly oligomerizes to form the C-complex, which in turn would interact with sequences at the stem-loop to form the O-complex. After a free 3'-OH primer-terminus appears as a product of the initiation reaction, Rep facilitates recruitment of the cellular replication factor C. This is most likely recruited as a complex with proliferating cell nuclear antigen (PCNA), the stimulatory factor of DNA polymerase delta which could be responsible for elongation in a processive manner.

The begomovirus origin consists of the stem-loop and sequences upstream, but not downstream, of this structure. Rep forms a high-affinity oligomeric complex in a region overlapping the c-sense promoter, which may be structurally and functionally similar to the C-complex identified in mastreviruses. Sequences mediating Rep binding consist of two copies of a 5 bp motif separated by a 3 bp spacer. In addition, three other DNA sequences, which act as auxiliary regions, are required for full activity of the begomovirus origin: the G box, the TATA box, and the CA motif. The viral REn protein stimulates DNA replication, and it has been proposed that this depends on the Rep-REn interaction, since REn would bind simultaneously to the stem-loop and to the upstream Rep-DNA complex. This mechanism would facilitate access of the high-affinity Rep complex to the initiation site at the loop and, in general terms, is similar to that of mastreviruses. Since the latter do not encode a REn protein, Rep itself, perhaps in collaboration with RepA, would play such a role.

Caulimoviridae

The replication cycle of CaMV and RTBV has been studied in detail. All members of the family *Caulimoviridae* seem to follow similar replication strategies that consist of two stages (see Figure 7.4). First, after the genomic dsDNA is transported to the nucleus by an unknown mechanism, the ssDNA discontinuities are sealed, and ds DNA is transcribed by the host RNA polymerase II. Then, the genomic 35 S RNA transcript is transported to the cytoplasm, where, in addition to serving as a template for translation of viral products, it acts as a template for reverse transcription, a process mediated by the virus-encoded reverse transcriptase (the CaMV ORF V). This second stage includes the elimination of the RNA strand and the conversion of the remaining ssDNA circle into a dsDNA genomic DNA. In the case of CaMV this replication step occurs in the cytoplasmic inclusion bodies, where the ORF VI gene product is important.

Two types of replication intermediates are formed—one with the typical nucleic acid nature and another that resembles encapsidated DNA. An early replication intermediate product consists of a 600 bp-long DNA fragment of (–)-strand polarity which maps to a region of the CaMV genome located between the 5' end of the 35S RNA and the (–)-strand discontinuity. This DNA fragment contains in its 5' end a covalently attached 75 nt long RNA molecule that was identified as tRNA^{met}. This represents the remainder of the priming event for reverse transcription, a complex and unique process where the 3'-OH for the reverse transcriptase is provided by the tRNA^{met} molecule. While the DNA strand is synthesized, the process requires an associated RNase H activity, which resides in the gene V product, to degrade the RNA template. Then, synthesis of the second DNA strand is discontinuous, thus allowing the appearance of the site-specific nicks.

Localization studies indicate that the inclusion bodies are the cellular compartments where CaMV DNA synthesis occurs. Such large replication complexes include, in addition to the inclusion bodies' matrix protein, the CP, the (–)-strand primer and the 35S RNA template. The interactions among these components facilitate the positioning of the RNA-dependent DNA polymerase moiety of gene V product in the right place to initiate (–)-strand synthesis and elongate the primer.

In broad outline, the RTBV replication cycle resembles that of CaMV, including the two steps (transcription and reverse transcription) and the use of similar replication intermediates, e.g., the small DNA molecule linked to the tRNA^{met} molecule. However, there is no evidence (for RTBV and badnaviruses) of cytoplasmic inclusion bodies functioning as replication factories.

REGULATION OF DNA REPLICATION/ VIRUS-HOST INTERACTIONS

The requirements for host cell functions are quite different in pararetroviruses and in ssDNA viruses (geminivirus and members of the *Nanoviridae*).

This is dictated mainly by the kind of proteins encoded by the virus and those required from the host cell. In the case of pararetroviruses, the replication cycle involves a host factor, the RNA polymerase II, ubiquitously present in all cells, and the virus-encoded reverse transcriptase. In the second case, however, viral genome replication factors are not encoded by the virus and, except for the initiator Rep protein, viral genome replication relies entirely on host cell proteins. Most, if not all of them, participate in host chromosomal replication, a process that takes place exclusively in proliferating cells, and are not functional in resting, differentiated cells. However, geminivirus DNA replication is excluded from meristems where proliferating cells are located—a situation suggesting that geminivirus infection somehow must promote a cellular state permissive for viral DNA replication, as animal oncoviruses do. This view was supported by early revealing experiments showing that (1) PCNA, an accessory factor of DNA polymerase delta, accumulates in TGMV-infected, Rep-expressing, differentiated cells and (2) WDV RepA interacts with members of the retinoblastoma (RB) tumor suppressor family, both of human and plant origin. In higher eukaryotes, cell cycle reactivation of quiescent cells and the G1 to S transition in proliferating cells is controlled by the RB family of proteins (known as RB-related [RBR] in plants). They modulate the expression of genes required in later stages of the cell cycle and are regulated by the E2F/DP family of transcription factors. In any case, studies with different geminiviruses clearly point to a rather complex situation regarding the interference with cell cycle regulatory components, where more than one mechanism seems to be operative. It remains to be determined whether the different mechanisms (see the following section) have evolved in different viruses or they are redundant for a particular virus.

Mastrevirus RepA interacts with RBR via a LxCxE amino acid motif, which is important for WDV DNA replication in cultured wheat cells. Other mastreviruses, for instance, MSV and BeYDV, also use a similar RBR interaction motif but, intriguingly, RepA mutations that impair RBR binding have no major effects on virus infectivity in plants. Begomoviruses do not encode any RepA homologous protein, but it is their Rep protein that is responsible for interaction with RBR via a non-LxCxE motif. In this case, and contrary to the situation in mastreviruses, point mutations that destroy RBR binding highly impair accumulation of viral DNA and lead to an altered tissue-specific pattern. In fact, begomovirus Rep-mediated accumulation of PCNA depends on a complex effect on the PCNA promoter activity, largely overcoming E2F-mediated repression. Begomovirus REn is another viral protein that may affect the cell cycle, since it also can bind RBR and Rep. Moreover, the REn-binding and the RBR-binding domains of Rep are closed and partially overlap, suggesting that competition between these viral proteins may occur. Clearly, more work is needed in this area to obtain a coherent and global picture of the interplay between geminivirus DNA replication and host cell proliferation.

The other group of ssDNA viruses belonging to the *Nanoviridae* also seems to interfere with RBR while using a Rep-independent mechanism. Although the amount of information is still rather limited, it is clear that the nanovirus FBNYV encodes a 20 kDa LxCxE-containing protein, named Clink (for cell-cycle link) which can bind RBR. Furthermore, Clink also contains an F-box that mediates its binding to the Skp1 protein, an integral component of SCF complexes involved in recruiting target proteins for specific degradation by the proteasome. These complexes, which are named after three of their components, Skp1, Cullin 1, and F-box, enzymtically add ubiquitil residues to certain proteins, which have been targeted to these complexes through their specific interaction with a particular F-box protein. The functional significance of the interactions of nanovirus Clink with these components of the proteolytic machinery of the cell during nanovirus DNA replication awaits further study.

VIRUSES WITH INCOMPLETE GENOMES

A common feature of many RNA viruses, both of plant and animal origin, is the presence of the so-called satellite or defective genomes. These are generally characterized by the presence of the minimal *cis*-acting sequences required for genome replication and encoding a variable number of viral ORFs, as well as by being maintained in association with the full-length viral genomes. Recently, the identification of smaller-than-genomic DNA molecules, in the case of geminivirus infection, has expanded our view on the presence of defective genomes in plant DNA viruses also. We are just starting to gain information on this topic and very little is known about the significance of these associated small genomes. As a general rule, they all contain the stem-loop structure with the invariant 9 nt present in all geminiviruses. In some cases, they interfere with replication of full-length genomes and, consequently, can be considered as defective interfering particles similar to those associated with many animal RNA viruses. In other cases, they seem to have an effect during virus infection. For example, the presence of defective molecules, known as DNA beta, are associated with an enhancement of symptom development, as has been demonstrated for yellow vein disease in *Ageratum* and in cotton leaf curl disease. The mechanism behind such effects is not fully understood but, at least in the case of cotton leaf curl disease, it seems that DNA beta may act as a suppressor of host gene silencing, a host defense response, thus allowing an increase of viral DNA. Another observation that increases the complexity of this field is the presense of defective DNA molecules (known as DNA 1) associated with the begomoviruses (i.e., *Ageratum yellow vein virus* and *Cotton leaf curl virus* cultures). In this case, the small circles replicate autonomously but instead of having the invariant 9 nt sequence they contain a sequence that is present in the nanovirus genomic circles. Since they encode a nanovirus Replike protein, the origin of such molecules is not clear.

BIBLIOGRAPHY

Aronson MN, Complainville A, Clerot D, Alcalde H, Katul L, Vetten HJ, Gronenborn B, Timchenko T (2002). In planta protein-protein interactions assessed using a nanovirus-based replication and expression system. *Plant Journal* 31: 767-775.

Boulton MI (2002). Functions and interactions of mastrevirus gene products. *Physiological and Molecular Plant Pathology* 60: 243-255.

Egelkrout EM, Mariconti L, Settlage SB, Cella R, Robertson D, Hanley-Bowdoin L (2002). Two E2F elements regulate the proliferating cell nuclear antigen promoter differently during leaf development. *Plant Cell* 14: 3225-3236.

Gutierrez C (1999). Geminivirus DNA replication. *Cellular and Molecular Life Sciences* 56: 313-329.

Gutierrez C (2000). DNA replication and cell cycle in plants: learning from geminiviruses. *EMBO Journal* 19: 792-799.

Gutierrez, C (2002). Strategies for geminivirus DNA replication and cell cycle interference. *Physiological and Molecular Plant Pathology* 60:219-230.

Hanley-Bowdoin L, Settlage SB, Orozco BM, Nagar S, Robertson D (1999). Geminiviruses: Models for plant DNA replication, transcription and cell cycle regulation. *Critical Reviews in Plant Sciences* 18: 71-106.

Harrison BD, Robinson DJ (2002). Green shoots of geminivirology. *Physiological and Molecular Plant Pathology* 60: 215-218.

Kobayashi K, Tsuge S, Stavolone LK, Hohn T (2002). The cauliflower mosaic virus virion-associated protein is dispensable for viral replication in single cells. *Journal of Virology* 76: 9457-9464.

Kong LJ, Hanley-Bowdoin L (2002). A geminivirus replication protein interacts with a protein kinase and a motor protein that display different expression patterns during plant development and infection. *Plant Cell* 14: 1817-1832.

Liu Y, Robinson DJ, Harrison BD (1998). Defective forms of cotton leaf curl virus DNA-A that have different combinations of sequence deletion, duplication, inversion and rearrangement. *Journal of General Virology* 79: 1501-1508.

Luque A, Sanz-Burgos AP, Ramirez-Parra E, Castellano MM, Gutierrez C (2002). Interaction of geminivirus Rep protein with replication factor C and its potential role during geminivirus DNA replication. *Virology* 302: 83-94.

Nagar S, Hanley-Bowdoin L, Robertson D (2002). Host DNA replication is induced by geminivirus infection of differentiated plant cells. *Plant Cell* 14: 2995-3007.

Palmer KE, Rybicki EP (1998). The molecular biology of mastreviruses. *Advances in Virus Research* 50: 183-234.

Ryabova LA, Pooggin MM, Hohn T (2002). Viral strategies of translation initiation: Ribosomal shunt and reinitiation. *Progress Nucleic Acids Research and Molecular Biology* 72: 1-39.

Turnage MA, Muangsan N, Peele CG, Robertson D (2002). Geminivirus-based vectors for gene silencing in *Arabidopsis*. *Plant Journal* 30: 107-114.

Chapter 8

Viroids

Ricardo Flores
Vicente Pallás

STRUCTURE

Primary Structure

Viroids are small (between 246 and 401 nucleotides [nt], excluding those with sequence repeats), single-stranded, circular RNAs without any apparent protein-coding capacity that are able to infect certain plants and to incite in most cases pathological alterations (see Table 8.1). Because of their minimal size, viroids can be regarded as the current lowest step of the biological scale. The 29 viroid species sequenced so far (Table 8.1) have been grouped within two families. Members of the family *Pospiviroidae,* whose type species is *Potato spindle tuber viroid* (PSTVd), are characterized by a central conserved region (CCR) and either a terminal conserved region (TCR) or a terminal conserved hairpin (TCH), but they lack hammerhead ribozymes (see Figure 8.1). Conversely, members of the family *Avsunviroidae,* whose type species is *Avocado sunblotch viroid* (ASBVd), lack any of these three conserved motifs but can adopt hammerhead structures in both polarity strands and self-cleave in vitro and in vivo accordingly (see also Replication). Within the family *Pospiviroidae,* the conserved motifs, particularly the type of CCR that is formed by two sets of conserved nucleotides flanked by an imperfect inverted repeat in the upper strand, serve to classify viroid species into five genera. The four known species of the family *Avsunviroidae* are allocated to three genera according to their base composition, global secondary structure and hammerhead architecture.

Secondary Structure

Convincing evidence of a different nature supports the view that PSTVd, and by extension the other species of its family, adopts in vitro a rodlike (or

TABLE 8.1. Viroid taxonomy with abbreviations and size of typical sequence variants.

Family	Genus	Species	Abbreviation	Size (nt)
Pospiviroidae	Pospiviroid	Potato spindle tuber[a]	PSTVd	356, 359-360
		Tomato chlorotic dwarf	TCDVd	360
		Mexican papita	MPVd	359-360
		Tomato planta macho	TPMVd	360
		Citrus exocortis[b]	CEVd	370-375, 463
		Chrysanthemum stunt	CSVd	354, 356
		Tomato apical stunt	TASVd	360, 363
		Iresine 1	IrVd-1	370
		Columnea latent	CLVd	370, 372
	Hostuviroid	Hop stunt[c]	HSVd	295-303
	Cocadviroid	Coconut cadang-cadang	CCCVd	246-247, 287-301
		Coconut tinangaja	CTiVd	254
		Hop latent	HLVd	256
		Citrus IV	CVd-IV	284
	Apscaviroid	Apple scar skin[d]	ASSVd	329-330
		Citrus dwarfing	CDVd	294, 297
		Apple dimple fruit	ADFVd	306-307
		Grapevine yellow speckle 1	GYSVd-1	366-368
		Grapevine yellow speckle 2	GYSVd -2	363
		Citrus bent leaf	CBLVd	318
		Pear blister canker	PBCVd	315-316
		Australian grapevine	AGVd	369
	Coleviroid	Coleus blumei 1	CbVd-1	248, 250-251
		Coleus blumei 2	CbVd-2	301-302

Family	Genus	Species	Abbreviation	Size (nt)
		Coleus blumei 3	CbVd-3	361-362, 364
Avsunviroidae	*Avsunviroid*	*Avocado sun-blotch*	ASBVd	246-250
	Pelamoviroid	*Peach latent mosaic*	PLMVd	335-338
		Chrysanthe-mum chlorotic mottle	CChMVd	398-401
	Flaviroid	*Eggplant latent*	ELVd	332-335

[a]The complete name of each species is followed by the word *viroid* (e.g., *Potato spindle tuber viroid*).
[b]Agent also of Indian tomato bunchy top.
[c]Agent also of cucumber pale fruit, plum dapple, peach dapple, and citrus cachexia.
[d]Agent also of dapple apple, pear rusty skin, and Japanese pear fruit dimple.

quasi-rodlike) secondary structure formed by alternating short double-stranded regions and single-stranded loops. This class of secondary structure most likely also exists in vivo, at least during certain stages of the biological cycle, because the repetitions and deletions observed in some viroids occur in such a way that the rodlike structure is preserved. From an analysis of sequence similarities, the rodlike structure of members of the family *Pospiviroidae* has been divided into five structural domains: central (C), pathogenic (P), variable (V) and terminal left and right (T_L and T_R, respectively) (see Figure 8.1, part A). The C domain contains the CCR, whereas the TCR and TCH are located in the T_L domain. Some of these structural domains have been involved in determining specific functions (e.g., the C domain in certain replication steps and the P domain in pathogenicity in PSTVd and closely related viroids). However, the situation is probably more intricate, with some functional roles being regulated by discrete determinants residing in different domains.

In the family *Avsunviroidae*, ASBVd also appears to adopt a quasi-rodlike secondary structure, but *Peach latent mosaic viroid* (PLMVd) and *Chrysanthemum chlorotic mottle viroid* (CChMVd) fold into a branched structure (see Figure 8.1, part B). This complex structure, composed of hairpins and double-stranded regions separated by internal loops, most likely also exists in vivo because the sequence heterogeneity found in a high number of natural variants (particularly CChMVd) does not affect the branched conformation; the changes map at the loops or, when located in stems, co-variations or compensatory mutations preserve their stability. Therefore,

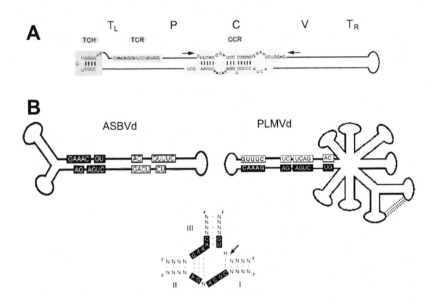

FIGURE 8.1. Structural models for viroids. (A) Rodlike secondary structure proposed for members of family *Pospiviroidae*. The approximate locations of the five structural domains C (central), P (pathogenic), V (variable), and T_L and T_R (terminal left and right, respectively) are indicated. The sequence and location of the TCH (terminal conserved hairpin, present in genera *Hostuviroid* and *Cocadviroid*), TCR (terminal conserved region, present in *Pospiviroid* and *Apscaviroid* genera, and in CbVd2 and CbVd3) and CCR (central conserved region, here presented in that of genus *Pospiviroid*) are shown. Arrows denote flanking sequences that, along with the core nucleotides of the upper CCR strand, form imperfect inverted repeats. (B) Quasi-rodlike and branched secondary structures proposed for *Avocado sunblotch viroid* (ASBVd) and *Peach latent mosaic viroid* (PLMVd) of family *Avsunviroidae,* with nucleotide residues conserved in most natural hammerhead structures shown within boxes with black and white backgrounds for plus and minus polarities, respectively. Broken lines in PLMVd indicate a pseudoknot interaction between two loops. Below, consensus hammerhead structure with nucleotide residues conserved in most natural hammerhead structures shown within boxes with black background and the arrowhead marking the self-cleavage site. H indicates A, C, or U, and N any residue; continuous and broken lines denote Watson-Crick and non-Watson-Crick base pairs, respectively. The central core is flanked by three helixes I, II and III; in most natural hammerhead structures helixes I and II are closed by short loops.

the paradigm that considered the rodlike or quasi-rodlike secondary structure a universal property of all viroids is no longer tenable.

Other Elements of Higher Order Structure

In vitro, viroids display structural transitions during thermal denaturation, with the most stable conformation being unfolded and other metastable conformations appearing. This is the case of PSTVd, in which the formation of a series of three hairpins (I, II, and III) is well documented. Some of these hairpins most likely also form in vivo, particularly during replication, and play an important role. In fact, hairpin I, which involves the conserved nucleotides of the CCR upper strand and the flanking imperfect inverted repeat (Figure 8.1, part A), is structurally conserved in all members of the family *Pospiviroidae*. Non-Watson-Crick interactions are also present in viroid RNAs. A first example is a special loop containing nonstandard interactions termed loop E, initially identified in 5S rRNA, which is part of the binding site of two proteins. This loop has been mapped at the CCR of PSTVd and proposed to have a role in the final ligation step of the replication cycle that leads to the progeny of circular viroid molecules. A second example is illustrated by members of the family *Avsunviroidae,* which are able to adopt hammerhead structures also stabilized by a complex array of non-Watson-Crick interactions (Figure 8.1, part B). Finally, pseudoknot elements (Watson-Crick interactions between a hairpin loop and a single-stranded region that can be the loop of another hairpin, which are then termed kissing loops), have also been identified in PLMVd and CChMVd, where they may contribute to the stabilization of the branched conformation of these two viroids (Figure 8.1, part B).

REPLICATION

Cellular Biology

Increasing evidence supports the view that the nucleus and the chloroplast are the accumulation sites for viroids of the family *Pospiviroidae* and *Avsunviroidae,* respectively. Initial studies using cell-fractionation techniques, particularly differential centrifugation, later confirmed by in situ hybridization combined with confocal laser scanning and transmission electron microscopy, showed that PSTVd accumulates in the nucleus and ASBVd in the chloroplast. Results also indicate that the other members of the two families behave in a manner similar to their corresponding type species. Moreover, additional data on the localization of the viroid complementary strands that are synthesized during replication (see sections addressing molecular biology and movement) support the notion that PSTVd replicates in the nucleus and ASBVd in the chloroplast (see Figure 8.2) and that this is also likely the case with the other members of both families.

Molecular Biology

As with any other replicating nucleic acid, viroid replication can be viewed from two main perspectives (although they are not mutually independent): the templates that are transcribed and the enzymes that catalyze these reactions, as well as others dealing with the subsequent processing of the primary transcripts. Viroids replicate through a rolling-circle mechanism in which the infecting monomeric circular RNA (arbitrarily referred to as having the plus polarity) is transcribed by a DNA-dependent RNA polymerase (subverted to act on an RNA template) into head-to-tail (–) multimers that, on their turn, serve as the template for a second RNA-RNA transcription step. The resulting head-to-tail (+) multimers are then precisely cleaved into unit-length strands and subsequently ligated to give rise to the final progeny of monomeric (+) circular RNAs; these two steps are catalyzed by RNase and RNA ligase activities, respectively, some of them with unexpected characteristics. This is the asymmetrical version of the rolling-circle mechanism that is followed by PSTVd and the other members of the family *Pospiviroidae*. In contrast, ASBVd and the other members of the family *Avsunviroidae* follow an alternative symmetrical version in which the head-to-tail (–) multimers are processed to their monomeric circular counterparts that then serve as the template for the second part of the replication cycle, which is symmetric to the first (Figure 8.2). Therefore, the demarcating difference between both versions is the nature of the (–) template: whereas the existence of the monomeric (–) circular RNA has been detected in ASBVd-infected avocado, attempts to identify this form in PSTVd-infected tissues have been unsuccessful.

Since viroids do not code for proteins of their own, the enzymology of viroid replication is determined by the cell compartment where this process occurs. On the basis of the effects of α-amanitin in in vivo and in vitro systems, the enzyme catalyzing the elongation of RNA strands of PSTVd and closely related viroids appears to be the nuclear RNA polymerase II. Results obtained with a different experimental approach, the use of a monoclonal antibody against a conserved domain of the major subunit of RNA polymerase II, also support this view. Data about members of the family *Avsunviroidae* are more recent and limited and suggest the involvement of a nuclear-encoded chloroplastic RNA polymerase (or another polymerase

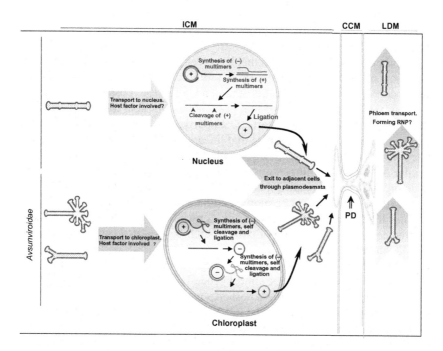

FIGURE 8.2. Schematic representation of the different steps occurring during viroid infection. The incoming viroid RNA, which adopts a rodlike (or quasi-rod-like) secondary structure in species of the family *Pospiviroidae* and in *Avocado sunblotch viroid* or a branched structure in the other two species of the family *Avsunviroidae*, travels to the replication site: the nucleus or the chloroplast. Members of the family *Pospiviroidae* replicate through the asymmetric pathway of the rolling-circle mechanism, whereas members of the family *Avsunviroidae* follow the alternative symmetric pathway in which (+) and (−) multimers self-cleave to unit length by hammerhead ribozymes. The demarcating difference between both pathways is whether or not the monomeric (−) circular template exists. The resulting progeny moves to the cell periphery in a cytoskeleton-independent manner (intracellular movement, ICM), and then, through plasmo-desmata (PD), to the adjacent cells (cell-to-cell movement, CCM). In the final step, the viroid RNA reaches the vascular tissue, through, which it is trans-located to the distal plant parts (long-distance movement, LDM). This latter class of movement is thought to occur through the formation of ribonucleoprotein complexes (RNP) with phloem proteins.

resistant to the inhibitor tagetitoxin) in ASBVd, and of a plastid-encoded RNA polymerase in PLMVd. In any case, it should be noted that under normal physiological conditions these enzymes transcribe DNA templates, and are redirected by viroids to accept RNA templates; how they manage to facilitate this template switch is an intriguing issue that remains unsolved.

Another interesting question that has recently been addressed is whether initiation of viroid RNA synthesis is promoter-driven and, consequently, site-specific, or whether it occurs at random, a feasible possibility given the circular nature of the template that permits its complete transcription irrespective of where synthesis starts. Data derived from in vitro capping and RNase protection assays indicate that linear ASBVd (+) and (–) RNAs isolated from infected avocado begin with a UAAAA sequence that maps at similar A+U–rich terminal loops in their predicted quasi-rodlike secondary structures. However, results obtained with an in vitro system in which a nuclear extract from potato was supplemented with the PSTVd monomeric (+) circular RNA have identified two initiation sites for the PSTVd (–) strands located 15 to 16 nt upstream of two GC-rich base-paired boxes in the rodlike secondary structure.

Regarding the RNase activity catalyzing the cleavage of the multimers of (+) or both polarity strands, the almost general view assumes that in members of the family *Pospiviroidae* it is a host enzyme whose site-specificity is determined by the spatial folding of the RNA substrate. In members of the family *Avsunviroidae,* however, the RNase activity is not an enzyme but a ribozyme (a catalytic RNA) embedded in the (+) and (–) strands. These ribozymes belong to the class termed hammerhead structures—a small RNA motif that self-cleaves in the presence of Mg^{2+} at a specific phosphodiester bond, producing 5'-hydroxyl and 2',3'-cyclic phosphodiester termini—which are composed of a central core of several conserved sequences flanked by three double-stranded regions with loose sequence conservation closed by loops (Figure 8.1). X-ray crystallography has revealed a complex array of non-Watson-Crick interactions between the residues of the central core that explains why they are strictly conserved in natural hammerhead structures. Strong evidence supports the involvement of hammerhead structures in the in vivo self-cleavage of multimeric viroid RNAs containing these catalytic domains. Recent results indicate that this processing is stimulated by certain host proteins that behave as RNA chaperones (proteins that facilitate proper RNA folding).

Data on the third catalytic activity, the RNA ligase mediating the circularization of the monomeric linear strands, are scarce. Whereas a host nuclear enzyme with similar properties to the wheat germ RNA ligase appears to be involved in PSTVd, two possibilities have been advanced for PLMVd: a wheat germ RNA ligaselike chloroplastic enzyme, considering that such an enzyme requires the 5'-hydroxyl and 2',3'-cyclic phosphodiester termini produced by hammerhead ribozymes or, alternatively, an autocatalytic self-ligation leading to unusual 2',5'-phosphodiester bonds at the ligation sites.

MOVEMENT

To systemically invade plants, viroids must be able to move from the initially infected cells to the surrounding ones and then to the vascular system, through which they reach the most distal parts. Three different types of movement can be considered: intracellular, cell to cell, and long distance (Figure 8.2).

Intracellular Movement

The incoming viroid RNA, when reaching the first cells that eventually become infected, must first move to the replication site; the resulting progeny, in turn, has to move then to the cell periphery. Studies with permeabilized protoplasts have demonstrated the import of PSTVd into the nucleus, but not of two viroids that replicate and accumulate in the chloroplast. These studies have also revealed that the nuclear import of PSTVd is a cytoskeleton-independent process mediated by a specific and saturable receptor. The mechanism by which members of the family *Avsunviroidae* are transported into the chloroplast is unknown.

Cell-to-Cell Movement

As an intermediate step to establish a systemic infection, viroids must move from one cell to the adjacent ones. This intercellular transport occurs through the plasmodesmata (PD), the structural units that connect plant cells and allow the traffic of macromolecules such as nucleic acids and proteins (Figure 8.2). Microinjection of PSTVd into symplastically connected mesophyll cells has shown that the viroid moves rapidly from cell to cell, whereas a 1,400 nt RNA containing only vector sequences is unable to move out of the injected mesophyll cells. Interestingly, a fusion of PSTVd to a nonmobile RNA was able to confer PD transport upon the latter, strongly suggesting that the viroid molecule possesses a specific sequence or structural motif for PD transport.

Long-Distance Movement

This third kind of movement permits infection of the distal plant parts and occurs through the vascular system; more specifically, via the phloem. Viroids, as most plant viruses, move following the net flow of photosynthetic products in the phloem (i.e., from fully expanded leaves upward to young developing leaves and to the shoot tip, and downward to the roots). The (–) polarity strand of PSTVd has been detected in the phloem, strongly suggesting that the long-distance transport of this viroid is likely sustained by replication in the phloem. In addition, PSTVd trafficking is governed by plant developmental and cellular factors.

Recently, it has been demonstrated that *Hop stunt viroid* (HSVd) can form a ribonucleoprotein complex with a phloem protein from cucumber in vitro and in vivo. This protein, the phloem protein 2 (PP2), is a dimeric lectin able to move from cell to cell through PD and toward sink tissues with the assimilate stream. These properties of PP2, together with its newly reported RNA-binding ability, makes it an excellent candidate to facilitate the systemic movement of viroids and, possibly, of other RNAs in vivo. Supporting this view, recent results indicate that PP2 is able to translocate HSVd through intergeneric grafts.

PATHOGENESIS

One of the most intriguing aspects of viroids is how these small noncoding RNAs are able to elicit a pathogenic effect on certain plant hosts. The underlying mechanisms must be in principle diverse, considering that members of both viroid families replicate (and accumulate) in different subcellular sites. Dissecting the influence of the two components of the viroid-host interaction is significantly easier from the viroid side due to its minimal genomic size. The substitution of only one nucleotide has been reported to alter the host range of PSTVd, and sequence changes of 3 to 4 nucleotides are sufficient to modify the phenotypic effects, transforming a severe strain into a latent one. These changes, however small, may have important effects on the spatial conformation of the viroid RNA. Mapping the regions where these changes occur has led to the characterization of a virulence-modulating region in PSTVd and of a pathogenicity determinant in CChMVd. Since these changes do not seem to influence replication, because viroid titers remain essentially the same, it appears more likely therefore that they might affect the interaction of the viroid RNA with a host factor, either a cellular RNA or a protein. Although the first possibility cannot be discarded, the second appears more probable. In this context it has been proposed that viroids could exert their pathogenic effects by interfering with the normal functioning of the plant RNA polymerases needed for their replication. However, some viroids can reach high accumulation levels without triggering a visible pathogenic alteration, whereas others accumulating to low titers cause notorious effects. Sequestering or altering the role of a regulatory host protein appears, then, a more attractive alternative. The nature of this protein remains to be determined, but data indicate that PSTVd can activate in vitro a mammalian protein kinase involved in regulating the initiation of protein synthesis; this suggests that perhaps viroids could also activate a homologous plant enzyme and trigger a signal transduction pathway ultimately leading to the onset of symptoms. Interestingly, the activation effect of a PSTVd severe variant is significantly greater than that of a mild one, thus providing an explanation for the different pathogenicity of the two variants. More recently, viroids have been proposed to exert their pathogenic effects via RNA silencing, which could be involved in cross-protection (see the section Diseases).

DISEASES

As inferred from their names (Table 8.1), most viroids incite specific diseases. Historically, it was the search for the etiological agents of certain diseases, initially presumed to be caused by viruses, which led to viroid discovery. Some of these diseases have devastating consequences, such as that caused by *Coconut cadang-cadang viroid* (CCCVd), which has destroyed millions of coconut palms in Southeast Asia. However, some other viroids replicate without inducing any visible pathological disorder in their hosts. Certain viroids have a relatively wide host range, whereas others are restricted to their natural hosts and closely related species. Plants can be coinfected by several viroids, as illustrated by the cases of old grapevine and citrus cultivars, although there are restrictions resulting from cross-protection phenomena that occur between two coinfecting strains of a viroid or even between two different viroids sharing extensive sequence similarities. Viroids are pathogens of warm climates because viroid accumulation and symptom expression are characteristically stimulated by high temperatures (30°C and even higher). Consequently, thermotherapy treatments aimed at obtaining pathogen-free plants are less efficient for viroids than for viruses, with some viroids being totally recalcitrant. Certain viroids can be transmitted by seed and pollen and even by aphids in one specific case, but, in general, transmission results from the use of contaminated pruning tools and of infected material for vegetative propagation. Therefore, the principal control measure relies on using viroid-free propagative sources.

DIAGNOSIS AND IDENTIFICATION

Serological techniques cannot be applied to viroid diagnosis because of the lack of viroid-encoded proteins. Therefore, detection of these pathogens has to rely on bioassays on indicator plants or on direct detection of the genomic viroid RNA (or a derivative thereof). Bioassays, despite their great sensitivity, are inappropriate for screening large numbers of samples because they are time-consuming and require considerable space. Likewise, special gel electrophoresis techniques based on the differential mobility of the circular viroid molecules are also unsuitable when dealing with numerous samples. The extraordinary progress made on nucleic acid research over the past few years has led to the development of new diagnostic methods for viroids, prominent among which are molecular hybridization and reverse transcription (RT) combined with the polymerase chain reaction (RT-PCR).

Molecular Hybridization

This methodology is based on the formation of a specific and stable hybrid between part (or the totality) of the viroid RNA and a labeled complementary probe. In the most common format, dot-blot, a nucleic acid preparation from the sample to be tested is directly applied to a nitrocellulose or a nylon membrane, which is then hybridized with a specific probe. Four main steps can be considered in this methodology: synthesis of the labeled probe, sample preparation, hybridization, and detection.

Labeling of nucleic acids has been performed for many years with radioactive ^{32}P. With the advent of nonradioactive labels, based on biotin and digoxigenin, molecular hybridization has become more accessible for routine diagnosis. Radioactive and nonradioactive labels are incorporated into RNA probes (riboprobes) by in vitro transcription from viroid cDNA clones. DNA probes are synthesized by random-primed polymerization or by PCR amplification with specific primers. Riboprobes are preferred because of the higher stability of the RNA-RNA hybrids.

Most methods for viroid extraction use phenol or other toxic organic solvents, making them undesirable for diagnostic laboratories. To circumvent this problem, a nonphenolic method previously described for isolating plant genomic DNA has been adapted for viroid extraction. The total nucleic acid preparations usually require an additional clarification to reduce the high background due to nonspecific hybridization. This clarification is performed by chromatography on nonionic cellulose and, in those tissues particularly rich in polysaccharides, by further extraction with methoxyethanol. For routine analysis in certification programs, sample manipulation has to be kept to a minimum. This can be achieved by using the tissue printing technique that avoids sample extraction and requires only the direct transfer to the membrane of the nucleic acids from a fresh stem, leaf, or fruit section. However, some viroids accumulating to low levels may escape detection by tissue printing.

Prior to hybridization, nucleic acids have to be fixed to the membrane. This can be done by baking or, preferably, by UV irradiation, which significantly enhances the hybridization signal. A previous denaturation step with formaldehyde also usually improves the final signal intensity. Hybridization depends on several factors, including the concentration of the target and the probe, mismatches between both polynucleotides, temperature, and buffer composition. The stringency of the hybridization conditions and the intrinsic stability of the hybrids determine their specificity. For viroids, the use of riboprobes, high temperatures, low-salt buffers, and the presence of formamide increase the specificity and lower the background.

After removing the nonspecific hybrids with washes of different stringency, the membranes are revealed immediately or stored dried for later detection. The labeled hybrids are detected by autoradiography when using radioactive probes or by digoxigenin- or biotin-specific antibodies coupled to a reporter enzyme, usually alkaline phosphatase, and a substrate, which upon reaction leads to color development or to chemiluminiscence.

RT-PCR Amplification

This methodology is based on the exponential amplification of specific DNA sequences catalyzed by a thermostable DNA polymerase. This goal is achieved through multiple cycles of three steps performed at different temperatures: denaturation of the DNA, annealing of two specific primers to the DNA single-strands, and primer extension mediated by a thermostable DNA polymerase to synthesize the final product. In the case of viroids, a previous reverse transcription step must be performed to transcribe the target RNA into a cDNA, which is then amplified. The sensitivity of the technique theoretically allows the amplification of a low number of molecules in a complex mixture.

PCR specificity is mainly provided by the primers and the annealing temperature. Primers are designed to have lengths between 18 and 25 nt, a G+C content of about 50 percent, and lack of an intra- and intermolecular complementarity, especially in their 3' termini. The extensive self-complementarity existing between regions of viroid RNAs may impair primer annealing or impede proper DNA elongation, particularly in the RT step. To overcome this problem, longer primers and higher temperatures, compatible with the reverse transcriptase activity, are recommended. To optimize detection, primers can be devised covering the motifs conserved in members of the family *Pospiviroidae* and the hammerhead structures in members of the family *Avsunviroidae*. Although it is not necessary to amplify the complete viroid genome, its small size makes it a common practice. In addition to primer design, the concentration of salts (particularly $MgCl_2$) in the RT and PCR buffers affect the specificity and yield of the amplified products. Therefore, a preliminary optimization should be carried out. An additional point to consider is the presence in the extracts of inhibitors of the RT-PCR enzymes that, otherwise, would provide false negatives. Woody plants, in particular, contain substances (polysaccharides and polyphenols) that inhibit DNA polymerases. Removal of the inhibitors can be achieved by using synthetic polymeric resins (i.e., Genereleaser) or polyvinylpyrrolidone. In some cases, a simple serial dilution of the extract can considerably mitigate the problem. Special caution must be taken to minimize contaminations that are a common source of false positives. To this end, "one-use" aliquots should be prepared from stocks of primers, reagents, and enzymes. Parallel positive and negative controls must be included in each experiment series. Recently this methodology has een extended to simultaneously detect several viroids (multiplex RT-PCR).

BIBLIOGRAPHY

Branch AD, Robertson HD (1984). A replication cycle for viroids and other small infectious RNAs. *Science* 223: 450-455.

Diener TO (2001). The viroid: Biological oddity or evolutionary fossil? *Advances in Virus Research* 57: 137-184.

Flores R, Hernández C, Martínex de Alba E, Darós JA, Di Serio F (2005). Viroids and viroid-host interactions. *Annual Review of Phytopathology* 43: 117-139.

Flores R, Randles JW, Owens RA, Bar-Joseph M, Diener TO (2005). Viroidae. In Fauquet CM, Mayo MA, Maniloff J, Desselberger I, Ball AL (eds.), *Virus Taxonomy, Eighth Report of the International Committee on Taxonomy of Viruses* (pp. 1145-1159). London: Elsevier/Academic Press.

Riesner D (1991). Viroids: From thermodynamics to cellular structure and function. *Molecular Plant-Microbe Interactions* 4: 122-131.

Symons RH (1997). Plant pathogenic RNAs and RNA catalysis. *Nucleic Acids Research* 25: 2683-2689.

Chapter 9

Transmission of Plant Viruses by Arthropods

Dick Peters

INTRODUCTION

Because of the rigidity of plant cell walls, most plant-infecting viruses have to be introduced into cells with the help of vector organisms, such as invertebrates and fungi (see also Chapter 10). A few viruses do not make use of vector organisms and are mechanically transmitted through wounds (see also Chapter 11). When arthropods (insects, mites) are virus vectors, transmission results from a cascade of activities related to probing and feeding on a plant. During the latter processes, the viruses are acquired from infected plants, retained (with or without propagation in the vector), released, and inoculated to virus-free plants. Different modes of transmission are distinguished as a consequence of the biology and the anatomy of the vectors, the outer structure of the viruses, the intrinsic properties of viral coat proteins, and the genomics of the viruses. The transmission of various viruses by aphids is well described. A nonpersistent and a persistent mode were distinguished in the initial studies made on the transmission of plant viruses by aphids. These terms reflected the length of the period during which a vector remains viruliferous. Aphids that transmit a virus nonpersistently lose their infectivity rapidly and acquire and transmit the virus in short-duration probing. Aphids, transmitting viruses persistently, acquire the virus after a long feeding period on the infected plant in a so-called acquisition access period (AAP), and they can transmit the virus to a plant during the inoculation access period (IAP) after a latency period (LP). They usually remain infective for their whole life span. A few viruses transmitted by aphids are optimally acquired and transmitted in longer periods than encountered in nonpersistent transmission, but usually shorter than in persistent transmission. The infectivity of these viruses is lost after a molt, but is retained for several days in adults. This mode is known as semipersistent transmission. Nonpersistently and semipersistently transmitted viruses are carried at the stylet tips or on the cuticular linings of the anterior alimentary canal; they are, therefore, preferentially referred to as stylet-borne or as foregut-borne viruses, respectively. The persistently transmitted viruses circulate in their vector and some of them even replicate during this circulation and are, therefore, referred to as circulative and circulative/propagative viruses, respectively. Although nonpersistently and semipersistently transmitted viruses do not circulate in their vector (even though peculiar differences exist), and persistently transmitted viruses circulate in the vector (with or without propagation), a distinction between noncirculative and circulative transmission modes will be more appropriate (see Table 9.1).

THE VECTORS

Arthropods transmitting plant viruses mainly belong to the families of the Aleyrodidae (whiteflies), Aphididae (aphids), Cicadellidae (leafhoppers), Fulgoroidae (planthoppers), and Pseudococcidae (mealybugs), all of which are large families of the order Homoptera (see Table 9.1).

Aphids and two whitefly species are mainly involved in noncirculative and circulative/nonpropagative transmission, while hoppers are specialized in circulative/propagative transmission (Table 9.1). Nine species of thrips (order Thysanoptera: family Thripidae) have also been recorded as vectors. The vectors belonging to all of these families ingest their food after penetrating the plant with their stylets. More than 60 beetle species from the families Chrysomelidae, Coccinellidae, Curculionidae, and Meloidae of the order Coleoptera transmit viruses. So far no vectors have been reported in the order Collembola, whose members also thrive on plants.

THE BIOLOGY OF APHIDS

The transmission of a virus is a function of the biology of its vector and is often related to its stage of development. The biology of some aphid species is well understood and as a consequence, the aphid vector and virus relations have been studied longer and in more detail than any other vector-virus system.

Aphids feed by inserting their slender mouthparts, a bundle of four stylets, into plant cells. The outer part of this bundle consists of two mandibular stylets and the inner part has two maxillary stylets (Figure 9.1). Intertwining grooves and ridges leaving two channels hold the maxillary stylets together. These channels are the food channel and the saliva channel, respectively. The stylets are inserted in the plant by a process of protracting first the mandibular stylets and then the maxillary stylets. During penetration

TABLE 9.1. The mode by which plant viruses belonging to the various genera are transmitted by their arthropod vectors.

Mode of transmission	Relation to vector tissue	Virus family	Genus	Vector
Noncirculative transmitted viruses	Stylet-borne	*Bromoviridae*	*Alfamovirus*	Aphids
			Cucumovirus	Aphids
			BMV[a]	Aphids
		Comoviridae	*Fabavirus*	Aphids
		Potyviridae	*Macluravirus*	Aphids
			Potyvirus	Aphids
			Rymovirus	Mites
		Flexiviridae	*Carlavirus*	Aphids
			CPMMV[b]	Whiteflies
			PAMV[c]	Aphids
	Foregut-borne	*Caulimoviridae*	*Badnavirus*	Aphids, leafhoppers, mealybugs
			Caulimovirus	Aphids
			Tungrovirus	Leafhoppers
		Sequiviridae	*Sequivirus*	Aphids
			Waikavirus	Aphids, leafhoppers
		Closteroviridae	*Closterovirus*	Aphids, mealybugs, pseudococcids, whiteflies, mites
			Crinivirus	Whiteflies
		Flexiviridae	*Capillovirus*	No vector known
			Trichovirus	No vector known
			Vitivirus	Aphids, mealybugs
			Allexivirus	Mites
Circulative transmitted viruses	Circulative	*Geminiviridae*	*Begomovirus*	Whiteflies (2 species?)
			Curtovirus	Leafhopper
			Mastrevirus	Leafhopper
			Topocuvirus	Treehopper (one species)
		Luteoviridae	*Enamovirus*	Aphids
			Luteovirus	Aphids

TABLE 9.1 *(continued)*

Mode of transmission	Relation to vector tissue	Virus family	Genus	Vector
			Polerovirus	Aphids
	*		*Umbravirus*	Aphids
Circulative/		*Bunyaviridae*	*Tospovirus*	Thrips
propagative	*		*Tenuivirus*	Planthoppers
	*		Double membranelike particles	Mites
		Rhabdoviridae	*Cytorhabdo-virus*	Aphids, leafhoppers, planthoppers
			Nucleorhabdo-virus	Aphids, lacebugs, leafhoppers, planthoppers
	*		Coffee ringspot[d]	Mites
		Reoviridae	*Phytoreovirus*	Leafhoppers
			Fijivirus	Leafhoppers
			Oryzavirus	Leafhoppers
		Tymoviridae	*Marafivirus*	Leafhoppers
Noncirculative/ propagative	*		*Sobemovirus*	
		Comoviridae	*Comovirus*	Beetles
		Tombusviridae	*Carmovirus*	Beetles?
		Tymoviridae	*Machlomovirus*	Beetles
			Tymovirus	Beetles

* Unassigned genus.
**Not placed in any taxon.
[a]BMV = *Brome mosaic virus.*
[b]CPMMV = *Cowpea mild mottle virus.*
[c]PAMV = *Potato aucuba mosaic virus.*
[d]This virus and other viruses were often considered to be rhabdoviruses. Recent studies show that they may be members of a new genus in the family of the *Rhabdoviridae.*

of the tissues of a plant, the stylets follow a path between the cells, while occasionally making an insertion in a cell or rarely passing through a cell, until they reach the sieve tubes. This penetration in the plant tissue is accompanied by the formation of a saliva sheath around the stylets. The most important function of this sheath is probably to provide rigidity to the flexible stylets. Before penetrating plant tissue, aphids make a short-duration

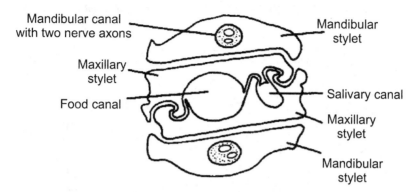

FIGURE 9.1. Diagram of a transversal section through the stylet bundle of an aphid. Figure reprinted with permission from A. F. G. Dixon (1973) *Biology of Aphids.* London: Edward Arnold.

probe (not longer than 20 sec) to test the suitability of the plant as host. On accepting the host, the aphid penetrates into the direction of phloem and reaches this tissue within 10 min to 1 h or possibly longer.

Aphids have a complex life cycle in the temperate regions. Most aphid species usually colonize on one host plant species. In contrast, some polyphagous aphids such as *Aphis gossypii, Aphis fabae, Myzus persicae,* and *Rhopalosiphum padi* live on more than one plant species. Their life cycles are characterized by an alternation between a primary host (winter host) and several secondary hosts (summer hosts). The eggs of the black bean aphid, *A. fabae,* produce fundatrices in the spring. The offspring of this generation and the next generation on the winter host are designated fundatrigeniae. After two to four apterous (wingless) generations a generation of alatae (winged aphids) develops. These alatae, also called spring migrants, migrate to various plant species, including bean, beet, and several weeds. The aphid then passes through different generations of apterous specimens (virginoparae) on these plants. When overcrowding is imminent or food supply deteriorates, the aphids produce alate virginoparae, which fly off to other hosts. In autumn, alate gynoparae fly to the winter host on which they produce oviparae. Up to this stage, all generations have been reproduced parthenogenetically (i.e., viviparous larvae were produced). The oviparae produce eggs after mating with alate males produced in the same period. The eggs are laid around the buds and in crevices in the bark of the spindle tree, *Euonymus europaeus.* The fundatrigeniae migrating from the winter to the summer hosts are the main spreaders of nonpersistently transmitted viruses (Figure 9.2). The population that develops on summer hosts is mainly

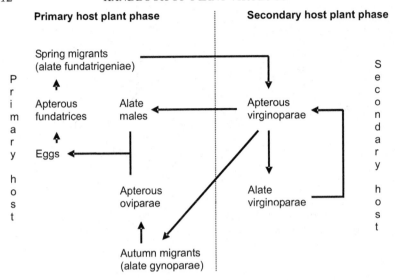

FIGURE 9.2. Schematic presentation of the annual life cycle of a holocyclic aphid species such as *Myzus persicae* or *Aphis fabae*.

involved in the spread of the semipersistently and persistently transmitted viruses.

In contrast to this holocycle, another cycle, called anholocycle, may exist in which the aphids do not produce sexually, but reproduce parthenogenetically. This cycle is typical of most aphid species in the tropics and subtropics and in greenhouses in temperate regions.

NONCIRCULATIVE TRANSMISSION OF VIRUSES BY APHIDS

All viruses of the genera *Alfamovirus* and *Cucumovirus* (family *Bromoviridae*), *Fabavirus* (family *Comoviridae*), *Potyvirus* (family *Potyviridae*), one or two carlaviruses (members of the family *Flexiviridae*), and several closteroviruses, caulimoviruses, and sequiviruses are transmitted in a noncirculative mode by aphids (Table 9.1). This type of transmission has been best studied in potyviruses. Aphids acquire viruses while probing on an infected plant. The aphid will transmit the virus during subsequent visits to other plants in short-duration probes or in phloem-seeking probings. The aphids remain infective for a couple of hours. However, their infectivity drops rapidly in the first hour after acquisition and then at a lower rate in the following few hours. Moreover, the aphid loses its infectivity after one or a few probes and in every molt. The virus is acquired from epidermis or mesophyll cells and is transmitted to these cells or lower-positioned cells. The acquisition and subsequent transmission efficiency are enhanced when the aphid is prevented from feeding for one or two hours (starvation period) prior to acquisition.

The nonpersistently transmitted viruses are transmitted to hosts as well as nonhosts of the aphid species. Aphids, probing on a nonhost, will move to another plant in search of a suitable host, hence visiting several plants. As a result of this host-searching behavior, spread by aphids may be even higher after probing on nonhosts than on hosts. Control of the spread of these viruses by insecticide application is often not useful, as the aphid will not acquire a killing dose during these short probes. Spread of the potyviruses and other nonpersistently transmitted viruses can be considerably reduced by the application of mineral or plant oils on plants. The effect of oil may partly be explained by some behavioral changes, but other, nonbehavioral factors must be mainly responsible for its inhibitory effect on transmission.

The molecular basis of potyvirus transmission has been elucidated during the past decade. Frequent mechanical transmission of potyviruses resulted in isolates that were no longer transmissible by aphids. Complementation between aphid-transmissible (AT) isolates and non-aphid-transmissible (NAT) isolates led to the discovery of the helper component (HC).

It is now well established that two viral-encoded proteins, namely the capsid protein (CP) and the HC, play a role in the transmission. The role of the CP depends on the presence of the DAG motif (the amino acids aspartic acid-alanine-glycine) in the surface-exposed N-terminus. Mutations in this motif can result either in a decrease of transmission efficiency or make the virus nontransmissible. Amino acids flanking this motif in *Tobacco etch virus* and *Tobacco vein-mottling virus* also seem to affect aphid transmission, as shown by substitution of residues upstream and downstream of the DAG motif.

The involvement of an HC component in the transmission has been demonstrated in transmission studies. Comparative studies of wild type and nontransmissible strains of several potyviruses have revealed the existence of a KITC motif (the amino acids lysine-isoleucine-threonine-cysteine) in the HC proteins. Analysis of a helper-deficient strain of *Zucchini yellow mosaic virus* showed the existence of a second motif (PTK, the amino acids proline-threonine-lysine) next to the intact KITC box.

Two models are proposed to explain the role of the HC in the transmission of potyviruses. In one model (the bridge model) it is assumed that the

N-terminal region of the HC (presumably the KITC motif) is recognized by an unknown receptor at the inner cuticle of aphid's stylet. Another domain at the HC may specifically interact with the DAG motif on the viral CP, by which the virus is retained at appropriate sites in the vector. This theory is supported by the fact that acquisition of the HC results in the acquisition of the virus. In the other model (the binding model), the HC is supposed to bind directly to the DAG motif of the CP, followed by a conformational change resulting in a direct binding of the virus particle to a putative receptor. In both models the virus has to be released either from the HC protein or the cuticle. The virus may be released by some proteolytic activity provided by the HC at the latter's C-terminus region. To connote its transmission and proteolytic function HC is nowadays termed HC-Pro.

The HC-mediated transmission as established in the potyviruses has not been found in *Cucumber mosaic virus* (CMV), another nonpersistently transmitted virus. Aphids, which feed through membranes on solutions of purified CMV, can acquire and transmit the virus.

Presently, two hypotheses have been proposed to explain the non-persistent virus transmission. The first is the ingestion-egestion mechanism, which assumes that the virus is ingested with small amounts of plant sap and is egested thereafter in the next probe. The second one is the ingestion-salivation theory, which is based on the assumption that watery saliva flushes the acquired virus particles after their release from the binding site or complex into the cytoplasm of a cell.

VIRUSES TRANSMITTED IN THE SEMIPERSISTENT WAY

Semipersistent transmission has to be considered in context with nonpersistent and persistent transmission. Some properties, such as the loss of the virus at molting, the absence of the circulation of the virus in the insect, and absence of an LP, resemble nonpersistent transmission. However, semipersistent transmission differs from nonpersistent transmission in that viruliferous aphids can inoculate three to five plants as against only one or two plants in nonpersistent transmission. The length of the acquisition period required to get optimal transmission suggests a persistent type of transmission. The probability that a plant becomes infected is positively correlated with the length of the AAP and IAP.

Viruses belonging to five genera are transmitted in the semipersistent mode (Table 9.1), but they do not show uniform transmission properties. The closteroviruses are mainly restricted to the phloem, whereas the caulimoviruses are found in most cell types. Helper components play a role in the transmission of caulimoviruses. Involvement of helper agents, proteins, or virus particles has been suggested in the transmission of other semi-persistent viruses. *Parsnip yellow fleck virus,* a sequivirus, can be acquired from plants by aphids already carrying *Anthriscus yellows virus* (AYV), another sequivirus. This suggests that the AYV CP may function as a helper. Helper components probably also play a role in the transmission of closteroviruses and carlaviruses. One or more proteins forming the rattlesnake tail at the 5' end of the closterovirus particles probably act as HCs. A unique relation exists between *Heracleum latent virus* (HLV), a vitivirus, and *Heracleum virus 6* (HV-6), a closterovirus, in which the transmission of HLV depends on that of HV-6.

Caulimoviruses are transmitted in a noncirculative fashion with properties similar to those of nonpersistent and semipersistent transmission. These viruses are acquired preferentially from the phloem, and at a low rate from nonphloem tissue. It is likely that these viruses could be transmitted in a nonpersistent as well as in a semipersistent fashion (so-called bimodal transmission). Loss of the virus at molting excludes the possibility that the virus circulates in its vectors.

The transmission of caulimoviruses, like potyviruses, also follows a helper strategy. The simple bridge hypothesis explaining the transmission of the potyviruses did not suffice for caulimoviruses, as two viral non-structural proteins seem to fulfill the function assigned to the HC-Pro of potyviruses. One protein (P2) may form a bridge between another viral protein (P3) and the aphid's cuticle, and the virus particle may bind to P3. In contrast to potyviruses, the structure of *Cauliflower mosaic virus* (CaMV) is such that it offers only one or a few contact sites for binding to the aphid's cuticle. It is not known how CaMV and potyviruses are released from the vector's cuticle.

CIRCULATIVE TRANSMISSION OF PLANT VIRUSES

The circulatively transmitted viruses can be divided into viruses circulating through the vector, such as members of the family *Luteoviridae* (transmitted by aphids) and those of the family *Geminiviridae* (transmitted by the whitefly species, *Bemisia tabaci,* or some leafhoppers). Some viruses belonging to this category circulate through the vector while they undergo replication (Table 9.1). Most circulatively transmitted viruses are usually transmitted to the host on which the vector can colonize, with a few exceptions.

The different members of the family *Luteoviridae* are specifically transmitted by one or a few vector species. The aphid *M. persicae* forms one exception in transmitting different members of the genus *Polerovirus.* These viruses can be acquired and inoculated within several minutes, but the efficiency increases considerably with the length of the AAP and IAP. Maximal acquisition and transmission is often obtained in one or more days. Following the acquisition, the vector becomes viruliferous after an LP, during which the virus circulates through the vector. Several studies have been made to determine the minimal or maximal LP. The median latency period (LP_{50}) is defined as the period between the start of the AAP and the moment at which 50 percent of the vectors have made a first inoculation. The LP_{50}, the frequency by which a vector can make a successful inoculation, and the percentage of transmitting individuals are probably the most useful parameters to estimate the competency of a vector population to transmit a virus. The LP_{50} of a circulative virus is shorter than that of the circulative/ propagative viruses. Median latency periods of *Barley yellow dwarf virus* (BYDV), a luteovirus, and *Potato leafroll virus* (PLRV), a polerovirus, are approximately 30 h, whereas an LP_{50} of 10-11 h has been reported for *Pea enation mosaic virus-1,* an enamovirus, in the pea aphid *Acyrthosiphum pisum.*

The pathway of members of the family *Luteoviridae* in the vector involves a cascade of several steps. After ingestion of phloem sap from infected plants, the virus particles move through the alimentary canal, enter the hemocoel, are transported to the accessory salivary glands (ASG) and, after their release into the salivary duct, are transmitted to the plant with saliva. To enter the hemocoel, the viruses belonging to genera *Luteovirus* and *Polerovirus* induce the formation of coated pits on the apical plasmalemma of the epithelial cells of the hindgut or the posterior part of the midgut, respectively, in a process reminiscent of receptor-mediated endocytosis (see Figure 9.3). The virus particles are then enclosed in these coated pits, which fuse to a larger structure, the receptosomes. Smooth, tubelike vesicles containing a few virus particles bud off from the receptosomes and are believed to deliver the virus in a process of exocytosis into the hemocoel. Following their release in the hemocoel the particles may diffuse in the hemocoel, and those that encounter the ASG will be taken up in a process that runs in a reverse direction to the process by which the virus is taken up into the hemocoel (see Figure 9.4). Before the virus can enter the cytoplasm of the ASG cells, the virus particles have to pass through the extracellular basal lamina and then come in contact with the plasmalemma of the highly invaginated secretory cell. The particles, enclosed in tubelike structures, accumulate in these invaginations. The coated vesicles, budding from these structures, release the virus via the apical plasmalemma into the salivary duct by exocytosis (Figure 9.4).

During migration of the virus in hemocoel the virus particles form a complex with a protein, called symbionin, a GroEL homologue produced

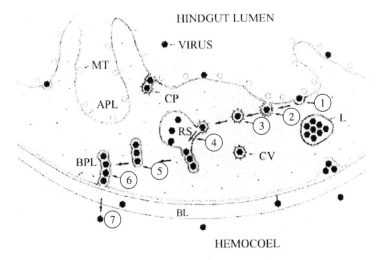

FIGURE 9.3. Transport of a luteovirus across a gut epithelial cell and its release in the hemocoel. APL: apical plasmalemma; BL: basal lamina; BPL: basal plasmalemma; CP: coated pit; CV: coated vesicle; L: larger smooth vesicles; RS: receptosome; 1 to 6: successive steps in the transport of a virus particle from the midgut lumen to the hemocoel. Figure reprinted with permission from F. E. Gildow (Penn State University, University Park, PA).

by the endosymbiont *Buchnera* living in the cells of the mycetocytes in the hemocoel. This protein is not restricted to the symbiont cytosol, but also occurs in large amounts in the hemolymph. Feeding antibiotics to aphids resulted in a decreased level of symbionts in the hemolymph and a markedly decreased transmission of PLRV. These data suggest that some interaction between the symbionin and virus is essential for retention of the infectivity of the aphid.

The luteoviruses display considerably high host plant and vector specificity. *Myzus persicae* is the most important vector infecting dicots, followed by *A. gossypii*. Luteoviruses infecting plants of the family Poaceae (grasses) are all transmitted by aphid species of which the summer migrants colonize on grasses. These viruses form a complex of different viruses, but all are known as BYDV. They can specifically be distinguished by the vector species, which preferentially transmits them (see Table 9.2). BYDV-MAV, which is transmitted by *Macrosiphum avenae* (= *Sitobion avenae*), is rarely transmitted by *Rhopalosiphum padi*. However, both viruses are efficiently transmitted by *R. padi* when oat plants are infected with both BYDV-MAV and *Cereal yellow dwarf virus* (CYDV-RPV[*Rhopalosiphum padi* virus]). These viruses are also transmitted by *R. padi* after the addition

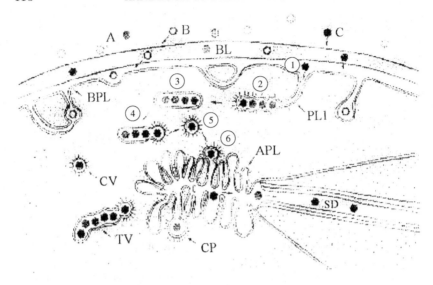

FIGURE 9.4. Passage of a luteovirus from the hemocoel through the accessory salivary gland cell and to the salivary duct. APL: apical plasmalemma; BL: basal lamina; BPL: basal plasmalemma; CP: coated pit; CV: coated vesicle; PLI: plasmalemma invaginations; SD: salivary duct; TV: tubular vesicle; A: particles that are prevented from entering the BL; B: particles whose circulation is restricted to the BL, BPL, and PLI; C: particles that make a complete circulation through the aphid; 1-6: successive steps in the transport of a virus particle from the hemocoel to the salivary duct. Figure reprinted with permission from F. E. Gildow (Penn State University, University Park, PA).

TABLE 9.2. The preferential transmission of viruses belonging to the complex of *Barley yellow dwarf virus* (BYDV) by their respective vectors.

Original name	Vector species	Acronym (virus-vector)
Barley yellow dwarf virus	*Sitabion avenae*	BYDV-MAV
Barley yellow dwarf virus	*S. (Macrosiphum) avenae*	BYDV-MAV
Cereal yellow dwarf virus (Barley yellow dwarf virus)	*Rhopalosiphum padi*	CYDV-RPV
Barley yellow dwarf virus	*Rhopalosiphum maidis*	BYDV-RMV
Barley yellow dwarf virus	*Schizaphis graminum*	BYDV-SGV

of antiserum to BYDV-MAV to a purified preparation from doubly infected plants, whereas this aphid species does not transmit when CYDV-RPV antiserum is added. These experiments show that the BYDV-MAV genome is encapsidated in the CP of CYDV-RPV, and thus protected from neutralization by antiserum to BYDV-MAV. This phenomenon is known as heterologous encapsidation, transcapsidation, or genomic masking.

The specificities between the different BYDV viruses are based on the differential uptake of the virus in the ASG and its movement through these glands, rather than on the uptake of the virus in the gut epithelial cells and release into the hemocoel of the aphids. All BYDV viruses can be detected in the hemocoel of vectors as well as in that of nonvectors after feeding on infected plants, with the exception of CYDV-RPV. This virus can be found in the hemocoel of all nonvectors, except in that of *Metopolophium dirhodum,* suggesting that this species lacks the receptors in its midgut or hindgut necessary for recognition of CYDV-RPV and invasion of its hemocoel.

TRANSMISSION OF GEMINIVIRUSES BY WHITEFLIES AND PLANTHOPPERS

The members of the *Geminiviridae,* like those of the *Luteoviridae,* are transmitted in a circulative manner. The geminiviruses, characterized by their geminate particles, are placed in the genera *Begomovirus, Curtovirus, Mastrevirus,* and *Topocuvirus.* During the past two or three decades, the begomoviruses rocketed to probably the world's most serious cause of diseases in several crops. They are transmitted by the whitefly *Bemisia tabaci.* Two biotypes have been distinguished in this vector species. One, called the "B" or silverleaf biotype, is also known as *B. argentifolii.* The curtovviruses and mastreviruses are in most cases transmitted by leafhoppers (cicadellids). The treehopper *Micrutalis malleifora* (a membracid) transmits the virus *Tomato pseudo-curly top virus,* the only member of the genus *Topocuvirus.*

The circulation of *Tomato yellow leaf curl virus* (TYLCV), a begomovirus, has been analyzed by amplifying the TYLCV-DNA in a polymerase chain reaction (PCR) on dissected organs of *B. tabaci.* This virus can be acquired in 5 to 10 min and can reach the midgut in one hour. It can be detected after 1.5 h in the hemolymph and 7 h later in the salivary glands. The minimal LP is about 8 h.

A symboninlike protein, a product of a C-type coccoid bacterium in *B. tabaci,* also safeguards transmission of the begomoviruses by the whiteflies. Whiteflies that ingested antibodies produced to the *Buchnera* symbionin isolated from *M. persicae* reduced TYLCV transmission by more than 80 percent.

A rapid circulation of the curtovirus *Beet mild curly top virus* (BMCTV) has also been demonstrated in its vector *Circulifer tenellus*. This virus was detected in the hemolymph after an AAP of 3 h and in the salivary glands after an AAP of 4 h. The results with TYLCV and BMCTV are consistent with the circulative/propagative mode of transmission. It is generally accepted that the members of the families *Luteoviridae* and *Geminiviridae* do not replicate in their vectors. Also, no evidence suggests that these viruses are transovarially transmitted.

CIRCULATIVE/PROPAGATIVE TRANSMISSION OF PLANT VIRUSES

The transmission of these viruses is limited to members of the families *Reoviridae* and *Rhabdoviridae;* to those of the genus *Marafivirus,* belonging to the family *Tymoviridae;* and to the genus *Tospovirus* of the *Bunyaviridae* and the genus *Tenuivirus* of the *Bunyaviridae*. Transmission of the last two genera will be covered in a separate section. Members of genera *Phytovirus, Fijivirus,* and *Oryzavirus* of the *Reoviridae* infect plants. The viruses of the first genus are transmitted by leafhoppers, and those of last two genera by planthoppers. The plant-infecting rhabdoviruses are restricted to the genera *Cytorhabdovirus* and *Nucleorhabdovirus*. They are transmitted by aphids, leafhoppers, and planthoppers. The transmission of *Beet leaf curl virus,* a nucleorhabdovirus, is transmitted by four lacewing species. Transmission of the reoviruses and rhabdoviruses is rather similar to that of the circulative viruses, except that they are replicated during the circulation through the vector. Circumstantial evidence that plant viruses can replicate in their vector was obtained when transovarial transmission of *Rice dwarf virus* (RDV), a phytoreovirus, was shown to occur in its leafhopper vector *Nephotettix cincticeps*. The offspring of noninfected males and infected females was partly able to transmit this virus. Evidence for the replication has been confirmed in several studies applying serial injection and by localization and accumulation of the virus in the vector, and analysis of growth curves of the virus in the vector and cell cultures derived from the vector.

Two proteins are involved in the transmission of RDV. Treatment of the virus particles with carbon tetrachloride resulted in the loss of P2, a minor outer CP. Only after administering these particles to the vector by injection, but not by feeding, did the latter become infective. Transmission of this virus is blocked by incubating virus preparations with antiserum to P8, a major CP.

Replication of some rhabdoviruses in insects is not restricted to their vectors. *Sowthistle yellow vein virus* can be found in various tissues of nonvectors and can even accumulate in their salivary glands without being transmitted.

THRIPS TRANSMISSION OF TOSPOVIRUSES

The transmission of tospoviruses is one of the most unique modes among the circulatively and propagatively transmitted viruses. These are the only viruses transmitted by thrips (family Thripidae), although some pollen-borne viruses are occasionally transmitted by thrips, probably in a mechanical fashion. Thrips are minute insects feeding on fungi or plants. Their life cycle consists of an egg stage, two larval stages, a prepupal and a pupal stage, and an adult stage. The larvae and the adults feed on plants, while the pupal stages probably do not feed at all (Figure 9.5).

The mean AAP and IAP last approximately 1-2 h, while the LP_{50} covers almost the whole larval developmental period. The viruses are successfully acquired only by young larvae, when acquisition is defined as ingestion of the virus resulting in replication and transmission. Recent studies have shown that the efficiency of acquisition drops rapidly with the age of the larvae and is almost zero when the larvae become second instars. The virus replicates in second instars and also in adults when they ingest the virus, but it cannot be transmitted. However, a high percentage of second instar larvae that had acquired the virus as newborn larvae transmit the virus just before pupation. During the LP, the virus replicates in epithelial cells of the anterior midgut. The infection then spreads to the visceral and longitudinal muscle cells and subsequently to the foregut and posterior parts of the midgut. The epithelium is sloughed during pupation. The virus must reach the salivary glands before pupation. No evidence suggests that the virus released in the hemocoel infects salivary glands. Electron microscopic studies have failed to detect the virus in the hemocoel, and its injection into the hemocoel did not render the thrips viruliferous. It has been proposed that infection of the salivary glands is due to contact between the lobed salivary glands and the visceral muscle cells of the midgut. During the larval period, the lobed salivary glands come into tight contact with the visceral cells of the midgut for several hours. During this contact the virus released by the infected midgut cells enters the salivary glands. In the later second instar larvae, the brain reposites into the head and the salivary glands lose their tight contact with the muscle cells. The development of the pterothorax and

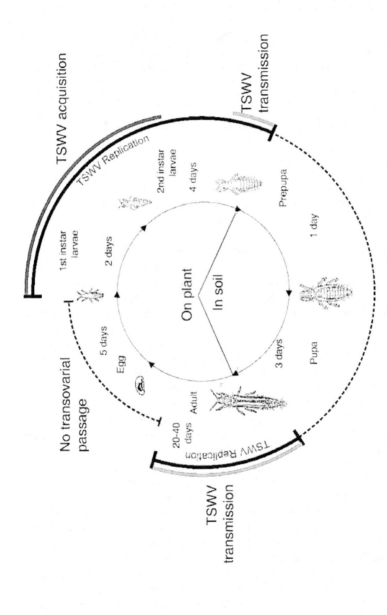

FIGURE 9.5. Life cycle of *Frankliniella occidentalis* and the stages in which the virus is acquired, replicated, and transmitted.

the wing muscles separates the salivary glands and the midgut and blocks further virus uptake. The loss of this contact may explain why second-stage larvae and adults do not transmit *Tomato spotted wilt virus* (TSWV) after ingestion and replication in the vector.

The various populations of *Frankliniella occidentalis* transmit TSWV at efficiencies varying from 12 to 80 percent. *Thrips tabaci* populations infecting tobacco seem to be efficient transmitters, whereas the arrhenotokous populations (consisting of males and females) from onion and other hosts seem to be poor transmitters and the thelotokous populations (consisting of females only) from onion fail to transmit.

NONCIRCULATIVE/CIRCULATIVE TRANSMISSION OF PLANT VIRUSES BY BEETLES

Viruses belonging to the genera *Bromovirus, Carmovirus, Comovirus, Machlomovirus, Tymovirus,* and probably most of the viruses belonging to the genus *Sobemovirus* are transmitted by beetles (Table 9.1). The transmission of these viruses is characterized by short AAP and IAP. The beetles become viruliferous and transmit in a single bite, but a longer AAP and IAP make the probability of transmission higher. The viruses can persist for shorter or longer periods in beetles. Evidence of an LP has not been found. Depending on the beetle and host plant, the virus remains restricted to the alimentary tract, whereas in other models a hemocoel phase is known. *Southern bean mosaic virus,* a sobemovirus, can be detected in the hemocoel of viruliferous bean leaf beetles, but not in that of the spotted cucumber beetles. This virus cannot be found in the hemocoel of the Mexican leaf beetle, whereas *Bean pod mottle virus,* a comovirus, is present in the hemocoel of this beetle. A beetle becomes immediately viruliferous after acquisition and the virus can enter the hemocoel, which acts as virus reservoir, suggesting that the transmission of viruses by beetles is noncirculative as well as circulative.

No proof has been found that the virus circulates through the salivary glands. Circumstantial evidence shows that the virus is transmitted when the beetle regurgitates during feeding. This regurtitant is probably a mixture of the secretion products of the maxillary and mandibular glands, the gnathal glands, and contents of the foregut. The virus is not transported from the hemocoel to these glands and, hence, is not added to their secretion products.

VIRUS TRANSMISSION BY MITES

Several economically important diseases, some of which are without a known etiology, are transmitted by phytophagous mites. The mites belong to three subfamilies of the Eriophyidae. The viruses transmitted by mites are divided into three groups. One group of viruses infects monocots. These viruses with long filamentous particles are known to induce cylindrical or "pinwheel" inclusions like the potyviruses and are placed in the genus *Rymovirus* of the *Potyviridae*. A second group consists of closteroviruslike particles. The quasi-spherical particles of a third, but interesting group, are bound by a double membrane particle (DMP) and have a diameter of 100 to 200 nm.

The transmission of *Wheat streak mosaic virus* (WSMV), a rymovirus, is probably the best-studied mite-transmitted virus. This virus, as well as some other rymoviruses and DMP, are transmitted by *Aceria tosichella* (formerly *A. tulipae*). Both nymphal stages and adults transmit WSMV, but only the nymphal stages acquire the virus from infected plants in such a way that they become infective. However, it has been shown that adults can ingest the virus. In transfers of one individual to a plant, 40 to 67 percent of the plants become infected. However, the acquisition and transmission efficiencies are very low. Nymphs having access for 10 to 15 min on infected plants may transmit WMSV to less than one percent of the plants. With an AAP and IAP, each of 16 h, about 50 percent of the plants tested became infected. The virus survives molting of the nymphs and persists for at least a few days. Transmission is found after keeping viruliferous mites on immune plants for seven days at 23-28°C or for 61 days at 3°C. The virus was found to be localized to the posterior part of the midgut and to the hindgut. The virus is not transovarially transmitted. Since there is no evidence for the existence of an LP, WSMV may not circulate in the vector. Therefore, it seems likely that this virus, like beetle-transmitted viruses, is transmitted in a noncirculative manner. The DMP particles, now considered to be virus particles, may be transmitted in a circulative/propagative fashion like all other membrane-bound viruses.

BIBLIOGRAPHY

Czosnek H, Morin S, Rubinstein G, Fridman V, Zeidan M, Ghanim, M (2001). Tomato yellow leaf curl virus: A disease sexually transmitted by whiteflies. In Harris KF, Smith OP, Duffus JE (eds.), *Virus-insect-plant interactions* (pp. 1-27). San Diego, CA: Academic Press.

Fereres A, Collar JL (2001). Analysis of non-circulative transmission by electrical penetration graphs. In Harris KF, Smith OP, Duffus JE (eds.), *Virus-insect-plant interactions* (pp. 87-109). San Diego, CA: Academic Press.

Gergerich RC (2001). Mechanism of virus transmission by leaf-feeding beetles. In Harris KF, Smith OP, Duffus JE (eds.), *Virus-insect-plant interactions* (pp. 133-142). San Diego, CA: Academic Press.

Harris KF, Harris LJ (2001). Ingestion-egestion theory of cuticula-borne virus transmission. In Harris KF, Smith OP, Duffus JE (eds.), *Virus-insect-plant interactions* (pp. 111-132). San Diego, CA: Academic Press.

Kim KS, Ahn KK, Gergerich RC, Kim SB (2001). Possible etiology of eriophyd mite-borne pathogens associated with double membrane-bound particles. In Harris KF, Smith OP, Duffus JE (eds.), *Virus-insect-plant interactions* (pp. 29-50). San Diego, CA: Academic Press.

Laval Kumar P, Teifion Jones A, Reddy DVR (2003). A novel mite-transmitted virus with a divided RNA genome closely associated with pigeonpea sterility mosaic disease. *Phytopathology* 93: 71-81.

Moritz G (2002). The biology of thrips is not the biology of their adults: A developmental view. In Marullo R, Mound L (eds.), *Thrips and Tospoviruses: Proceedings of the 7th International Symposium on Thysanoptera* (pp. 259-267). Reggio Calabria: Università degli studi Mediterranea di Reggio Calabria.

Nagata T, Peters D (2001). An anatomical perspective of tospovirus transmission. In Harris KF, Smith OP, Duffus JE (eds.), *Virus-insect-plant interactions* (pp. 51-67). San Diego, CA: Academic Press.

Raccah B, Huet H, Blanc S (2001). Potyviruses. In Harris KF, Smith OP, Duffus JE (eds.), *Virus-insect-plant interactions* (pp. 181-206). San Diego, CA: Academic Press.

Soto MJ, Gilbertson RL (2003). Distribution and rate of movement of the curtovirus *Beet mild curly top virus* (family *Geminiviridae*) in the beet leafhopper. *Phytopathology* 93: 478-484.

Chapter 10

Plant Virus Transmission:
Fungi, Nematodes, and Seeds

Jeanne Dijkstra
Jawaid A. Khan

TRANSMISSION BY FUNGI

Already in 1958 it was demonstrated that the fungus *Olpidium brassicae* played a role in the transmission of the big-vein disease of lettuce. Since that discovery, several soilborne viruses belonging to three different genera have been shown to be transmitted by fungi.

All virus-transmitting fungi are obligate, zoospore-forming endoparasites of plants. They belong to the Chytridiomycota (chytrids; genus: *Olpidium*) or the Plasmodiophoromycota (plasmodiophorids; genera: *Polymyxa* and *Spongospora*). The species *O. bornovanus* (= *O. radicale* = *O. cucurbitacearum*), *O. brassicae, P. betae, P. graminis,* and *Spongospora subterranea* are natural virus vectors.

The fungi in the three genera have comparable life cycles.

Thick-walled resting spores develop in the cells of the roots or young tubers of the host plant. The type of resting spores depends on the fungus: the chytrids have single resting spores, whereas the plasmodiophorids are characterized by clusters of resting spores (spore balls or cystosori). The resting spores are released when the infected roots or tubers decay in the soil. When soil conditions are favorable (high moisture content) resting spores germinate and release motile primary zoospores which are uniflagellate in chytrids and biflagellate in plasmodiophorids.

The zoospores move to the roots and attach themselves to root hairs or epidermal cells. Thereafter, they reach the encystment phase, during which the flagellae are withdrawn and a cyst wall is secreted.

Chytrids and plasmodiophorids penetrate host cells in a different way. The cyst of chytrids dissolves a small pore in the cell wall and its released protoplast passes through it. Plasmodiophorids penetrate the cell wall with a stylet in a tube formed as soon as the cyst has settled down on roothairs or epidermal cells. After evagination of the tube an adhesorium is formed and the host cell is punctured with the stylet. Subsequently, the stylet and the protoplast of the cyst are released into the cytoplasm of the host cell.

In both types of vectors, the fungal protoplast develops into a plasmodium that is enveloped in a thin thallus membrane (immature thallus). Thereafter, the thallus forms sporangia, from which the secondary spores are released through an exit tube into the water surrounding the root.

Later in the cycle, the mature thallus with its thicker membrane develops into thick-walled resting spores or resting sporangia, which may remain viable in decaying root material in the soil for a long time: in the case of *P. betae* for more than ten years.

CONTACT BETWEEN VIRUS AND FUNGUS

Viruses are acquired by the fungus in two different ways and are, therefore, divided into two categories.

In the case of chytrid-borne viruses belonging to the genera *Tombusvirus, Carmovirus, Necrovirus,* and *Dianthovirus,* the isometric virions are adsorbed on the zoospores outside the host plant (externally borne viruses).

During the encystment phase, the virus particles are taken into the cytoplasm of the zoospore and introduced into the host cell after penetration of the latter by the cyst. As soon as the fungus forms a thallus, virus particles escape from the thallus into the host cell.

Transmission can take place only when virus particles bind to the zoospore, which is a highly specific process. For example, particles of *Tobacco necrosis virus,* a necrovirus, were found to adsorb on zoospores of a lettuce isolate of *O. brassicae,* but not to those of a mustard isolate of this fungal species.

In contrast to these externally borne viruses, those belonging to the second category are already present in the zoospores when they are released into the soil from viruliferous resting spores. Viruses thus acquired by zoospores and resting spores inside the host plant (internally borne viruses) are found in the genera *Furovirus, Pomovirus, Pecluvirus, Benyvirus, Bymovirus, Ophiovirus,* and *Varicosavirus.* Acquisition of the rod-shaped or filamentous virions by the fungus takes place as follows.

When virus-free zoospores penetrate the rootlets of a virus-infected host plant, they form a plasmodiumlike thallus. At that stage the virus is thought to pass through the still-thin membrane of the thallus, in which it accumulates but does not multiply. When zoosporangia or resting spores are formed in the mature, thicker-walled thallus, the virus is incorporated in them so that in the subsequently formed zoospores the virus will also be present.

Inside the resting spores, which can survive in dry root material, the virus may remain infectious for a very long time, often for years.

The relationship between virus and fungal vector is highly specific.

MOLECULAR INTERACTIONS
BETWEEN VIRUS AND FUNGUS

Externally Borne Viruses

The capsid protein (CP) of *Cucumber necrosis virus* (CNV), a tombusvirus, has been shown to play a role in the uptake of the virus by the zoospore. In vitro mutagenesis studies with CNV revealed that the substitution of glutamic acid to lysine in the S domain of the CP resulted in reduced transmissibility by its vector *O. bornovanus* and less binding of virus particles to the zoospores.

Recently, more amino acids exposed on the surface of the virus particle have been found to be essential for fungus transmission. It is suggested that a certain region of the capsid links the virus particles to putative receptors in the zoospore plasmalemma.

Internally Borne Viruses

Spontaneous mutants of BNYVV with deletions in the RNA2 of this virus produced smaller readthrough proteins of the CP and could no longer be transmitted by *P. betae*. The readthrough domain of the fusion protein contains a KTER (lysine-threonine-glutamic acid-arginine) motif known to be a determinant of vector transmission. Isolates of BNYVV maintained by serial mechanical transmissions in the laboratory for a long time, showed deletions in the KTER-encoded domain and were no longer fungus-transmissible.

Potato mop-top virus (PMTV), a pomovirus, also lost its ability to be transmitted by the fungus *S. subterranea* after prolonged mechanical transmission. In this case, deletions were detected in the readthrough-protein-encoding region of RNA2 (previously referred to as RNA3). However, as transmissible isolates of PMTV do not possess a KTER motif in their readthrough domain, other regions in this domain may act as determinants for interaction with the vector.

An isolate of *Barley mild mosaic virus,* a bymovirus, transmitted mechanically for many years, was no longer transmissible by its vector, *P. graminis*. This isolate was shown to contain a deletion in the P2 protein of its RNA2, leading to the production of a truncated protein. Presence of a truncated protein has also been reported for an isolate of another bymovirus, *Barley yellow mosaic virus,* which was no longer transmissible by its vector, *P. graminis.* Therefore, it is likely that such truncated proteins are responsible for the lack of fungus transmission.

TRANSMISSION BY NEMATODES

Grapevine fanleaf virus (GFLV), a nepovirus, was the first virus shown to be transmitted by an ectoparasitic soil-inhabiting nematode (eelworm). At present, a number of viruses belonging to the genera *Nepovirus* and *Tobravirus* are known to be transmitted by nematodes belonging to the family Longidoridae (longidorids) in the order Triplonchida, and the family Trichodoridae (trichodorids) in the order Dorylaimida, respectively.

Longidorids belonging to the genera *Longidorus* and *Xiphinema* are slender, 4-8 mm long nematodes with relatively long (60-250 μm) hollow stylets. The stylet consists of an 8-140 μm long odontostyle surrounded by a stylet guide sheath for penetration of root tip cells and a stylet extension (odontophore) with nervous tissue and protractor muscles. The odontophore passes into the esophagus, which connects the stylet to a muscular pump for sucking up the cell contents, and the esophageal bulb in which large gland cells secrete saliva for liquefaction of the cytoplasm. The ingested cell contents are forced into the gut.

Trichodorids belonging to the genera *Trichodorus* and *Paratrichodorus* are small rather plump, 0.5-5 mm long nematodes with a 20 to 80 μm long, slightly curved stylet (onchiostyle). They feed on root tips and also on growing tips of stem parts in the soil. In contrast to longidorids, the short, toothlike stylet of trichodorids does not penetrate cells, but damages the cell walls so that the cell contents can be sucked in. Thereafter, the ingested food passes through the pharynx and esophagus into the gut.

CONTACT BETWEEN VIRUS AND NEMATODE

Ingested virus particles are adsorbed to the inner surface of the odontostyle, or between the odontostyle and the guide sheath in *Longidorus* spp., to the cuticular lining of the lumen of the odontophore and the esophagus of *Xiphinema* spp., or to the cuticular lining of the lumina of the pharynx and esophagus in *Paratrichodorus* and *Trichodorus* spp.

The adsorbed virus particles can be retained by the nematode for a long time, varying from months to years, without loss of infectivity.

Unlike ingestion, retention is a specific process occurring only in nematodes that can transmit the virus. Surface structures on the cuticle of the alimentary tract of the nematode may determine the interaction between the vector and the virus with its genetic determinants for nematode transmission.

Virus transmission may occur when a virus-containing nematode starts feeding on another plant. The adsorbed virus particles dissociate from the retention site and are transferred, most likely together with saliva, to the new plant.

No evidence has been found that the virus replicates in the nematode.

Transmission is in a semipersistent manner. Both larvae and adults can transmit, but the larvae lose the virus after molting.

MOLECULAR INTERACTIONS
BETWEEN VIRUS AND NEMATODE

Nepoviruses

Based on the amino acid similarities in that part of the RNA2-encoded polyprotein which is situated near the N-terminal region of the CP, nepoviruses have been placed in two groups. Viruses in one group are transmitted solely by *Longidorus* spp., whereas the vectors of the other group, to which, among other things, GFLV belongs, are restricted to *Xiphinema* spp.

The specificity of transmission of GFLV by its vector proved to be determined by a domain located within the 513 C-terminal residues of the RNA2-encoded polyprotein.

Tobraviruses

Experiments with *Tobacco rattle virus* (TRV) and *Pea early-browning virus* (PEBV) have proved that in addition to the CP, viral nonstructural proteins are also essential for nematode transmission.

The nonstructural 40 kilodalton (kDa) protein encoded by the 2b gene of TRV has been shown to be required for transmission by both *Paratrichodorus pachydermus* and *P. anemones*. In PEBV, a difference of two amino acids in the 29.6 kDa nonstructural protein was responsible for the lack of nematode transmission, but other nonstructural proteins encoded by its RNA2, such as the 23 and 9 kDa proteins, were also shown to be involved in vector transmission.

Nonstructural virus-encoded proteins have also been shown to play a role in transmission by other vectors of viruses, possibly by forming a link between the CP and putative receptor sites in the food canal of the vector. A well-known example is the helper component-protease (HC-Pro) of potyviruses found to be essential for their transmission by aphids.

Besides nonstructural proteins, certain carbohydrates identified at the site of retention of the virus in the vector nematode may also be involved by linking the virus-protein complex to the cuticle of the alimentary tract.

SEED TRANSMISSION

Bean common mosaic virus (BCMV), a potyvirus, was the first virus shown to be transmitted through seed. At present, more than 100 viruses are known to be seed-transmitted. Seed transmission offers an attractive strategy for the survival of particularly those viruses that have restricted host range and follow a nonpersistent mode of transmission through insect vectors. Seed transmission, even with low frequency, may be crucial for the spread of viruses and has epidemiological impact. The transmission frequency, varying from nearly 0 to 100 percent, depends on the genetics of both host and virus and is influenced by environmental factors. For instance, a seed-transmissible virus may have nontransmissible strains or different genotypes of the same host may show completely different transmission efficiency of a given virus. Since the virus hardly undergoes any changes during the transmission process, it is the host, which changes physiologically and genetically (from haploid to diploid, and back to haploid). Thus, the transmission process is complex and coordinated by several determinants.

CONTACT BETWEEN VIRUS AND SEEDS

All seedborne viruses infect the embryo and survive in it during the maturation and storage of seeds. An exception is *Tobacco mosaic virus,* a tobamovirus, as it is present on the surface of the seed coat but does not infect the embryo. Seed transmission requires that the virus should be able to infect floral meristems at an early stage of development, resulting in infected pollen and ovaries, or the embryo during embryogenesis. In ovule-based transmission, the virus infects the maternal meristematic tissues before their differentiation into the ovule and the physical separation of the embryo from the mother plant. While in pollen transmission, the virus should be able to infect the floral meristems and pollen mother cells before the appearance of the callose layer. The embryo infection takes place either before fertilization (indirect embryo infection) or after fertilization (direct embryo infection). In the former case, the virus is able to infect floral meristems before embryogenesis and the embryo is infected as a result of the fusion of gametes during fertilization. Examples include *Tobacco ringspot virus* (TRSV, a nepovirus), BCMV, and *Barley stripe mosaic virus* (BSMV, a hordeivirus). The involved mechanisms are not yet known in detail. TRSV is transmitted (predominantly) through the megagametophyte. It is pollen transmission in BCMV that is more effective than ovule transmission. Infection of megaspore mother cells or pollen mother cells is believed to play a role in BSMV transmission.

The fact that the embryo is physically separated from the maternal plant tissues makes its direct invasion a complex process. In order to accomplish it, the virus should be able to invade the embryo at an early stage (i.e., before the maternal tissues are separated from those of a progeny). This may explain why infection of a mother plant by a virus after flowering does not lead to seed-mediated transmission. The virus pathway has been well studied in pea in combination with *Pea seed-borne mosaic virus* (PSbMV), a potyvirus. It was shown that during an early developmental stage of embryo, PSbMV entered into the embryo of a PSbMV-transmissible pea cultivar through the suspensor (a transient structure that supplies nutrition to the developing embryo). It was suggested that transport of PSbMV via the suspensor is possible because of the presence of plasmodesmatal connections between the suspensor cells and the embryo during early stages. In the PSbMV-nontransmissible cultivar, the virus cannot reach the suspensor due to blockage of its replication and movement. As a result, PSbMV is not able to cross the boundary between the maternal and progeny tissues. In seed transmissible isolates, the virus infects the embryo, multiplies in the embryonic tissues, and persists during seed maturation. The involved mechanism is, however, a complex one, as PSbMV resistance genes, namely *sbm-1, sbm-3,* and *sbm-4,* located on chromosome 6 of the pea genotypes, are also suggested to influence replication of PSbMV pathotypes. Further, multiple maternal genes regulate the ability of the virus to replicate and spread in the nonvascular tissues, thereby blocking the virus transmission in the nonseed-transmissible pea cultivars.

MOLECULAR INTERACTIONS
BETWEEN VIRUS AND SEED

Availability of full-length, infectious cDNA (complementary DNA) clones of a few viruses has illustrated the mechanism of seed transmission. Interestingly, mutagenesis studies have shown that the viral genes/products influence transmission mainly by regulating the replication capacity and their movement in the reproductive tissues of infected hosts, though participation of host factors cannot be ruled out. Several viral/host determinants participate in seed transmission.

Pseudorecombinants

A possible role of pseudorecombinants of some bi/multipartite RNA viruses has been studied. The RNA1 component of both *Cucumber mosaic virus,* a cucumovirus, with a tripartite genome, and some of the bipartite nepoviruses has been shown to be the determinant of seed transmissibility. Though pseudorecombinants have lower seed transmissibility, other minor determinants are also suspected in reducing the rate of virus transmission.

Helper Component Protease (HC-Pro)

HC-Pro, a nonstructural gene present in potyviruses, is involved in aphid-mediated transmission, replication, and long-distance movement. Mutagenesis studies have illustrated that the HC-Pro regulates replication and movement of PSbMV in the reproductive tissues of pea plants, thus influencing invasion of the embryo. Besides the helper component, the N-terminal region of the capsid protein gene of PSbMV and a few maternal genes of the pea genotype also participate in this direction.

12 K Gene

Pea early browning virus (PEBV, a tobravirus) possesses a bipartite genome (RNA1, RNA2). RNA1 encodes four proteins of 141, 201, 30, and 12 kDa. Notably, the 12 kDa (12 K) gene resembles to the HC-Pro a potyvirus and the RNAγ gene of a hordeivirus. The latter two are associated with the seed transmission of PSbMV and BSMV, respectively. PEBV infects male as well as female reproductive tissues of pea. With mutation in the 12 K gene, PEBV is unable to infect the pollen grains, eggs, or ovules. Although infection of the gametes is necessary for the seed transmission, the 12 K gene controls the movement of PEBV to reach the reproductive tissues. The mutant causes necrotic symptoms in the vegetative tissues, while in the reproductive tissues the virus is eventually absent. It is believed that the 12 K gene influences virus transmission either by invading the gametes or by enabling replication of the virus in the reproductive tissues of pea cultivars.

RNAγ

BSMV contains three RNA species: RNAα, RNAβ, and RNAγ. Major seed determinants are located in the 5' untranslated region (UTR), and a 369 nt repeat present in the γa and γb genes. Mutations in these determinants significantly influence replication and translation of the virus. It was shown that seed transmission is completely blocked when the repeat sequence is present. This sequence is suggested to enhance the replication and movement of BSMV, while its mutation affects both the transmission and symptom expression in barley plant.

It is worth mentioning that the 12 K gene of tobraviruses, HC-Pro of potyviruses and the γb gene of hordeiviruses share cystein-rich domains that regulate the transmission process by affecting the replication and movement of the virus in the reproductive tissues of respective hosts, and thus influences the infection of the embryo.

TRANSMISSION THROUGH POLLEN GRAINS

Pollen-, and seed-mediated transmission are two overlapping phenomena. Many viruses transmitted through seed, such as ilar-, nepo-, sobemo-, and idaeoviruses are also transmitted through pollen. Though gametic infection of the embryo and direct infection of the mother plants are two suggested mechanisms, they remain unclear. A pollen transmission requires the ability of the virus to infect meristematic tissues (developing pollen mother cells). Pollen infection of healthy mother plants, which yield infected seeds, may contribute to the seed-mediated transmission of viruses. Self-pollinated infected plants yield a higher percentage of infected seeds than when only one gamete is from an infected plant. Sometimes, insects play an important role in pollen-mediated transmission. For instance, in ilarviruses and probably nepoviruses, insects with foraging habits carry virus-contaminated pollen to healthy plants. Similarly, thrips in the pollen receptors in flowers may cause wounds, which are helpful in mechanical transmission to nongametophytic tissues. Often, pollen grains are carried along with the virus in such a fashion that there are chances of mechanical transmission of the virus.

BIBLIOGRAPHY

Belin C, Schmitt C, Demangeat G, Komar V, Pinck L, Fuchs M (2001). Involvement of RNA2-encoded proteins in the specific transmission of *Grapevine fanleaf virus* by its nematode vector *Xiphinema index. Virology* 291: 161-171.

Dijkstra J, De Jager CP (1998). *Practical plant virology: protocols and exercises.* Berlin, Heidelberg, New York: Springer-Verlag.

Dijkstra J, Khan JA (2002). Virus transmission by fungal vectors. In Khan JA, Dijkstra J (eds.). *Plant viruses as molecular pathogens,* (pp. 77-104). Binghamton, NY: The Haworth Press, Inc.

Gao Z, Eyers S, Thomas C, Ellis N, Maule A (2004). Identification of markers tightly linked to *sbm* recessive genes for resistance to *Pea seed-borne mosaic virus. Theoritical and Applied Genetics* 109: 488-494.

Hull R (2002). *Matthews' plant virology.* San Diego: Academic Press.

Kakani K, Sgro J-Y, Rochon D'Ann (2001). Identification of specific cucumber necrosis virus coat protein amino acids affecting fungus transmission and zoospore attachment. *Journal of Virology* 75: 5576-5583.

MacFarlane SA, Brown DJF (1995). Sequence comparison of RNA2 of nematode-transmissible and nematode-non-transmissible isolates of *Pea early-browning virus* suggests that the gene encoding the 29 kDa protein may be involved in nematode transmission. *Journal of General Virology* 76: 1299-1304.

MacFarlane SA, Wallis CV, Brown DJF (1996). Multiple genes involved in the nematode transmission of *Pea early browning virus. Virology* 219: 417-422.

Roberts IM, Wang D, Thomas CL, Maule AJ (2003). *Pea seed-borne mosaic virus* seed transmission exploits novel symplastic pathways to infect the pea embryo and is, in part, dependent upon chance. *Protoplasma* 222: 31-43.

Sandgren M, Savenkov EI, Valkonen JPT (2001). The readthrough region of *Potato mop-top virus* (PMTV) coat protein encoding RNA, the second largest RNA of PMTV genome, undergoes structural changes in naturally infected and experimentally inoculated plants. *Archives of Virology* 146: 467-477.

Schmitt C, Mueller A-M, Mooney A, Brown D, MacFarlane S (1998). Immunological detection and mutational analysis of the RNA2-encoded nematode transmission protein of pea early browning virus. *Journal of General Virology* 79: 1281-1288.

Vassilakos N, Vellios EK, Brown EC, Brown DJF, MacFarlane SAA (2001). Tobravirus 2b protein acts *in trans* to facilitate transmission by nematodes. *Virology* 279: 478-487.

Chapter 11

Mechanical Transmission
of Plant Viruses

Jeanne Dijkstra
Jawaid A. Khan

In mechanical transmission a virus is introduced into a plant by mechanical means (i.e., by wounding of the plant). The wounds may be inflicted experimentally by mechanical inoculation (experimental mechanical transmission) or they may have been made in nature by accident (natural transmission).

EXPERIMENTAL MECHANICAL TRANSMISSION

In this type of transmission, also called (mechanical) inoculation or sap transmission, healthy test plants are exposed to a virus-containing suspension (inoculum) in such a way that infection may occur.

Experimental mechanical transmission is a valuable method and is commonly used for diagnosis of diseases, propagation and maintenance of viruses, quantitative virus assays (e.g., local-lesion assays), virus detection, and in studies on plant-virus interactions.

However, not all viruses are mechanically transmissible. Another drawback of mechanical inoculation of plants is that the efficiency is very low, as usually more than 500 virus particles are required to give rise to a visible infection. In addition, even when infection occurs, virus multiplication does not take place synchronously, as the infectious entity moves from cell to cell. Such asynchronous infection hampers studies of stages in the infection process. To overcome this problem, suspensions of protoplasts isolated from mesophyll cells may be inoculated instead of leaves or other plant parts.

Inoculation of Plants

The plant parts, usually leaves, to be inoculated are dusted with an abrasive, such as Carborundum powder (silicon carbide), and then rubbed with the inoculum by using the forefinger, foam plastic blocks, or glass spatulae. Instead of sprinkling the leaves prior to inoculation, Carborundum powder or Celite (diatomaceous earths), another abrasive, can also be mixed with the inoculum.

The wounds in the epidermal cells thus created should be small enough not to cause too much damage and big enough to give the virus particles access to the cells.

Another type of mechanical inoculation is slash inoculation, used for viruses of woody plants; for instance, *Citrus tristeza virus* (CTV), a closterovirus. In this technique, citrus trees are inoculated by slashing the bark of a healthy tree with a blade contaminated with CTV.

A more uniform infection of wheat plants with *Soil-borne wheat mosaic virus,* a furovirus, has been obtained by inoculation of young wheat leaves abraded with an artist's tool called a Nib prior to inoculation with a mixture of purified virus and carborundum powder. Unlike conventional inoculation with carborundum powder, no significant damage (necrosis, yellowing) occurred in the inoculated leaves.

For viruses that are difficult to transmit mechanically, alternative techniques have been developed. Some of the more successful ones are the vascular puncture technique and particle bombardment. The former technique has been used for, among other things, vector-obligated viruses of maize belonging to the genera *Fijivirus, Marafivirus, Nucleorhabdovirus,* and *Waikavirus.* This technique involves the puncturing of soaked maize kernels with one or more entomological minuten pins by pushing the pins through a drop of inoculum placed on the inoculum site of the kernel.

In particle bombardment (biolistic inoculation), microprojectiles consisting of 4 μm tungsten particles or 0.6 μm gold particles on which a virus or viral nucleic acid is adsorbed, are fired into target cells or tissues. This is achieved by means of a gunpowder charge, which accelerates a macroprojectile down the barrel of the device, or by a helium-powered acceleration device.

The success of mechanical inoculation depends on both the quality of inoculum (virus concentration, stability of the virus, presence of other constituents in the inoculum) and the susceptibility of the test plant.

Inoculum

In order to obtain a good inoculum, selecting the source plant and preparation of the inoculum are of great importance. In general, young leaves with clear symptoms contain the highest virus concentrations and are therefore preferred as the virus source. Depending on the stability of the virus, the infected plant material is ground in a mortar with a pestle or in a blender with deionized water or a buffer (e.g., 0.01 M phosphate buffer) and the slurry thus obtained is pressed through cheesecloth.

Many plants contain constituents that inhibit infection either by inactivating the virus or by making the test plant less susceptible. The following substances may act as inhibitors of infection.

Polyphenoloxidases present in plant sap oxidize polyphenols to *o*-quinones, which can inactivate viruses. The oxidation can be countered by adding reducing agents to the leaf maceration fluid (e.g., ascorbic acid, sodium sulphite, 2-mercaptoethanol, and so on) or copper-chelating agents such as sodiumdiethyldithiocarbamate (Na-DIECA).

Many woody plants and members of the family Rosaceae contain tannins. These secondary metabolites may bind irreversibly to the virus particles, thus making them noninfective. To prevent this, compounds known to compete with the virus particles for the tannins (e.g., caffeine, egg albumen, hide powder, and nicotine) are added to the maceration fluid.

Nucleases in plant sap constitute a problem when the source plant contains a virus devoid of a protein coat (e.g., in the case of a plant inoculated with only the long particles of *Tobacco rattle virus,* a tobravirus). Infectivity may be retained by adding bentonite clay to the maceration fluid to adsorb the nucleases.

The other type of inhibitors that affect the susceptibility of the test plant consists of certain proteinaceous substances present in members of the Amaranthaceae, Caryophyllaceae, Chenopodiaceae, and Phytolaccaceae. Such proteins compete with the virus particles for infectible sites in the epidermal cells of the test plant. Their action can be diminished by diluting the virus-containing sap, as dilution usually decreases the effect of the inhibitor more than it affects the infectivity of the virus.

SUSCEPTIBILITY OF THE TEST PLANT

Amaranthaceae, Chenopodiaceae, Cucurbitaceae, Leguminosae, and Solanaceae are known to be very susceptible to a large number of viruses. Host range studies are, therefore, usually carried out with plant species in these families.

The susceptibility of a plant depends on its genotype and physiological condition. As the latter is influenced by a number of factors, such as the temperature, light intensity, and humidity at which the plant has been grown, as well as its nutrition and age, it is clear that for successful inoculation these factors have to be taken into account. In general, leaves of young well-nourished plants raised under moderate light intensity are most susceptible. In cucurbitaceous and leguminous plants the cotyledons of the former and the primary leaves of the latter are usually more susceptible than the true leaves and trifoliolate leaves, respectively. In order to establish a host plant range, inoculated test plants not showing symptoms have to be back-inoculated (i.e., sap from leaves of such symptomless plants are inoculated onto a sensitive assay plant). In this way, latent infections can be established.

The susceptibility of a plant also varies with the time of the day and with the time of the year in the temperate zone.

INOCULATION OF ISOLATED PROTOPLASTS

Protoplasts are obtained by treating leaves from which the lower epidermis has been stripped with a mixture of pectinases and cellulases to dissolve the middle lamella and cell walls. The resulting protoplasts, mostly those from the palisade parenchyma cells, are spherical and possess only the plasma membrane (plasmalemma).

When the protoplast suspension is mixed with the inoculum, usually in the presence of a polycation, for example, poly-L-ornithine, or polyethylene glycol (PEG), virus particles penetrate into the protoplasts.

Another method used to introduce a virus or viral RNA into protoplasts is by electroporation. This method is less time-consuming than the previous one, as it eliminates the incubation period of protoplasts with the inoculum. In electroporation, a mixture of protoplasts and inoculum is exposed to a brief high-voltage pulse by which the protoplasts become transiently permeable to the virus or viral RNA.

Protoplasts can also be inoculated by particle bombardment.

NATURAL MECHANICAL TRANSMISSION

Stable viruses that occur in high concentrations in plants, such as *Tobacco mosaic virus* (TMV) and other tobamoviruses; *Tobacco necrosis virus*, a necrovirus; and *Potato virus X*, a potexvirus; and sobemoviruses can be readily transmitted in the field by contact. This may happen when leaves of healthy and infected plants rub together by wind, agricultural implements, or grazing animals. *Subterranean clover mottle virus*, a sobemovirus, is transmitted by grazing and treading of sheep in subterranean clover crops.

Rice yellow mottle virus, another sobemovirus, endemic in Africa, can efficiently be spread in fields by cows, donkeys, and the grass rat *(Arvicanthis niloticus)*; infectious virus was shown to be present in the saliva of grazing donkeys. Virus particles may also contaminate the feathers of birds and the fur of animals in the field when they brush past infected plants. Many infections are man-made, for instance, by using contaminated tools. Tobacco smokers have been shown to transmit TMV by their hands, as even cured tobacco, as present in cigarettes, may still contain infective virus.

Stable viruses may also be present in soil and water after having been released from decaying infected plant debris. Infection of cucumber plants with *Cucumber green mottle mosaic virus,* a tobamovirus, in a greenhouse could be attributed to the presence of this virus in the water used for irrigation. Such soilborne or waterborne viruses may infect plants mechanically, usually via slightly damaged rootlets, or by abrasion when soil particles come into contact with leaves.

Mechanical transmission can also occur by air. Infective TMV was present in air samples taken over an infected tobacco field. However, so far airborne transmission has not been shown to be of great importance.

The previously mentioned types of transmission are examples of so-called abiotic transmission; in other words, transmission without involvement of a biological vector (arthropod, fungus, or nematode).

TRANSMISSION BY GRAFTING

Grafting is used to transmit viruses that cannot be transmitted by other methods, as is the case with phloem-limited or very unstable viruses and viruses without known biological vectors.

It is the oldest method of transmitting infectious agents from diseased plants to healthy ones. In fact, the word "inoculation" has been adopted from the ancient horticultural practice of grafting, in which a bud, popularly called an "eye" (Lat. *oculus*) from one plant was inserted into another.

For a long time, grafting was the only way to establish whether an abnormality in a plant was caused by a virus.

Transmission by grafting may occur when a scion is grafted on a rootstock and one of the graft components is infected systemically with a virus. In general, graft transmission is successful when the cambial tissues of the grafted components have fused. Fusion takes place only when the tissues of the two plants are compatible, which implies that scion and stock should be from the same or related species.

In exceptional cases, however, even without proper union of the tissues, virus transmission may occur as a result of the transient intimate contact during which exchange of cell constituents may have taken place between cut surfaces of the two plants. However, such transmission is an example of mechanical transmission rather than graft transmission. This type of transmission may also result from leaf grafting, in which pieces of infected leaves are inserted under flaps of bark of healthy trees. It is, however, possible that a degree of transient tissue fusion may occur when the inserted leaf material is from the same species.

Graft inoculation is commonly used for diagnosis of virus diseases, detection of viruses in crop plants or propagation material (indexing), generation and production of virus-free plant material, and screening for virus resistance.

The most commonly used grafting techniques in plant virology are bud grafting, wedge-grafting, and bulb and tuber grafting.

Bud grafting is primarily used in fruit trees for indexing and is of two types, namely single-bud grafting and double-bud grafting. In single-bud grafting, a bud from the tree to be indexed is grafted onto an indicator rootstock. If the source tree is infected with a particular virus, the indicator will show the characteristic symptoms of that virus.

In double-bud grafting, buds from both source and indicator trees are grafted onto the same healthy rootstock with the bud of the indicator tree above that of the source tree.

In wedge-grafting, the scion is trimmed to a wedge shape and inserted into the longitudinally split stem of the rootstock. Wedge-grafting is used to transmit viruses in woody or herbaceous crop plants. Introduction of a virus by grafting may lead to symptoms that differ from those resulting from mechanical inoculation. For example, when a healthy plant of *Nicotiana glutinosa* is wedge-grafted onto a plant of *N. tabacum* cv. White Burley infected with TMV, the former plant, which is hypersensitive to TMV, develops a severe systemic necrosis and dies.

In bulb and tuber grafting, parts of the healthy and diseased bulbs or tubers are brought into contact. This is done by fixing cut parts together (as in bulbous crops) or by core-grafting of potato tubers, in which a core without eyes is inserted into a bore hole made in the virus-free tuber. In both cases, plants developing from the initially healthy bulb or tuber will be diseased. In nature, root grafting has been shown to occur in woody plants.

TRANSMISSION BY VEGETATIVE PROPAGATION

When a plant is systemically infected, usually all its tissues except, in some cases, those in the meristematic regions of roots and shoots, contain the virus. Hence both scions, rootstocks and budwood, and bulbs, corms, rooting cuttings, runners, suckers, and tubers taken from virus-infected plants will also be infected.

The difference between the transmission of the two categories of plant material is that the former leads to real transmission of virus from a diseased plant to a healthy one, whereas the latter is not strictly a real transmission, as the virus was already present in these plant parts from the beginning.

Vegetatively propagated plants are the main virus source for annuals.

Many economically important viruses have been spread and are still disseminated by vegetative propagation.

TRANSMISSION BY DODDER

Virus transmission from a plant to another unrelated plant by grafting is usually impossible, because incompatibility between the two plants prevents organic union. In such a case, dodder (*Cuscuta* spp.), a parasitic plant of the family Convolvulaceae, may be used to transport virus from the diseased plant to a healthy one.

When a dodder stem comes into contact with a stem of the host plant, its epidermis forms a projection, which attaches itself firmly inside the stem of the host so that a haustorium (sucker) is formed. Cells on the surface of the haustorium develop into hyphelike bundles which penetrate deep into the host plant. Some of these bundles make contact with the xylem of the host and differentiate into water-transporting elements, others make contact with phloem and act as metabolite-transporting elements. Some evidence suggests that plasmodesmata are present in the haustoria, which are in contact with the cortex cells of the host plant.

When dodder parasitizes a virus-diseased plant, its stems can be trained to a healthy plant. The parasite then acts as a sort of pipeline between the diseased plant and the healthy one through which the virus is transported in the flow of metabolites.

Also, detached virus-containing dodder stems are able to transmit virus when they have established themselves on a healthy test plant.

Dodder has proved to be especially useful for the transmission of viruses or phytoplasmas from trees to herbaceous plants.

In most cases, no evidence of multiplication of the virus in dodder. An exception is *Cucumber mosaic virus,* a cucumovirus, which has been found to multiply in dodder and to cause symptoms in it.

Transmission of a virus by dodder may sometimes be complicated by the presence of a seed-borne virus, dodder latent mosaic virus, in the parasite. This virus is symptomless in dodder, but causes symptoms in a number of unrelated plant species.

BIBLIOGRAPHY

Bennett CW (1944). Latent virus of dodder and its effect on sugar beet plants. *Phytopathology* 34: 77-91.

Desjardins PR, Drake RJ, French JV (1969). Transmission of citrus ringspot virus to citrus and non-citrus hosts by dodder *(Cuscuta subinclusa). Plant Disease Reporter* 53: 947-948.

Dijkstra J, De Jager CP (1998). *Practical plant virology: Protocols and exercises.* Berlin, Heidelberg, New York: Springer-Verlag.

Driskel BA, Hunger RM, Payton MC, Verchot-Lubicz J (2002). Response of hard red winter wheat to *Soilborne wheat mosaic virus* using novel inoculation methods. *Phytopathology* 92: 347-354.

Franz AWE, Van der Wilk F, Verbeek M, Dullemans AM, Van den Heuvel JFJM (1999). Faba bean necrotic yellows virus (genus *Nanovirus*) requires a helper factor for its aphid transmission. *Virology* 262: 210-219.

Klein TM, Wolf ED, Wu R, Sanford JC (1987). High-velocity microprojectiles for delivering nucleic acids into living cells. *Nature* 327: 70-73.

Louie R (1995). Vascular puncture of maize kernels for the mechanical transmission of maize white line mosaic virus and other viruses of maize. *Phytopathology* 85: 139-143.

Madriz-Ordeñana K, Rojas-Montero R, Lundsgaard T, Ramirez P, Thordal-Christensen H, Collinge DB (2000). Mechanical transmission of maize rayado fino marafivirus (MRFV) to maize and barley by means of the vascular puncture technique. *Phytopathology* 85: 139-143.

Sarra S, Peters D (2003). *Rice yellow mottle virus* is transmitted by cows, donkeys and grass rats in irrigated rice crops. *Plant Disease* 87: 804-808.

Chapter 12

Serology

Marc H. V. van Regenmortel

INTRODUCTION

When preparations of plant viruses are injected into animals such as rabbits and mice antibodies specific to the viruses appear in the blood of the animals within a few days. The collective term for those substances that give rise to the production of antibodies is antigens. After collecting blood from the injected animal, the serum containing the antiviral antibodies is obtained by removing the red blood cells by centrifugation. This serum, which is referred to as an antiserum specific to the virus used for immunization, can be used in a variety of serological tests for detecting the presence of the corresponding virus. Antibodies specific to a virus can also be obtained from other sources such as hybridoma cells, from the egg yolk derived from laying hens that were immunized, or by genetic engineering and recombination technology. Since such antibodies are not obtained from serum, the term serology is somewhat inappropriate for describing techniques that use such reagents. Instead of referring to serological methods, it is quite common to describe these methods as immunoassays or immunochemical techniques.

ANTIBODIES

Antibody molecules are glycoproteins known as immunoglobulins, which are found in serum. Immunoglobulin G (IgG), the most common type of immunoglobulin, consists of four polypeptide chains linked by disulphide bridges (see Figure 12.1). There are two identical heavy (H) chains of about 450 amino acid residues and two light (L) chains of about 220 residues. Each H and L chain consists of several homologous domains of about 110 residues. The sequence of the amino terminal domains of the H and L chains varies in antibodies of different specificity. Within each of these variable (V) domains (V_H and V_L), three segments exhibit hypervariability and

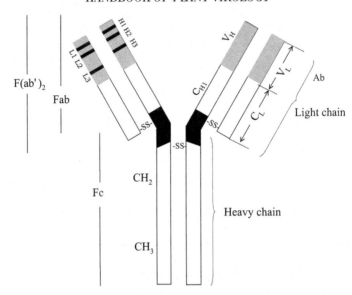

FIGURE 12.1. Schematic representation of an immunoglobulin G (IgG) molecule Ab, antigen-binding site. C: constant region; C_{H1}, C_{H2}, C_{H3}: constant regions on the heavy chain; CL: constant region on the light chain; F(ab')$_2$: fragments antigen-binding sites; Fc: fragment crystallizable; H: heavy chain; H_1, H_2, H_3: hypervariable (CDR) regions of the heavy chain; L: light chain; L_1, L_2, L_3: hypervariable (CDR) regions of the light chain; SS: disulphide bridge; V_H, V_L: variable domains on the heavy and light chain, respectively.

form the complementarity-determining regions (CDRs) of the immunoglobulin. The CDRs of the light chain are labeled L1, L2, and L3, and those of the heavy chain, H1, H2, and H3. The six CDRs, comprising in total about 50 amino acid residues, constitute the antigen-binding (Ab) site and are made up of six hypervariable loops located at the tip of the two variable domains of the Y-shaped IgG molecule.

The remaining domains in the H and L chains are invariant and are called constant (C) regions. The H chain contains three constant domains known as C_{H1}, C_{H2}, and C_{H3}, while the L chain contains a single constant domain called C_L.

Immunoglobulins can be cleaved into various fragments by the action of proteases. Treatment of IgG with papain cleaves the H chains at the N-terminal side of the disulphide bridges that keep the H chains together, thereby generating two Fab fragments (fragment antigen binding) and one Fc fragment (fragment crystallizable). Each Fab fragment contains one of the two identical combining sites of the antibody. Treatment of IgG with pepsin cleaves the H chains at the C-terminal side of the disulphide bridges, generating a single $F(ab')_2$ fragment that contains the two combining sites. These sites can be dissociated into two univalent Fab fragments after reduction of the disulphide bridge. Using pepsin digestion it is also possible to obtain Fv fragments, which consist of only the variable V_H and V_L regions that retain the antigen-binding capacity of the antibody. Using recombinant DNA technology it is nowadays straightforward to obtain Fv fragments or single-chain variable fragments (scFv) that consist of the V_L and V_H domains linked by a flexible peptide.

The V_L and V_H domains associate noncovalently in such a way that the six CDR loops come close together to form the binding site or paratope. Although the 50 or so residues that contribute to the CDRs may be viewed as a potential binding pocket, each individual paratope of a given antibody is made up of no more than 15 to 20 residues that are in contact with the specific antigen recognized by that antibody. This means that about two-thirds of the CDR residues are available for binding to other antigenic structures that may bear little or no resemblance to the first antigen. It is, therefore, not surprising that immunoglobulins are always multispecific (i.e., able to bind to a variety of antigens and not only to the one against which the antibody was elicited). The usual finding, however, is that the different subsites present in the immunoglobulin binding pocket partly overlap, and this will prevent two different antigens from being accommodated simultaneously. Although a given paratope tends to react with the homologous antigen used for raising the antibody with higher affinity than with heterologous cross-reacting antigens, it also sometimes happens that a paratope binds more strongly to a heterologous antigen. This phenomenon, known as heterospecific binding, can be put to practical use, for instance, when a variety of monoclonal antibodies reacting differentially with a number of viruses are selected following immunization with a single virus.

VIRAL ANTIGENS AND EPITOPES

A macromolecular assembly such as a virion or a dissociated viral protein subunit are both antigens because they are able, when injected into an animal, to elicit an immune response and to be recognized specifically by the antibodies induced by the immune response. Antigenic reactivity or antigenicity refers to the ability of the antigen to react with antibodies, while immunogenicity refers to the ability of the antigen to elicit an immune response.

Those parts of the antigen that react specifically with the paratopes of an antibody are known as antigenic determinants, or epitopes. Each epitope is delimited by the paratope footprint of about 800 Å^2: (3.2 × 2.5 nm) and consists of an array of atoms juxtaposed on the antigen's three-dimensional surface. The cluster of atoms forming an epitope can be identified only through its capacity to bind to a complementary paratope and not by intrinsic structural features that could be recognized independently of the epitope-paratope relationship.

Protein epitopes are usually classified as either continuous or discontinuous, depending on whether the amino acid residues in the epitope are contiguous in the polypeptide chain or not. In practice, the label "continuous epitope" is given to any linear peptide fragment of the protein that reacts with antibodies raised against the protein. Discontinuous epitopes, which correspond to the majority of epitopes in a protein, are made up of residues that are brought close together by the folding of the peptide chain. The distinction between continuous and discontinuous epitopes is fuzzy, since discontinuous epitopes usually contain several segments of a few contiguous residues that may be able on their own to bind to the paratope. Such segments may then be given the status of continuous epitope in spite of being part of a larger discontinuous epitope.

Since the quaternary structure of virions influences the properties of viral epitopes, it is useful to distinguish three types of epitopes, known as cryptotopes, neotopes, and metatopes. These different types of epitopes are illustrated in Figure 12.2 in the case of *Tobacco mosaic virus* (TMV), a tobamovirus. The virion consists of 2,130 identical capsid protein subunits arranged as a helix around an RNA molecule.

Cryptotopes are hidden epitopes buried inside the assembled virion that become accessible only after dissociation of the particle. Several cryptotopes have been located at the surface of the TMV capsid protein subunit by identifying which peptide fragments of the protein were able to bind to antibodies raised against the protein. Neotopes are found only on the surface of the assembled virion and are not present at the surface of dissociated subunits. They arise either through conformational changes in the monomers induced by intersubunit bonds or through the juxtaposition of residues from neighboring subunits. Metatopes are epitopes that are present in both dissociated and polymerized forms of the viral protein. The location of neotopes and metatopes on TMV particles has been visualized by immunoelectron microscopy using gold-labeled antibodies.

The surfaces of a viral subunit that harbor cryptotopes tend to have a chemical structure that is more conserved in the members of a given viral genus or family than the surface that harbors neotopes (part A in Figure 12.2). Since cryptotope-bearing surfaces interact during the polymerization

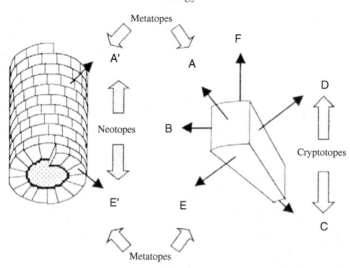

FIGURE 12.2. Schematic model of the protein subunits of *Tobacco mosaic virus* in monomeric form and in the virion. Neotopes are found on surfaces A' and E'. Metatopes are found on surfaces A, A', E, and E'. Cryptotopes are found on surfaces B, C, and D. The type of epitope present on surface F has not yet been defined. The allocation of epitopes was done by biosensor mapping. (*Source:* From M. H. V. van Regenmortel, 1999, Synthetic Peptides As Antigens, *Laboratory Techniques in Biochemistry and Molecular Biology,* Volume 28, page 51, figure 1.6. Reprinted with permission of Elsevier.)

of the subunits into virions, there are limits to the number of mutations they can accommodate while still retaining their functional activity. Antibodies directed to cryptotopes of a virus therefore tend to cross-react more widely with other members of the same virus species or genus than antibodies directed to neotopes. Antibodies raised against dissociated viral subunits or against the conserved core region of a viral protein frequently show broad-spectrum cross-reactivity. By immunizing animals with viral subunits or with synthetic peptides corresponding to conserved regions of the coat protein of potyviruses, it has been possible to obtain antibodies cross-reacting with a large number of distinct potyviruses.

ANTIBODY PRODUCTION

Polyclonal antisera against plant viruses are mostly obtained in rabbits, mice, or chickens by injecting the animals with purified suspensions of a virus or viral proteins. Usually a dose of 0.1 to 1.0 mg of purified antigen is administered intramuscularly at intervals of two to three weeks in the form of an emulsion with Freund's incomplete adjuvant, consisting of paraffin oil and emulsifier. Immunization with larger doses of antigen does not lead to proportionally higher antibody levels and has the disadvantage that contaminants are more likely to reach a level where they are also able to induce an immune response. Blood is collected from the animals after the second and subsequent injections. When laying hens are immunized, large quantities of antibody can be obtained on a daily basis from the egg yolks by precipitation with polyethylene glycol.

Antibodies obtained in this manner in rabbits or chickens are polyclonal (i.e., they correspond to heterogeneous mixtures of antibodies directed against the numerous epitopes present in a viral protein or a virion). Following the development of the hybridoma technology, monoclonal antibodies (MAbs) are increasingly replacing polyclonal antisera as the reagent of choice for identifying plant viruses. MAbs are obtained by cloning antibody-producing spleen cells from an immunized mouse after fusion of the spleen cells with myeloma cells. Methods for producing MAbs have been described in many textbooks, and procedures for obtaining MAbs to plant viruses have been summarized by Hampton and De Boer (1990) and Van Regenmortel and Dubs (1993). A critical step for obtaining suitable MAbs is the screening of culture supernatants after the cell-fusion step. Since MAbs are directed against a single epitope, it is possible to select antibodies that react only with the virus even when the mouse was injected with a preparation that contained contaminating plant antigens. MAbs directed against plant antigens can be identified by screening with extracts from healthy plants, and these can then be discarded. The screening of hybridoma is usually done by some form of enzyme-linked immunosorbent assay (ELISA), and it is important to use the same type of ELISA for screening as will be used in the subsequent work with the MAbs. The reason for this is that MAbs are not equally active in all types of ELISA. The experimental conditions used in different ELISAs (pH, temperature, adsorption step to the plastic surface, labeling of the antibody, and so on) may be detrimental to the activity of the Mab or may lead to the appearance or disappearance of cryptotopes or neotopes of the virus. For instance, MAbs selected for reacting with neotopes may fail to react in ELISA when the microtiter plates are coated with virus at pH 9.6, since this procedure tends to lead to the preferential adsorption of dissociated viral protein and to the disappearance of neotopes.

Advances in recombinant technology have made it possible to clone the genes encoding variable domains of MAbs and to express them in bacteria. The genes can be cloned from RNA prepared from hybridoma cells. The expressed proteins can be targeted to the periplasm of the bacteria, where proper folding of the antibodies will take place. The antibody genes can be cloned in different formats, for example, as scFv or Fab, and they can be genetically linked to peptide tags or enzymes in the form of fusion proteins. Recombinant antibodies against plant viruses obtained in this manner have been used in a variety of immunoassays.

The advent of the polymerase chain reaction and phage display technology has made it possible to obtain scFv antibody libraries. Polypeptides that correspond to antibody V_H and V_L regions are joined with a short linker sequence and fused to the N-terminus of the phage pIII minor coat protein (phagemid vector). The resultant ScFv-pIII fusion product is expressed on the surface of mature phage particles. The phage preparation is enriched for specific antigen-binding antibody fragments by a series of selection steps with viral antigen. Phages which carry antibodies that bind to the virus are separated from unbound phages, and the bound phages are then eluted from the antigen and used to infect bacterial cells. Phages are harvested from the infected bacterial cultures and used for further selection. This approach is particularly useful for obtaining scFv against viruses that are poorly immunogenic.

A major advantage of MAbs and recombinant antibodies over polyclonal reagents is the possibility of obtaining practically unlimited quantities of an antibody. These antibodies that recognize a single epitope also make it easier to differentiate between neotopes, metatopes, and cryptotopes, and they facilitate the mapping of epitopes at the surface of viral antigens.

The question is often asked whether a MAb or a polyclonal antiserum is the more specific reagent. As illustrated in Figure 12.3, the specificity of an antibody depends on its ability to discriminate between two antigens. If the aim is to distinguish between antigens 1 and 2, MAb anti-a would be considered nonspecific, since it reacts with both antigens, while MAbs anti-b and anti-d would be considered highly specific. For discriminating between the two antigens, a polyclonal antiserum containing a mixture of antibodies anti-a, -b, and -c would be more specific than MAb anti-a but less specific than MAb anti-b. It is in fact misleading to refer to a MAb as being specific for a protein antigen or a viral strain, since an antibody can be specific only for an epitope. In practice, antibodies are said to be specific if they achieve the discrimination that the investigator wishes to make. It is, indeed, possible either to select MAbs that do not recognize an antigenically related virus (because the epitope that binds to the MAb is not shared between the two viruses) or, on the contrary, to select MAbs that do not differentiate between the two viruses (because they recognize an identical epitope present in both).

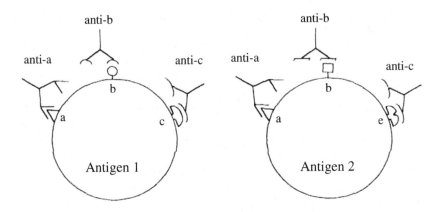

FIGURE 12.3. Potential cross-reactions between two antigens, 1 and 2. Since epitope a is present on both antigens, Mab anti-a will react in an identical fashion with antigens 1 and 2. MAbs anti-b and anti-d recognize unrelated epitopes and will show no cross-reactivity between the two antigens. MAb anti-c reacts strongly with the homologous epitope c and cross-reacts weakly with heterologous epitopes e. (*Source:* From M. H. V. van Regenmortel et al., 1998, From Absolute to Exquisite Specificity. Reflections on the Fuzzy Nature of Species, Specificity and Antigenic Sites, *Journal of Immunological Methods,* 216(1), 37-48. Reprinted with permission of Elsevier.)

PRECIPITATION TESTS

Precipitation occurs when sufficient quantities of antigen and antibody molecules are allowed to combine. The term precipitation is used when complexes of viral protein or virions become insoluble, whereas the term agglutination is used to describe the clumping of cells or particles of similar size. Quantitative precipitation or precipitin tests are performed in small tubes or in single drops of the mixed antigen and antiserum reagents deposited on a plastic or glass surface. Twofold dilutions of the reagents are usually mixed and incubated for 30 to 60 min before being examined for the presence of a visible precipitate.

In agglutination reactions, either the antigen or the antibody is first conjugated to the surface of red blood cells or of carrier particles such as latex. The large size of these conjugates will induce a visible serological clumping with antibody concentrations 100 to 1,000 times lower than is necessary for obtaining a visible precipitation. Detailed procedures have been described (see Bibliography).

IMMUNODIFFUSION TESTS

Immunodiffusion tests are precipitation tests carried out in gels instead of in free liquid. The most commonly used format is the double diffusion test, in which antigen and antibody diffuse toward each other into a gel that initially contains neither of them. As diffusion progresses, the two reactants meet and form a precipitation band.

Double diffusion tests are usually performed in petri dishes containing a gel of 0.5 to 1.5 percent agar or agarose. Wells are formed in the gel by positioning templates on the dish before pouring the agar or by using gel cutters after the agar has set. A commonly used well pattern consists of a central well of 4 to 7 mm diameter surrounded by six peripheral wells at a distance of 3 to 6 mm from the central well. After the reagents have been introduced in the wells, diffusion is allowed to proceed and precipitation bands are recorded. Evaporation can be prevented by pouring a layer of light mineral oil over the gel surface. The position and curvature of precipitation lines give an indication of the size and concentration of the antigen.

When two viral antigens diffuse from neighboring wells toward the same antibody well, three different precipitation patterns can develop at the position where the bands caused by the two antigens meet. These patterns, known as coalescence, spur formation, and crossing of lines indicate that the two antigens are serologically identical, related, or nonrelated. The main value of immunodiffusion tests lies in the ability of these patterns to provide a visible demonstration of the extent of similarity that exists between antigens.

Since many filamentous plant viruses do not diffuse readily in agar gel, it is customary to first dissociate such virus particles into protein subunits by the addition of various chemicals.

A procedure known as intragel cross-absorption is very useful for demonstrating the presence of distinct epitopes in related viruses. In this procedure, the cross-reacting antigen used for absorption is allowed to diffuse into the gel from a central well, thereby establishing a concentration gradient of antigen in the gel around the well. After 24 h, the antiserum is allowed to diffuse from the same well, which leads to a ring of precipitation around the well produced by the cross-reacting antibodies. Antibodies that are not precipitated by the cross-absorbing antigen will continue to diffuse freely and will form precipitation lines with the homologous antigen diffusing from a peripheral well; this indicates the presence of epitopes in the homologous antigen that are not present in the cross-reacting antigen.

ENZYME-LINKED IMMUNOSORBENT ASSAY

An ELISA is a solid-phase immunoassay in which each successive immunological reactant is immobilized on a plastic surface and the reaction is detected by means of enzyme-labeled antibodies. ELISA is the most commonly used immunological technique in plant virology. In direct ELISA procedures, the antivirus antibody is itself labeled with an enzyme, while in indirect ELISA procedures the enzyme conjugate is an anti-immunoglobulin reagent. The advantage of indirect ELISA is that a single, commercially available enzyme conjugate (for instance, an enzyme-labeled goat antirabbit immunoglobulin) can be used for detecting any number of viruses for which rabbit antibodies are available. This does away with the need to prepare separate enzyme conjugates for each virus. The precision of ELISA makes this assay suitable for quantitative measurements; for instance, for determining virus concentrations or the degree of antigenic relatedness between viruses. The maximum sensitivity of antigen detection by ELISA is of the order of 1 ng/ml.

ELISA is mostly performed in 96-well plastic microtiter plates. The first step consists in incubating antigen or antibody in the wells, usually for 2 h at 37°C or for 16 h at 4°C. After each step of adding reagents, the wells are washed with a buffer. After adsorption of the first layer of reagent on the plate, remaining sites on the plastic are saturated with 1 percent bovine serum albumin, ovalbumin, or defatted milk. Various ELISA procedures have been used in plant virology, and the most useful ones are listed in Exhibit 12.1. These methods have been described (see Bibliography). Parameters such as the incubation time, temperature, reagent concentration, and type of enzyme conjugate can be varied, and optimal conditions must be established empirically for each individual system.

Although ELISA is less sensitive than nucleic acid hybridization methods or biological assays on indicator plants, it is more suitable for routine, large-scale testing of field samples.

OTHER IMMUNOASSAYS

Various immunoblotting assays are commonly used in plant virology. In Western blotting, the components of a mixture are first fractionated by gel electrophoresis before they are transferred from the gel onto an overlayered membrane or nitrocellulose sheet by a second step of electrophoresis known as electroblotting. The different antigenic bands are then revealed with labeled antibodies.

EXHIBIT 12.1.
ELISA Procedures Used in Plant Virology

1. Ab^R, Ag, AB^R-E
2. Ag,Ab^R-E
3. Ag, AB^R, anti-R^G-E
4. Ab^C, Ag, Ab^R, anti-R^G-E
5. $F(ab')^R_2$, Ag, Ab^R, anti-Fc-E
6. $F(ab')^R_2$, Ag, Ab^R, PA-E
7. Ab^C, Ag, (Ab^R, anti R^G-E)
8. $F(ab')^R_2$, Ag, (Ab^R, anti-Fc-E)
9. Ab^C, Ag, Ab^M, anti-M^R, anti-R^G-E
10. PA, Ab^R, Ag, Ab^R, PA-E
11. $Ab^{C/M}$, Ag, $Ab^{R/M}$- biot, Avidin-E

Note: Reagents are listed as successive incubation steps.
Ab, antibody; Ag, antigen; R, rabbit; E, enzyme label; G, goat; C, chicken; M, mouse; PA, protein A; biot, bioten.

In dot-blot immunoassay, purified antigen or crude sap from infected plants is applied (2 μL drops) directly to a nitrocellulose membrane and air-dried. Antigen is detected by successive incubations with antivirus antibody, an anti-immunoglobulin enzyme conjugate and substrate. As little as 0.5 pg of virus can be detected by this technique.

In recent years, biosensors based on surface plasmon resonance have become the method of choice for analyzing antigen-antibody interactions in terms of affinity and kinetic constants. The affinity of antibodies determines to a considerable degree their suitability for use in different immunoassays.

Biosensor technology is also very useful for mapping epitopes at the surface of virions and viral proteins. Biosensor instruments make it possible to visualize on a computer screen the antigen-antibody binding process as a function of time, as one of the interacting partners binds to its ligand, immobilized on the surface of a sensor chip. None of the reactants needs to be labeled, which avoids the artifactual changes in binding properties that often result from labeling. The binding stoichiometry is easily calculated from biosensor data, and conformational changes that occur in a viral protein as a result of antibody binding can also be monitored. It is also possible to visualize each of the successive binding steps in a multilayered assay, and this allowed the unambiguous demonstration of conformational epitopes in TMV.

SEROLOGICAL DIFFERENTIATION INDEX

The degree of antigenic relatedness between two viruses is usually expressed in terms of a serological differentiation index (SDI), which corresponds to the average number of twofold dilution steps separating homologous and heterologous antiserum titers. The homologous titer refers to the titer of an antiserum with respect to the antigen used for immunizing the animal, while the heterologous titer refers to the titer with respect to another related antigen. If the homologous titer (the last twofold antiserum dilution at which a reaction is visible) is 1/1024 (2^{-10}) and the heterologous titer is 1/128 (2^{-7}) the SDI is 3. The SDI is a reliable measure of antigenic relatedness only if it represents the average value calculated from a large number of bleedings from different immunized animals. Average SDI values obtained in reciprocal tests with several antisera to the two viruses that are being compared have been shown to agree closely and to correlate with the degree of sequence similarity of the coat proteins of the two viruses. Since Mabs are specific for single epitopes, information on the degree of overall antigenic relatedness between two viral proteins that always possess many different epitopes can be obtained only from polyclonal antiserum. SDI values can be obtained from precipitin tests or from ELISA, which requires much smaller quantities of reagents. SDI values are calculated from ELISA results by comparing the antiserum dilutions that lead to the same absorbance (e.g., 0.5 or 1.0) for the homologous and heterologous antigens. Since ELISA titers are always much higher than precipitin titers for the same antiserum, ELISA is able to detect more distant antigenic relationships than precipitin tests. It is customary to consider that two viruses that differ by an SDI value larger than 4 to 5 correspond to separate virus species.

BIBLIOGRAPHY

Dijkstra J, De Jager CP (1998). *Practical plant virology: Protocols and exercises.* Berlin, Heidelberg, New York: Springer-Verlag.

Hampton R, Ball E, De Boer S (eds.) (1990). *Serological methods for detection and identification of viral and bacterial plant pathogens.* St Paul, MN: American Phytopathological Society Press.

Harper K, Ziegler A (eds.) (1999). *Recombinant antibodies: Applications in plant science and plant pathology.* London: Taylor & Francis.

Van Regenmortel MHV (1982). *Serology and immunochemistry of plant viruses.* New York: Academic Press.

Van Regenmortel MHV, Dubs MC (1993). Serological procedures. In Matthews REF (ed.), *Diagnosis of plant virus diseases* (pp. 159-214). Boca Raton, FL: CRC Press, Inc.

Chapter 13

Detection and Identification
of Plant Viruses and Disease Diagnosis

Francisco J. Morales

INTRODUCTION

Plant viruses remain one of the main biotic constraints to food and industrial crop production despite significant advances in the areas of molecular plant virology and genetic engineering. Moreover, the globalization of the economy and ensuing new world trade regulations, which discourage the application of strict plant quarantine measures, are expected to accelerate the international dissemination of plant viruses. Fortunately, highly sensitive molecular techniques are now available to detect and identify plant viruses. The adoption of these techniques greatly facilitates disease diagnosis, and it seems that the spectacular advances in plant virus characterization will further benefit in the near future from equally impressive advances in the field of electronics. Nevertheless, plant pathologists should keep in mind that plant disease diagnosis is largely dependent on the biological interactions between plant pathogens, their hosts, and the environment. Thus, adequate training of molecular plant pathologists in basic agricultural sciences remains an essential prerequisite for effective management of plant diseases in sustainable crop production systems.

DETECTION

To detect a plant virus is to establish the presence of the particular virus in plants, vectors, or inert media. The detection of plant viruses is greatly aided by the expression of disease symptoms in infected plants, but it is more critical in the case of plants that do not show visible symptoms (e.g., plant material used for certification programs or phytosanitary purposes). Also, some viruses are highly stable outside their plant hosts and may remain infectious on processed plant materials (e.g., tobacco) or in the environment (e.g., soil, water) for extended periods of time.

Physical Properties

The physical properties of the virion determine to a large extent the most reliable virus detection method. For instance, morphology and size may greatly facilitate or hinder the detection of plant viruses in infected plant material. Most rod-shaped, bacilliform and filamentous viruses are readily observed by simple negative-staining techniques using electron microscopy, whereas the majority of isometric viruses are not easily discerned from other morphologically similar plant components, unless the virus occurs in high concentration in the infected plant.

Chemical Properties

The chemical characteristics of virions are also important for virus detection. Most plant viruses are nucleoproteins made up of one or more protein subunits and either ribonucleic acid (RNA) or deoxyribonucleic acid (DNA). The properties of the viral nucleic acid are very useful for virus detection purposes. Considering that the majority of plant viruses contain RNA, and that RNA viruses form a double-stranded (ds) RNA molecule as an intermediate replication product, the extraction of dsRNAs from plants and their subsequent visualization by gel electrophoresis are valuable detection techniques for RNA plant viruses. The use of virus-free controls is important for this assay, as some symptomless plants may contain dsRNA molecules of diverse origin.

Nucleic Acids

Nucleic acid-based detection methods are essential for pathogens devoid of a capsid protein (CP), namely some viruses, satellite RNAs, defective and defective-interfering RNAs, and viroids. Nucleic acids can be detected by molecular techniques such as nucleic acid hybridization (NAH) and polymerase chain reaction (PCR). The NAH technique involves the binding of two strands of nucleic acids: two RNAs, two DNAs, or a DNA:RNA hybrid. To this effect, probes must be prepared according to the type of viral nucleic acid. In the case of dsDNA, nick-translation of the dsDNA is achieved with a restriction enzyme, followed by treatment of the ss-nicks with DNA polymerase to replace the single-stranded (ss) DNA with new nucleotides. One of the four new deoxyribonucleotides (dNTPs) is then labeled radioactively or nonradioactively to produce the labeled RNA probe. Single-stranded probes eliminate the possible reannealing problems of ds probes. Single-stranded DNA probes can be prepared by different methods, such as the production of complementary DNA (cDNA) reversely transcribed from viral RNA. For the preparation of ssRNA probes, a series of plasmid vectors and in vitro transcription systems are available. The most frequently used radioactive label is the isotope ^{32}P, which has a half life of about 14 days. For nonradioactive labeling, biotin and digoxigenin have been the labels of choice in recent years. These "labels" (often referred to as haptens), attach to nucleic acids and are then detected using a specific binding protein.

Sample preparation depends on the virus-host interaction (e.g., plant genotype, organ affected, virus location). In some cases, simple sap extracts might suffice. Different extraction buffers have been used in cases where stabilizing agents are required (e.g., antioxidants, ribonuclease inhibitors). However, for a large number of viruses and plant species, special nucleic acid extraction methods are required. Some of these methods are simple and rapid (e.g., phenol extraction and ethanol precipitation), but other methods may require more laborious and expensive procedures. The main sample support media for "dot-blot" or "spot hybridization" assays are nitrocellulose or nylon membranes. Nylon membranes are generally preferred due to the fragility of nitrocellulose. Samples are usually applied onto membranes as aliquots, although some researchers have squashed or printed plant tissue (e.g., leaf tissue) or plant organs (e.g., cut twigs or pods) onto membranes. RNA and DNA can be denatured before or after placing the samples on the membrane. In a prehybridization step, membranes are incubated in a solution to coat the sites on the membrane where the probe could bind nonspecifically. For hybridization, ss probes are added, adjusting the temperature up or down (37-75°C) depending on the stringency (high or low, respectively) of the test. The specificity of hybrid formation is also affected by the stringency of the final washing step. Increasing the temperature 5-15°C below the melting temperature of the hybrid molecule and lowering the salt concentration increases the specifity of hybrid formation. After washing, reactions on the membranes can be visualized by autoradiography (radioactive procedure) or colorimetrically (nonradioactive assays). Though the sensitivity of NAH depends on multiple factors, it is known to detect pg quantities of viral nucleic acid.

The PCR technique has been one of the major technological breakthroughs of modern science. The extreme sensitivity of PCR resides in its power to amplify specific DNA sequences in vitro. Basically, three different temperature cycles permit the melting, annealing, and extension of the oligonucleotide primers with a thermostable DNA polymerase, to produce large quantities of the target DNA fragments. PCR is at least 100 times more sensitive than most serological methods, which makes it a very useful technique to detect viruses that occur in very low concentration in infected plants. PCR is particularly suited for the detection of ss or dsDNA viruses, but RNA viral sequences can also be readily detected following their transformation into cDNA using reverse transcriptase (RT-PCR). The use of degenerate primers (oligonucleotides with only partial homology to the target nucleic acid) in the PCR technique permit the detection of conserved nucleotide sequences present in different virus species belonging to a given genus. These broad-spectrum or "universal" primers have been used to detect virus species belonging to at least 15 different virus genera. However, if virus detection must be accomplished at the species or strain level, highly specific primers can be designed based on less-conserved regions of the viral genome.

Proteins

Regarding the properties of viral proteins, the capsid protein of non-enveloped plant viruses consists of one or more protein subunits of Mr greater than 10,000, which possess important physical, chemical, and biological properties. For detection purposes, the CP of plant viruses absorbs ultraviolet radiation at wavelengths between 250 (minimum absorption) and 280 (maximum absorption) nm. Purified viral preparations consisting mainly of concentrated nucleoprotein show a characteristic curve with a smooth slope between 260 and 280 nm when analyzed in a UV spectrophotometer. In the absence of viral nucleic acids, the UV absorption curve would show a very steep slope with a sharp peak at 280 nm. Isometric viruses tend to show steeper slopes than most filamentous viruses, due to their greater nucleic acid:protein ratios. Viral CP subunits can be visualized by electrophoretic techniques in purified virus preparations and their molecular mass determined as a characteristic virus identification feature.

The CP is usually a good immunogen that can elicit the production of antibodies, particularly immunoglobulin G (IgG), in the immune system of vertebrate animals (see also Chapter 12). The availability of specific antisera to many plant viruses greatly facilitates plant virus detection. The majority of antisera produced to date are polyclonal antisera made up of different antibodies elicited in laboratory animals against the various epitopes or antigenic determinants present on viral CPs. Polyclonal antisera are not usually highly specific because different virus species belonging to the same genus may share common antigenic determinants, which may lead to cross-reactivity between closely related virus species. However, this limitation was overcome with the advent of monoclonal antibodies developed from single-antibody-producing cells fused to a mouse myeloma cell, to produce a hybrid antibody-producing cell (hybridoma) that can be grown and maintained in vitro. Monoclonal antibodies react only with a single antigenic determinant on a protein and, therefore, may be highly specific. However, this property has also been used to develop "broad spectrum" monoclonal antibodies to highly conserved antigenic determinants present in different virus species. These universal monoclonal antibodies are currently widely used for virus detection at the genus level.

Some plant virus genomes also code for nonstructural proteins that are necessary for the movement, transmission, and multiplication of viral pathogens. Some of these proteins can be detected in infected tissue using different methods, such as light or electron microscopy, electrophoresis, serology, blotting, and molecular techniques. For instance, the genome of potyviruses codes for a helicase, which forms cylindrical inclusions in the cytoplasm of infected hosts. These characteristic protein inclusions can be detected with the aid of the light microscope in infected tissues, and several workers have used protein inclusions to produce antisera.

Biological Properties

The biological properties of virions are equally important for the detection of plant viruses. Some susceptible plant species (e.g., tobacco, common bean) display characteristic symptoms when infected by different plant viruses. Establishing the host range of plant viruses helps identify those plant species that can be used to detect these pathogens following standard virus inoculation procedures. Some plant species react to plant viruses in a hypersensitive manner that localizes the pathogen at the inoculation site. These local-lesion assay plants can also be used to detect the presence of plant viruses, particularly in the case of latent infections. Unfortunately, not all viruses can be transmitted by manual or mechanical means (e.g., rubbing of infected plant extracts onto leaves of healthy test plants). Some viruses are transmitted only by vectors, such as insects, mites, fungi, and nematodes. In these cases, other techniques, such as grafting, can be used to detect plant viruses that are not mechanically transmissible (see also Transmission by Grafting in Chapter 11).

IDENTIFICATION

Once viruses are detected, they should be identified (i.e., their specific identity should be established). The identification of a viral pathogen implies that Koch's postulates have been satisfied (i.e., the virus must be isolated from the original host and transferred to a susceptible test plant, where disease symptoms must be reproduced and shown to be caused by the same virus that was inoculated onto the test plant). However, pathogens can be temporarily "associated" with a particular disease until their pathogenicity is demonstrated.

Physical Properties

For practical purposes, the physical properties of virions are only one of the various characteristics that must be taken into account for the identification of plant viruses. The shape of a virus reflects the limited number of ways in which CP subunits can be arranged around viral nucleic acid. Detailed examination of virus structure is possible and is usually performed by electron microscopy or X-ray crystallography.

Electron Microscopy

Electron microscopy usually provides basic information on the morphology of virions, making possible the tentative placement of a virus in a particular family or genus. Immunosorbent electron microscopy (ISEM) is a technique that combines the magnifying power of electron microscopy with the specificity of serology. This technique is recommended in the case of viruses that cannot be easily discerned in infected plant extracts because of their morphology (e.g., isometric viruses) or low concentration. Viruses possessing bacilliform, filamentous, or rod-shaped particles are easier to visualize and characterize according to their particle size. However, the shape and size of some of these viruses can be modified by the chemical composition of the suspending medium in which they are observed. Filamentous and rod-shaped viruses can also show considerable particle fragmentation or aggregation, which may drastically change their native morphological characteristics. Isometric viruses are best observed in partially purified (concentrated) suspensions. It is advisable to include particle size standards (microscopic particles of known size) in virus preparations used to determine virion size and measure a statistically significant number of particles (ideally 100 particles). The data collected can be analyzed using a histogram of particle length/diameter frequencies. This procedure is particularly recommended for multicomponent (multiparticulate) viruses, such as furoviruses, pecluviruses, benyviruses, hordeiviruses, and tobraviruses.

Chemical Properties

The chemical properties of virions and, more specifically, the properties of the viral nucleic acid constitute nowadays the ultimate criterion for identification of plant viruses, due to the rapid development of molecular techniques for the characterization of plant virus genomes. Approximately 5 percent of the plant viruses have DNA genomes and the rest (95 percent) have RNA genomes, including 3 percent with dsRNA genomes. The viral genomes can be

1. reverse transcribed (retroid) and encapsidate an RNA (retroviruses) or DNA (pararetroviruses) copy of the genome or
2. not reverse transcribed (nonretroid).

The nonretroid viruses can be ssDNA, dsRNA, negative/ambisense ssRNA, or messenger-sense ssRNA. Positive-sense means that the viral nucleic acid can act as messenger (m) RNA and, thus, code for a protein. Negative-sense means that a viral polymerase is necessary for the synthesis of a positive strand. Eukaryotic mRNAs are monocistronic (only the first open reading frame is translated). RNA viruses usually express more than one gene in their genomes following different strategies. For instance, the viral genome may be distributed over different genomic RNAs (multipartite genome). Some viruses follow an ambisense (ambiguous sense) transcription strategy. Ambisense viruses possess genes that are encoded on both positive and negative strands. Finally, some viral genes are transcribed from internal sites in the viral RNA into subgenomic (sg) RNAs. These sgRNAs are then accessible to ribosomes for translation. Understanding genome composition, structure, and function has made possible the development of reliable virus detection and identification methods. For virus identification purposes, it is the order or sequence of the nucleotides that make up the viral genome that ultimately establishes the identity and genomic organization of a plant virus.

In order to determine a nucleotide sequence, DNA fragments are generally produced with restriction endonucleases, and then amplified by inserting the DNA into a plasmid or bacteriophage. Plasmids are small circular dsDNA molecules found in bacteria, and bacteriophages are bacterial viruses. These cloning vectors integrate viral DNA fragments into their genomes. The resulting hybrid DNA molecules are reinserted into the bacterium and allowed to grow into colonies on culture media. Transformed bacterial colonies containing the viral DNA are selected based on the properties of their plasmid vectors (e.g., antibiotic resistance). As mentioned previously, bacteriophages can also be used to carry viral DNA fragments into a bacterial cell. The M13 phage, for instance, provides a ssDNA template, which facilitates cloning and sequencing procedures. Because most plant viruses contain RNA, a cDNA chain has to be synthesized on an RNA template. This step was possible after the discovery of the enzyme reverse transcriptase. The ss cDNA is then converted to a dsDNA molecule using a DNA polymerase, which can then be inserted into a plasmid to be cloned and sequenced. One of the most widely used among the sequencing methods has been the dideoxy chain termination method. Nowadays, however, sequencing is done using automatic sequencers or DNA analyzers designed for automated operation with minimum supervision, and very rapid results (minutes to a few hours, depending on the instrument).

The nucleotide sequence of selected regions of viral genomes can also be determined by PCR, using mixtures of different primers, followed by direct sequencing or cloning of the amplified PCR products into a plasmid vector.

Primer selection can be based on previously identified conserved regions of the viral genome for a particular virus species or genus. The use of degenerate oligonucleotide primers (i.e., a mixture of primers in which the position of some nucleotides vary according to a possible consensus sequence) permits the detection of plant viruses at the species, genus, or family level. The resulting PCR products can then be sequenced for increased specificity of detection. The existence of plant viruses with 3'-polyadenylated (polyA) tails further facilitates the use and design of degenerate primers for virus identification.

Partial sequencing of viral genomes is a very useful and rapid method to identify plant viruses. However, it must be kept in mind that plant viruses have evolved through mutation and recombination and that phenomena, such as pseudorecombination, frequently occur in nature and change viruses in specific regions of their genome. Hence, partial sequences may sometimes fail to recognize inherent or acquired differences in the organization of plant viral genomes. For instance, the genus *Begomovirus* contains an ever-increasing number of different species with two molecules of ssDNA (2,500 to 3,000 nucleotides). These molecules, designated as A and B, contain two divergent sets of genes separated by an intergenic region *(IR)* that includes a segment of 180 to 200 nucleotides, referred to as the common region *(CR)*. The *CR* is the region with the highest degree of homology between the two ssDNA components. The A component contains 4 to 5 genes, including the *CP* and the replication initiation protein *(rep)*, and the B component two genes. Although the complete sequence of the genomes of several bipartite begomoviruses is known, the identification of the numerous begomoviruses found in nature is usually done employing one or two sets of primers designed for target regions of the A component. The most widely used set of primers amplifies part of the *rep,* the complete *IR,* and part of the *CP.* However, most of the fragments obtained with this set of primers do not include the *IR.* The second set of primers amplifies the core region of the *CP.* Although the *CP* is the most conserved gene among begomoviruses both of New World and Old World origin, the latter set of primers can differentiate begomoviruses at the species cluster level. This limited sequence, however, does not detect mutations or recombinations in other parts of the viral genome. However, the first set of primers sometimes produces sequences that show a high degree of similarities or identities to different begomovirus species, depending on the gene compared *(rep* or *CP).* Moreover, none of the two sets of primers can detect pseudorecombination (the A component from one virus species and the B component from another begomovirus species).

In conclusion, the partial sequencing of plant virus genomes only permits a tentative identification of plant virus species. Molecular virus characterization techniques have practically replaced traditional methods used in the past to identify plant viruses. Yet, many of the virus species currently recognized by the International Committee on Taxonomy of Viruses remain to be characterized at the molecular level.

Biological Properties

With the advent of molecular virus identification techniques the biological properties of plant viruses have been relegated to a second place, even though the biological interaction between a plant virus and a plant genotype determines the importance of viruses as plant pathogens. For instance, Peanut stripe virus has been shown to be closely related to *Bean common mosaic virus* (BCMV), a potyvirus, at the molecular level and it is currently recognized as a strain of BCMV. However, from the biological point of view, Peanut stripe virus differs considerably in pathogenicity and virulence from known BCMV strains.

The biological interaction between plants and viruses also depends on the mode of transmission of these pathogens. *Tobacco mosaic virus* (TMV), a tobamovirus, one of the first viruses to be studied, was clearly a contagious agent transmitted by manual or mechanical means. Grafting was one of the first techniques used to transmit viruses, even before viruses were recognized as such. Systemic plant viruses are regularly transmitted through vegetative propagation material, such as tubers, corms, bulbs, and budwood. Transmission of plant viruses through seed can occur passively by contamination or infection of seed tissues outside the embryo. However, most seed transmission of plant viruses occurs in the embryo, before the production of gametes or cytoplasmic separation of embryonic tissue from the mother plant. Pollen may also become infected and result in the infection of the embryo.

A large number of plant virus species are transmitted by invertebrate animal vectors, namely insects and mites (see also Chapter 9). Insects with piercing and sucking mouthparts are by far the main vectors of plant viruses, particularly species in the order Homoptera, such as aphids, leafhoppers, planthoppers, and whiteflies. Some insects with chewing or rasping mouthparts, such as beetles and thrips, are also efficient virus vectors of economically important viruses. The mode of virus transmission by insects is primarily determined by the duration of retention (persistence) of the virus in the vector. Nonpersistent viruses can be acquired and transmitted within seconds, although some vectors of this type of virus can retain the virus for hours. Semipersistent viruses take longer to be acquired (about 5 min) and transmitted, and they can persist for a few days. However, as in the latter case, these viruses do not have a latency period and are lost when the insect molts. Persistent viruses are usually located in the vascular bundles and, hence, their insect vectors take approximately 15 min or more to acquire them from infected plants. Once acquired, persistent viruses must circulate within the insect (circulative viruses) and can become infective only after a latency period. Infectivity is not lost during molting, and the concentration of some of these viruses may increase inside the insect vector (propagative viruses). Some persistent viruses can establish a very close biological relationship with their insect vectors, becoming transovarially transmitted and sometimes pathogens of their invertebrate vectors. Transmission of plant viruses by soil-inhabiting vectors, such as nematodes and fungi, is also possible. Different species of nematodes and fungi are efficient vectors of several plant virus genera of economic importance (see also Chapter 10).

The biological transmission of plant viruses by vectors is a rather specific process, which contributes to the identification of plant virus genera. However, some of the recognized plant virus genera do not have a known vector, and they probably disseminate through contamination of water and soil media where susceptible plants grow.

DIAGNOSIS

Diagnosis is defined as "the identification of the nature of an illness or other problem by examination of the symptoms" (*Oxford English Dictionary,* 1999). However, although diagnosis of viral diseases is primarily dependent on symptom expression in infected plants, it must be borne in mind that many viruses do not incite symptoms in their plant hosts. Thus, reliable diagnosis of a virus disease also requires the previous identification of the causal viral pathogen (etiological diagnosis), including fulfillment of Koch's postulates.

Symptoms may be either "histological" or "morphological" (see also Chapter 3). Histological symptoms affect plant cells internally, and consequently require a cytopathological study; that is, the observation of changes in cell structure, content, or arrangement induced by the pathogen. In the case of plant viruses, these studies can be carried out with the aid of an optical microscope or, better yet, an electron microscope. The possibility of using ordinary light microscopes to diagnose plant virus infections was fully realized in the 1960s, when inclusion bodies or amorphous inclusions were consistently observed in plant cells infected by viruses such as *Bean yellow mosaic virus,* a potyvirus, or TMV. These inclusion bodies are now known to have a characteristic structure and composition according to the virus genus and, consequently, constitute a diagnostic characteristic. Some of these inclusion bodies have been shown by electron microscopy and other biochemical assays to consist of high concentrations of virus particles in regular crystalline arrays or dispersed in the cytoplasm. The amorphous inclusions often represent viroplasms (i.e., intracellular sites of virus synthesis) where immature virus particles form complexes with plant cell organelles, either in the cytoplasm or the nucleus. Plant viruses may also induce the formation of inclusion bodies consisting of proteins in characteristic crystalline arrays. For instance, potyviruses induce different types of cytoplasmic inclusions whose morphology does not vary with either the host or environmental conditions and, thus, are of diagnostic value.

In general terms, plant viruses may cause deleterious effects on nuclei, mitochondria, chloroplasts, cell walls, and other cell structures. Morphological symptoms consist of the morbid conditions displayed by plants as a result of the histological symptoms induced by the pathogen. For instance, rots, dwarfing, and plant malformations are morphological symptoms resulting from histological conditions causing cell necrosis and abnormal cell multiplication (hypoplasia or hyperplasia). Symptoms are either localized (e.g., local lesions) or systemic (e.g., variegation, yellowing, ringspots) and may be either primary (i.e., the direct damage caused by the pathogen, such as vascular necrosis) or secondary (i.e., the result of the primary symptom, in the latter case, wilting). Thus, symptom development is often the expression of the effect of a plant virus in a given organ or tissue of the plant host. For instance, phloem-limited viruses tend to affect the translocation of nutrients and, consequently, are often associated with systemic symptoms, such as yellowing and wilting (e.g., luteoviruses). Viruses replicating in the cytoplasm tend to cause leaf malformation (e.g., potyviruses), and viruses that replicate in the nucleus tend to disrupt the entire plant, causing plant malformation and significant yield losses (e.g., geminiviruses). Currently, some plant pathologists consider the diagnosis of plant virus diseases as an integral part of virus identification, which requires considerable laboratory support to complement the clinical observation of disease symptoms. However, once a virus has been adequately characterized (identified) and its biological properties studied, visual examination of disease symptoms may suffice for diagnosis of the disease.

CONCLUSION

DNA-based technologies have been the basis for molecular detection and identification of plant viruses in modern times. Genomics and biosystematics research are paving the way for the development of novel DNA detection techniques, such as the detection of single-nucleotide polymorphisms (SNPs), the DNA microarrays for multiplex pathogen detection, and electronic pathogen-detection methods. Companies providing diagnostic services based on automated DNA techniques will grow in number, and hand-held devices for plant disease diagnosis will probably become commonplace in the near future. Nevertheless, the etiological diagnosis of plant disease by well-trained plant pathologists should remain a valuable asset under field conditions, particularly in developing countries where modern diagnostic tools may not become widely adopted in the near future.

BIBLIOGRAPHY

Bos L (1999). *Plant viruses: Unique and intriguing pathogens—A textbook of plant virology*. Leiden: Backhuys Publishers.

Dietzgen RG (2002). Application of PCR in plant virology. In Khan JA, Dijkstra J (eds.), *Plant viruses as molecular pathogens* (pp. 471-500). Binghamton, NY: The Haworth Press, Inc.

Hari V, Das P (1998). Ultra microscopic detection of plant viruses and their gene products. In Hadidi A, Khetarpal RK, Koganezawa H (eds.), *Plant virus disease control* (pp. 417-427). St. Paul, MN: APS Press.

Padidam M, Beachy RN, Fauquet CM (1995). Classification and identification of geminiviruses using sequence comparisons. *Journal of General Virology* 76: 249-263.

Rojas MR, Gilbertson RL, Russel DR, Maxwell DP (1993). Use of degenerate primers in the polymerase chain reaction to detect whitefly-transmitted geminiviruses. *Plant Disease* 77: 340-347.

Sambrook J, Fritsch EF, Maniatis T (1989). *Molecular cloning: A laboratory manual,* Volumes 1-3, Second edition. Cold Spring Harbor, NY: Cold Spring Harbor Laboratory Press.

Sanger F, Nicklen S, Coulson AR (1977). DNA sequencing with chain-terminating inhibitors. *Proceedings of the National Academy of Sciences USA* 74: 5463.

Singh RP, Nie X (2002). Nucleic acid hybridization for plant virus and viroid detection. In Khan JA, Dijkstra J (eds.), *Plant viruses as molecular pathogens* (pp. 443-470). Binghamton, NY: The Haworth Press, Inc.

Chapter 14

Ecology and Epidemiology

Michael J. Jeger

INTRODUCTION

The word "ecology" was first used by the German zoologist Ernst Haeckel in 1869, broadly meaning the study of the natural environment and of the relations of organisms to one another and their surrounding environment (after the Greek *oikos,* meaning home). Since then more succinct definitions have emerged, stressing the factors and interactions (biotic and abiotic) that determine the distribution and abundance of organisms. At the level of the individual organism, ecology deals with how individuals affect and are affected by the biotic and abiotic environment. At the level of the population (consisting of individuals of the same species), ecology deals with the occurrence of particular species and their population dynamics. At the level of the community (consisting of more than one population), ecology deals with the composition, diversity, and functioning of the community. Ecology is equally concerned with relatively natural and human-made or -influenced environments.

Epidemiology is broadly defined as the study of disease in populations of hosts. Operationally, a distinction can be made between microparasites and macroparasites as disease-causing organisms. Microparasites multiply directly within their host, usually in host cells, macroparasites grow in their host but multiply by producing infective stages, which are released from the host to infect new hosts. In plant pathology, disease-causing agents are known as pathogens, and plant viruses are clearly microparasites which, to use the same epidemiological terminology, can be transmitted directly from host-to-host (mechanical, seed, pollen, soil) or indirectly through another species (the vector).

The author acknowledges Blackwell Publishing for permission to reproduce figures in this chapter.

These definitions of ecology and epidemiology raise clear questions in the study of plant viruses. What is the individual virion? What is the abiotic and biotic environment of the individual virion (the physico-chemical environment of the host cell or the vector)? Is the population definition sufficient to describe the genetically heterogenous collection of individuals within a virus species? Does the interaction of virus, host, and vector populations automatically mandate a community-level approach to virus epidemiology?

Plant pathologists have made much use of the so-called epidemiological triangle to stress the interactions between pathogens, hosts, and the environment. For most plant viruses this concept must be extended to include the vector. In the 1930s, the American entomologist Walter Carter developed the concept of the ecological trinity of viruses, hosts, and vectors within a particular environment. Accordingly, an ecological approach considering the interactions of viruses and vectors with crops, weeds, and other wild or volunteer hosts and the influence of the environment and cropping practices is necessary to understand the conditions in which crop epidemics occur. In 1980, an ecological approach to the epidemiology of plant virus diseases was presented by Thresh, who also introduced the concept of a dynamic equilibrium in which neither host nor virus gains permanent ascendancy and epidemics occur as a consequence of perturbation to previously stable situations caused by new crops, cropping practices, and intensification of cultivation.

SOURCES OF INFECTION

For a virus to persist in a given environment it must have ecological competence to invade and multiply rapidly when suitable habitats are present. There must also be an effective means of survival between growing seasons. For example, *Bean common mosaic virus,* a potyvirus, is seedborne; *Peanut clump virus,* a pecluvirus, persists in the long-lived resting spores of its soil-inhabiting fungal vector; and *Maize dwarf mosaic virus,* a potyvirus, infects wild perennial grasses that occur commonly in or near maize fields. Sources of infection can be internal or external to the field of interest and influence the subsequent development of disease. Sometimes this is unrelated to geographical distribution and activities of vectors. With *Zucchini yellow mosaic virus* (ZYMV), a potyvirus, for example, there can be equally high numbers of aphid vectors at different sites, but infection at only one site. Even when plants showing symptoms are routinely destroyed, thus eradicating secondary infection sources, new plants become infected over the whole season. ZYMV-infected plants as primary internal virus sources play a prominent role in starting epidemics of the disease in the field.

Weed species and wild hosts are major sources of inoculum for *Cucumber mosaic virus* (CMV), a cucumovirus, within and external to field-grown cucumber. Seed infection by CMV in lupins is a primary source for subsequent virus spread by aphids. Higher seed infection levels, better establishment of seed-infected plants, and early aphid arrival favor greater virus spread, yield loss, and infection in the harvested seed. Recommendations in Western Australia for the management of CMV infection in lupin include sowing seed with less than 0.5 percent infection to minimize infection sources and sowing at a high seeding rate to remove seed-infected, usually stunted plants through early canopy formation and shading by neighboring healthy plants.

VECTORS AND MODE OF SPREAD

Viruses must be able to move from infected to healthy plants if they are to persist in host populations. Viruses can be transmitted directly through seed (or vegetative propagules), pollen, or physical contact, but in most cases transmission is indirect, through vectors, including humans. Vectors of plant viruses include arthropods (insects and mites), nematodes, and fungi. There is often a marked specificity; for example, aphid-borne viruses are not transmitted by insects other than aphids. Vectors are able to disseminate viruses over short and greater distances. Local spread enables viruses to exploit habitats already colonized, whereas long-distance spread can lead to the colonization of new habitats. Many insect vectors have diverse life cycles and characteristics and are active at one or more stages, emphasizing the key epidemiological role of these active stages. A knowledge of the means of virus spread by vectors is essential to understanding epidemiology and developing effective control measures; for example, through the use of virus-free planting material, crop rotations, insecticides, natural enemies, and cultural practices, including intercropping.

For any virus-vector combination, information on transmission is basic to an understanding of disease epidemiology. Some examples illustrate the type of information necessary. *Beet western yellows virus* (BWYV), a polerovirus (family *Luteoviridae*), is persistently transmitted by aphids to many cultivated crops, including Chenopodiaceae, Cruciferae, and Asteraceae. Several aphid species have been identified as vectors, with *Myzus persicae* and *Lipaphis pseudobrassicae* being the most efficient. Minimal durations of acquisition and inoculation access periods for BWYV transmission by *M. persicae* are about 1.0 and 0.5 h, respectively. Aphid transmission efficiency varies with temperature, with the highest transmission rates found in the range of 20 to 25°C. The retention period of BWYV in *M. persicae* is at least two weeks. The virus does not pass on to the progeny of the vector. *Myzus persicae* is also one of many aphid species that transmits *Potato leafroll virus,* a polerovirus, in a persistent manner. The aphid readily colonizes solanaceous weeds in potato crops. In Washington State, population numbers can typically be higher on weed hosts than adjacent potato plants, and transmission to potato plants readily occurs by *M. persicae* from symptomatic *Solanum sarrachoides.* The same and other aphid species transmit virus in a nonpersistent manner. For example, all isolates of ZYMV are transmitted in a nonpersistent manner by *Aphis gossypii* and *M. persicae.*

Tomato spotted wilt virus (TSWV), a tospovirus belonging to the family *Bunyaviridae,* a family of animal viruses, is transmitted by several species of thrips. *Frankliniella fusca* is an abundant TSWV vector species in North Carolina. A wide range of common annual, biennial, and perennial plant species act as reproduction sites for *F. fusca* and serve as acquisition sources for the vector to spread TSWV to susceptible crops. Elsewhere, weeds and greenhouse crops, such as chrysanthemum, play an important role in the survival of TSWV and its thrips vectors. TSWV is also transmitted by *F. occidentalis* and can be detected in this species by double-antibody sandwich ELISA. *Frankliniella occidentalis* acquires TSWV with varying efficiency from different virus-infected plants and can spread TSWV from weeds to cultivated crop hosts and vice versa. Colonizing adult thrips have been reported to prefer landing and feeding on TSWV-infected plants with larval populations significantly greater on diseased plants. This reinforces the impact of the presence of virus-infected plants on the viruliferous thrips population.

Rice yellow mottle virus (RYMV), a sobemovirus, causes a damaging disease of rice in sub-Saharan Africa. It is transmitted by a range of coleopteran and orthopteran insects, but also mechanically by walking and feeding cattle and other animals. The vectors are leaf feeders with biting mouthparts and may be present in great abundance in rice and grasses in irrigated lowland fields. The dynamics of disease spread and the role of vectors is poorly understood. Information on the distribution, biology, host range, and damage relationships is required to develop integrated management of the vectors.

The planthopper *Laodelphax striatellus* is a vector of *Northern cereal mosaic virus,* a cytorhabdovirus, which overwinters in wheat stubble underground. Overwintered viruliferous nymphs emerge and are responsible for early infection of spring-sown wheat. *Barley yellow striate mosaic virus,* a cytorhabdovirus, *Maize rough dwarf virus,* a fijivirus, and *Wheat dwarf virus,* a mastrevirus, are also plant-/leafhopper-borne viruses. The first two are transmitted in a persistent circulative-propagative manner by *L. stria-tella,* the third in a persistent circulative (not propagative) manner by the leafhopper *Psammotettix alcencis.* Transovarial transmission to progeny occurs in *L. striatella.*

EPIDEMIOLOGICAL CYCLES OF INFECTION

When a healthy susceptible plant is exposed to virus inoculum, the plant may or may not become infected (establishment and replication of the virus). If infection occurs and the virus multiplies and becomes systemic, then the plant becomes infectious and serves as a source of inoculum for virus acquisition and further spread through vectors. The durations of the latent period (time from infection to infectiousness) and the infectious period (time the host plant remains infectious) depend on properties of both the host plant and virus strain and are of key epidemiological significance in determining the number of infection cycles that can occur during a crop cycle (time from planting to harvest). Mathematical models of these cycles (see Figure 14.1) have been developed to analyze the effects of control practices such as roguing, or removal of diseased plants.

A fundamental requirement for understanding the ecology and epidemiology of a plant virus disease is a basic description of the virus-host plant interaction. With many of the economically important virus diseases, these properties have been well worked out and attention by plant virologists has turned largely to molecular and genomic attributes of the virus. The importance of a basic description of virus and host plant properties becomes apparent when a relatively new and poorly described virus disease increases in the range and extent of damage it causes.

The virus causing citrus leprosis disease, Citrus leprosis virus (CiLV), an unassigned species in the family *Rhabdoviridae,* is a good example of this. The disease, although not confirmed as caused by CiLV, had severe effects on sweet oranges in Florida until the 1920s, after which it became rare. It appeared in South America in the 1930s and is now the major virus disease of citrus in Brazil. Citrus-growing areas in Latin America, including Panama and Costa Rica, are threatened by the disease. Using mechanical transmission to herbaceous test plants (e.g., *Chenopodium, Gomphrena,* and *Tetragonia*) in an appropriate environment, it is possible to diagnose CiLV in three to four days. Field assessment scales are available for evaluation of CiLV, suitable for epidemiological studies, germplasm screening, and evaluation of disease-yield loss relationships. The virus does not infect susceptible citrus systemically, but local symptoms are induced with mite

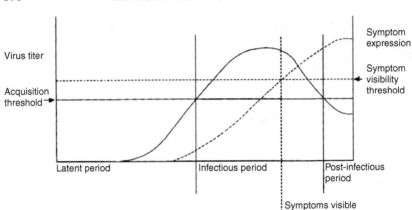

FIGURE 14.1. Diagram showing changes in virus titer and symptom expression with time defining the categories of latent, infectious, and postinfectious disease (Chan and Jeger, 1994). The solid curve shows the change in virus titer; the dashed curve shows changes in symptom expression. The solid horizontal line shows the virus acquisition threshold, which is used to define the latent, infectious, and postinfectious periods (solid vertical lines). The symptom expression threshold is represented by the horizontal dashed line. In the case shown, symptoms become visible (vertical dashed line) after the beginning of the infectious period. (*Source:* From Chan MS, Jeger MJ, 1994, An Analytical Model of Plant Virus Disease Dynamics with Roguing, *Journal of Applied Ecology,* 31, 413-427, figure 1. Reproduced with permission of Blackwell Publishing.)

transmission. Sweet orange fruits can also be infected. Transmission is persistent circulative-propagative with virus particles readily observed in the body of the mite, but there is no transovarial transmission (i.e., transmission through the eggs to the progeny of viruliferous females). Feeding for four days on symptomless but infected citrus leaves is sufficient for the mite to acquire the virus and subsequently transmit to healthy plants. The time for symptom development has been evaluated under different conditions.

QUANTIFYING VIRUS SPREAD

Virus spread occurs both spatially and temporally. A feature of many virus diseases is that there are clear trends in incidence with distance from sources or foci of primary infection. Such trends are known as disease gradients and, when disease is plotted against distance, often give curvilinear and concave relationships. These can be linearized by transforming both disease (expressed as a proportion or percentage of plants diseased) and distance using logarithms. The slope of the resulting straight line can be calculated using linear regression. The steeper the slope, the more restricted the spread from the source of infection. With time, disease gradients tend to become blurred with new satellite foci forming, which can ultimately coalesce to form areas of mass infection. The potential for establishing new healthy plantings in such areas has been modeled in the case of *Cacao swollen shoot virus,* a badnavirus, based on the model structure shown in Exhibit 14.1. The success of such plantings depends largely on the extent to which

---

EXHIBIT 14.1.
Model Structure Used to Analyze Reinfection
of Replanted Healthy Cocoa by Swollen
Shoot Virus in Areas of Mass Infection

Variables

Radius of the circular area unaffected by radial spread	*(R)*
Number of trees within the replanted area infected by jump spread	*(N)*
Proportion of perimeter invaded by the mealybug vector	*(P)*

Parameters

Tree density within replanted area	*(d)*
Radius of the circular area originally cleared	*(R$_c$)*
Rate of inward invasion (radial spread from the periphery)	*(k$_1$)*
Rate of jump spread into the noninvaded area	*(k$_2$)*
Width of cordon-sanitaire established around the replanted area	*(w)*

Model structure

$$\frac{dR}{dt} = f'(R,N,P;k_1,w,d,R_c)$$

$$\frac{dN}{dt} = g(R,N,P;k_1,k_2,w,d,R_c)$$

$$\frac{dP}{dt} = h(P;k_1,w,R_c)$$

f[1], g, and h are functions of the variables and parameters in parentheses and their form depends on the assumptions made in developing the model (Jeger and Thresh, 1993).

"jump" spread by windborne vectors is important relative to radial spread over short distances from the periphery, and whether routine inspection and roguing of diseased trees in the replanted area is practiced.

With repeated cycles of infection and dissemination from sources of infection, disease incidence within a crop increases with time. The pattern of temporal increase can readily be visualized by plotting disease progress curves of incidence against time. These often give sigmoid (S-shaped) curves, which can be linearized by using the logit transformation. The slopes of the resulting straight line can be calculated using linear regression. The steeper the slope, the faster the rate of disease increases. In most cases the amount of disease levels off at less than 1 (or 100 percent), and this parameter can also be estimated from disease-progress data. Temporal analysis of virus epidemics has since extended from these simple descriptive models to models based firmly on vector-virus transmission mechanisms and parameters estimated for the different transmission classes (see Table 14.1).

CULTURAL PRACTICES

Cultural practices can have a marked influence on the incidence and spread of virus diseases. They also are often the only means available to control virus diseases in many developing countries of the tropics. In particular, the transition from traditional to modern methods of crop production (Table 14.2) may have affected the incidence of virus diseases and attempts to control them. A major challenge for the future is to develop modern productive cropping systems that have the apparent stability and resilience of traditional systems with respect to virus diseases.

TABLE 14.1. Relevant parameter values for the four plant virus transmission classes.

Parameter[a]	Nonpersistently transmitted	Semipersistently transmitted	Persistently transmitted	
			Circulative	Propagative
Acquisition period	0.021 (0.5 h)	0.083 (2 h)	0.5	2.0
Inoculation period	0.021 (0.5 h)	0.083 (2 h)	0.5	2.0
Latent period	0	0	1	20
Infectious period	0.25 (6 h)	4	20	∞

[a]Parameters refer to the plant virus interaction with the insect vector. Units shown are days (with hours in parentheses for some short times).

TABLE 14.2. Contrasting features of traditional and modern methods of crop production.

Feature	Traditional	Modern
Fields	Small, irregular	Large, regular
Crop species	Often intermixed	Usually single
Cultivars	Often intermixed	Usually grown singly
	Usually landraces	Usually specially bred
Propagules	Own-grown or produced locally	Usually specially bred
		Usually purchased
		Seldom produced locally
Inorganic fertilizers	Seldom used	Used routinely
Herbicides/pesticides	Seldom used	Often used
Rotations	Much use of bush fallow	Limited use of fallow
Traction	Mainly human/animal	Mechanical

Source: Adapted from Thresh (1991).

Viruses cause severe disease problems in many annual crops. Integrated management of virus diseases, including the use of cultural practices adapted to disease risk, varietal characteristics including resistance, and location-specific factors, is often the only means of preventing, delaying, or reducing the rate of increase of an epidemic. For many vegetable crops, the use of virus-free seed in a clean environment can limit early infections, and these are often the most damaging in the field. In rice about 25 viruses are known to have a direct economic impact in rice production. Intensification of rice cultivation is based on the introduction of high-yielding varieties; increased use of irrigation, fertilizers, and pesticides; crop monoculture; and mechanization. Despite improved productivity, the more efficient cultural practices adopted in Asia, the Americas, and to a lesser extent in Africa, may lead to epidemics of virus diseases requiring integrated management practices for control.

With the virus complex in white clover (consisting of *Alfalfa mosaic virus*, an alfamovirus, *Clover yellow vein virus*, a potyvirus, and *Peanut stunt virus*, a cucumovirus), then cultural practices, including harvest date, affected both the spatial pattern of disease and its management. A statistical analysis showed that virus resistance altered the spatial pattern of disease (Table 14.3). Epidemics in the susceptible cultivar Regal were characterized by a stronger aggregation of diseased individuals and more numerous and larger clusters of diseased plants than with virus-resistant germplasm.

TABLE 14.3. Spatial statistics generated by two-dimensional distance class analysis of the virus complex in white clover in experimental plots of cv. Regal (virus-susceptible) and Southern Regional Virus Resistant germplasm.

Cultivar/ germplasm	Random (%)	Nonrandom (%)	Strongly non-random (% of non-random)	1990 Cluster size[a]	1990 Cluster no.	1991 Cluster size	1991 Cluster no.
SRVR	0	100	40	2.7 (1-4)	1.7 (1-2)	1.0 (1)	1.0 (1)
Regal	10	90	89	3.0 (1-7)	2.8 (1-5)	5.6 (1-19)	2.4 (1-3)

Source: Adapted from Nelson and Campbell, 1993.
[a]Cluster size and number are defined when at least one discrete cluster of diseased plants is present (range in parentheses).

ENVIRONMENTAL CONDITIONS

In many countries, intricate relationships between climate, cropping season, and intercrop periods and the effects on virus incidence are poorly understood. Climate influences virus disease outbreaks, the rate of development and activity of virus vectors and their migration, and the phenology of crops, weeds, and wild hosts that serve as infection sources.

In West Africa rainfall, temperature, and wind were identified as key weather variables affecting virus diseases in cereal, vegetable, and tuber crop production. These variables will be important in determining the timing and period of any crop protection measures. Recently, potato tuber necrotic ringspot virus, a new strain of *Potato virus Y,* a potyvirus, has spread through Europe, with its appearance determined by varietal susceptibility and weather conditions. Mild and drought conditions in winter-spring followed by a very warm summer in 1997 proved ideal for transmission of this new strain by vectors and for symptom development in the north of Italy, where the disease was first reported. *Maize streak virus* (MSV), a mastrevirus, is transmitted by leafhopper (*Cicadulina* spp.) vectors in Africa. Indigenous grasses act as reservoirs of both virus and vectors, although not all grasses perpetuate epidemiologically competent strains on maize. Peak populations of *Cicadulina* spp. are observed before the rains end in savanna zones but after the rains in the forest zones. Off-season survival of MSV and vectors is dependent on water availability in riverine environments and soils during dry seasons. MSV epidemics occur only when weather conditions allow vector survival and population increase, and where maize-competent strains are present in grass hosts. Whitefly *(Bemisia tabaci)* population numbers are influenced by weather. The temporal disease progress of *African cassava mosaic virus,* a begomovirus, was shown to be strongly related to fluctuations in temperature and radiation, but in tropical humid climates with a short dry season rain-related variables are unlikely to be a limiting factor.

FORECASTING DISEASE DEVELOPMENT

Mathematical models and decision-support schemes have been developed for forecasting and control of virus diseases. A simulation model for *Barley yellow dwarf virus* (BYDV), a luteovirus, was successfully validated in barley and wheat crops during the period from 1978 to 1989 (see Figure 14.2) and indicated the value of practical and reliable forecasting of BYDV and the need for control measures. A decision support system is now sponsored by the United Kingdom's Home Grown Cereals Authority, which provides an effective, user-friendly and field-specific system for determining the need to control the aphid vectors. The decision support system encapsulates the effect of weather on the development, reproduction, movement, and survival of aphids and is used to predict how much secondary spread of the virus from the initial foci of infection has occurred. Temperature following initial infection was found to be critical in determining final disease levels, irrespective of crop growth stage at infection. Over the range of temperatures encountered in the United Kingdom, autumn and winter temperatures were important at all stages in the virus transmission cycle. Forecasting models for virus yellows in sugar beet caused by *Beet yellows virus,* a closterorvirus, and/or *Beet mild yellowing virus,* a luteovirus, have also been developed in the United Kingdom based on abundance of winged and wingless *M. persicae* vectors and on regional differences in weather and overwintering hosts, viruses, and vector. The model was used to analyze changes in the epidemiology of virus yellows over the period from 1965 to 1996 associated with improvements in pest management practices, providing a strategic extension to disease forecasting.

The epidemiology of rice tungro virus disease, caused by *Rice tungro bacilliform virus,* a badnavirus, and *Rice tungro spherical virus,* a waikavirus, and its leafhopper vectors, *Nephottettix virescens* and *N. nigropictus,* is affected by climate and weather. A requirement for a threshold level of viruliferous leafhoppers in the field for epidemic outbreaks has been suggested. From this follows the prospect of forecasting vector numbers on the basis of rainfall and forecasting disease incidence on the basis of viruliferous vectors at the threshold level. In West Bengal peak numbers of *Nephotettix* spp. are to a large extent predictable some 65 days after the peak monsoon rainfall.

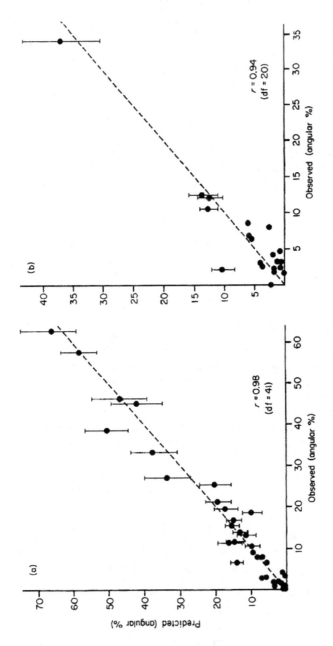

FIGURE 14.2. Agreement between observed and predicted infection at growth stage 32-33 in (a) winter barley and (b) winter wheat. The error bars show 95 percent confidence limits of predicted values and the broken line indicates equality of observed and predicted values. (*Source:* From Kendall DA, Brain P, Chin NE, 1992, A Simulation Model of the Epidemiology of barley yellow dwarf virus in Winter Sown Cereals and its Application to Forecasting, *Journal of Applied Ecology,* 29, 414-426, figure 3. Reproduced with permission of Blackwell Publishing.)

Finally, infection cycles are not always correlated with specific weather events. This was found in studies of the population dynamics of *Frankliniella* spp. and the progress of spotted wilt disease in tomato fields. The increase and spread of disease was largely related to numbers of immigrating thrips from hosts outside the crop rather than number of thrips generations, which is largely dependent on environmental conditions. Thus, the emphasis on forecasting spotted wilt disease moves to the need to monitor or model immigration/emigration of the vector.

BIBLIOGRAPHY

Atiri G, Njukeng AP, Ekpo EJA (2000). Climate in relation to plant virus epidemiology and sustainable disease management in West Africa. *Journal of Sustainable Agriculture* 16: 17-30.

Bosque-Perez NA, Buddenhagen IW (1999). Biology of *Cicadulina* leafhoppers and epidemiology of maize streak virus disease in West Africa. *South African Journal of Plant and Soil* 16: 50-55.

Chan MS, Jeger MJ (1994). An analytical model of plant virus disease dynamics with rouging. *Journal of Applied Ecology* 31: 413-427.

Fargette D, Jeger MJ, Fauquet C, Fishpool LDC (1994). Analysis of temporal disease progress of *African mosaic virus*. *Phytopathology* 84: 91-98.

Jeger MJ, Thresh JM (1993). Modeling reinfection of replanted cocoa by *Swollen shoot virus* in pandemically diseased areas. *Journal of Applied Ecology* 30: 187-196.

Kendall DA, Brain P, Chinn NE (1992). A simulation model of the epidemiology of *Barley yellow dwarf virus* in winter sown cereals and its application to forecasting. *Journal of Applied Ecology* 29: 414-426.

Madden LV, Jeger MJ, Vandenbosch F (2000). A theoretical assessment of the effects of vector-virus transmission mechanism on plant virus disease epidemics. *Phytopathology* 90: 576-594.

Nelson SC, Campbell CL (1993). Comparative spatial analysis of foliar epidemics on white clover caused by viruses, fungi, and a bacterium. *Phytopathology* 83: 288-301.

Thresh JM (1980). An ecological approach to the epidemiology of plant virus disease. In Palti J, Kranz J (eds.), *Comparative epidemiology* (pp. 57-70). Wageningen, the Netherlands: PUDOC.

Thresh JM (1991). The ecology of tropical plant viruses. *Plant Pathology* 40: 324-339.

Chapter 15

Recombination in Plant Viruses

Chikara Masuta
Masashi Suzuki

INTRODUCTION

Natural recombination and mutations are the two main driving forces leading to the evolution of plant viruses. Such events as the natural occurrence of mutations and recombination between plant viruses have been clearly detected from the sequence analysis of genomes of virus isolates of RNA as well as DNA viruses. This chapter mainly describes the recombination in RNA viruses, the latter part briefly deals with DNA viruses.

The majority of plant viruses are positive-sense RNA viruses and depend upon the virus-encoded, RNA-dependent RNA polymerase (replicase) for their viral multiplication. RNA viruses undergo rapid genetic changes due to the high error rate of viral replication. Due to this inherent error, heterogeneous populations of related RNA molecules are generated. RNA recombination takes place during viral replication via a replicase-mediated template-switching mechanism, which is the most accepted model for recombination in RNA viruses. Viral replicases sometimes pause before a recombination signal (i.e., secondary structure), detach themselves from the template, and subsequently reinitiate copying on a different position (or another RNA) to the end of the molecule. Ultimately, this template-switching mechanism results in the formation of truncated or chimeric molecules. A detailed explanation follows.

CLASSIFICATION OF RNA RECOMBINATION

Lai's System

Two systems are generally followed for the classification of RNA recombination. In Lai's system, RNA recombination is generally categorized into three types (see Figure 15.1). The first type, homologous recombination

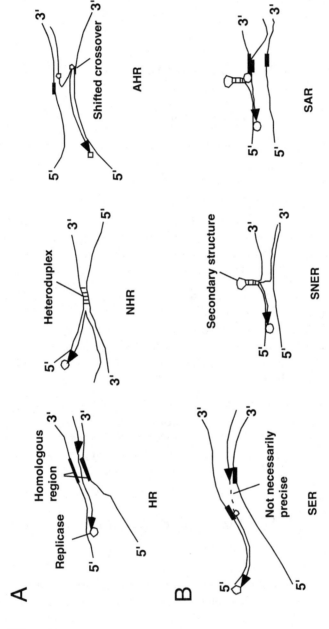

FIGURE 15.1. Classification of RNA recombination, based on systems proposed by Lai, 1992 (A) and Nagy and Simon, 1997 (B). AHR: aberrant homologous recombination; NHR: nonhomologous recombination; SAR: similarity-assisted recombination; SER: similarity-essential recombination; SNER: similarity-nonessential recombination.

(HR), takes place between two similar RNA molecules and crossovers occur at precisely matched sites. The second type, nonhomologous recombination (NHR), occurs between two molecules at noncorresponding sites. The third type, aberrant homologous recombination (AHR), represents somewhat of an intermediate form of the previous two types. AHR requires two similar molecules; however, crossovers do not occur at corresponding sites. The characteristics of AHR involve insertions, deletions, and even unrelated sequences at crossover sites. However, Lai's classification system creates some confusion. Since the definition of homologous is not clear, it is not possible to properly classify recombination associated with a very short sequence similarity.

Nagy and Simon's System

Considering the inconvenience in Lai's system, an alternative method was proposed by Nagy and Simon (1997) based upon the recombination mechanism and the features of the recombinant products (Figure 15.1). Their classification also contains three categories. Similarity-essential recombination (SER) does not necessarily require precise recombination, while sequence similarity between two molecules is essential; thus it is a hybrid of the HR and AHR described in Lai's system. Similarity-nonessential recombination (SNER) involves two unrelated molecules, but recombination is triggered by another factor(s) such as secondary structure, subgenomic promoter, heteroduplex formation, and so on. Similarity-assisted recombination (SAR) is an intermediate between SER and SNER. It occurs when an additional factor besides sequence similarity is required. Despite the fact that this new classification scheme appears to be more understandable, Lai's system is more generally accepted.

NATURAL RECOMBINATION

When RNA recombination is believed to occur under little selection pressure, it is regarded as natural recombination.

Natural Recombination in Viruses

By utilizing sequence comparisons followed by appropriate computer analyses, we can now speculate on past recombination events in plant RNA viruses. For example, recombination has been found in tobraviruses. The 3' end of RNA2 of a *Tobacco rattle virus* (TRV) isolate was shown to be corresponding to the 3' end of RNA1 of the same virus. Recombination was also documented to occur between two tobraviruses. Specifically, the RNA1 (or RNA2) of some TRV isolates was found to be a recombinant molecule with *Pea early browning virus* sequences. Recently, using computer programs Phylpro and SiScan, it has been demonstrated that *Turnip mosaic virus* (TuMV; a potyvirus) populations collected from around the world contain many recombinants between two pathotypes, *Brassica* and *Brassica-Raphanus*. The TuMV recombinants accounted for 6 percent of the population.

Natural Recombination in Subviral Species

Recombination in Viroids

Recombination does not appear to be restricted to viruses and occurs in some subviral RNA molecules such as viroids and satellite RNAs. Natural viroid chimeras between different viroids have been demonstrated by pairwise comparisons of viroid sequences. For example, *Tomato apical stunt viroid*, a pospoviroid, appears to be a recombinant between a *Potato spindle tuber viroid*-like viroid and a *Citrus exocortis viroid*-like viroid. *Australian grapevine viroid*, an apscaviroid, appears to have been generated by extensive RNA recombination among these four viroids.

Recombination in Satellite RNAs

In contrast with viroid, little evidence supports the notion that natural satellite RNAs actually originated from RNA recombination. The satellite RNAs that have been extensively studied are those of the *Turnip crinkle virus* (TCV), a carmovirus. TCV contains satellite RNAs of various sizes, which are dependent upon TCV for their replication. Apparently, a satellite (sat C) is derived from recombination events between another satellite (sat D) and the genomic RNA of TCV. Furthermore, new satellites are easily generated between sat C and sat D in experimental coinfections with sat C, sat D, and TCV.

Generation of Defective Interfering RNAs (DI RNAs)

DI RNAs are often associated with and dependent upon some RNA viruses for their replication. They represent a deleted form of a viral genome, usually compete with the intact viral genome, and cause attenuated symptoms. Their synthesis is believed to originate due to replication errors of viral replicase via the template-switching mechanism proposed for RNA recombination. The mechanism of DI RNA formation has been extensively studied in tombusviruses using an in vitro RNA-dependent RNA polymerase system.

RECOMBINATION BETWEEN VIRAL RNA AND TRANSGENES

Transgenic Plant Expressing Viral Sequence

The first transgenic plant that expressed a viral sequence was produced in 1985, and one year later overexpression of the capsid protein (CP) gene of *Tobacco mosaic virus* (TMV), a tobamovirus, resulted in the acquisition of resistance to TMV (pathogen-mediated resistance) in transgenic tobacco plants. Does a transgenic plant that produces a viral genome have the potential to generate novel viruses via RNA recombination during viral infection? This question has been addressed numerous times since a large number of plants with viral sequences were created to give viral resistance. Furthermore, the plants were monitored under field conditions.

Evidence of Recombination with Host Transgenes

To obtain sound evidence for RNA recombination with host transgenes, a system was developed that utilized *Cowpea chlorotic mottle virus,* a bromovirus, and a transgenic plant. A deletion mutant which lacked the 3' one-third of the CP gene was inoculated onto transgenic plants expressing the 3' two-thirds of the CP gene. As a result, a wild-type virus was recovered in the systemically infected plants, thus suggesting that RNA recombination with the transgene actually had taken place. The significance of such recombination events in relation to viral evolution will be described elsewhere.

PSEUDORECOMBINATION (REASSORTMENT)

Multipartite Viruses and Pseudorecombination

In a phenomenon called pseudorecombination (or reassortment), multipartite (segmented) viruses, which contain more than one RNA segment, can exchange their genomes between different isolates and sometimes between different (but related) viruses. Pseudorecombination is a drastic method to exchange genetic materials and is believed to be important for the evolution of multipartite viruses. Although no clear rule about how to use the synonyms exists, the term "reassortment" is often used for animal viruses.

Pseudorecombination in Cucumoviruses

Cucumoviruses have a tripartite genome and under experimental conditions many artificial pseudorecombinants were created to investigate the functions of viral proteins. For *Cucumber mosaic virus* (CMV), a cucumovirus, phylogenetic estimation revealed that pseudorecombination has indeed given rise to present forms. To this date, however, no distinct isolate generated between two subgroups by pseudorecombination has been demonstrated. However, between two cucumoviruses, it was shown that a natural cucumovirus isolate contained genomic segments from both CMV and *Peanut stunt virus.* It has been reported that RNAs 1 and 2 were not exchangeable between two cucumoviruses, while RNA 3 may be exchanged, thereby giving rise to pseudorecombinants. This incompatibility of the heterogeneous combination of RNAs 1 and 2 creates the three separate cucumovirus species.

MECHANISM OF RNA RECOMBINATION

Although some hypotheses on the mechanism of RNA recombination are still discussed, template-switching mechanism is the most widely accepted.

In Vitro Recombination

RNA Breakage and Ligation Mechanism
(Breakage and Rejoining Mechanism)

The breakage and rejoining mechanism operates mainly in doublestranded DNA. However, it is conceivable that the same mechanism could operate in single-stranded RNA viruses through double-stranded replicative intermediate RNAs. Moreover, RNA breakage and ligation that was mediated by ribozyme has been demonstrated for group II introns in vitro. Previously, in vitro studies that used Q beta phage replicase with some model templates revealed that recombination occurred at the 3' end of the acceptor strand by the breakage and ligation mechanism. Even though it does not exclude the template-switching mechanism, it may be difficult to assume that the RNA breakage and ligation mechanism is a natural recombination mechanism.

In Vivo Recombination

Currently, the most accepted mechanism of RNA recombination is the replicase-driven template-switching mechanism. This hypothesis has been commonly acknowledged and used to explain both homologous and nonhomologous recombination. For in vivo recombination, it has been noted that after an initial recombination event, survival of the recombinants depends upon the cellular factors that impose selection for or against a particular recombinant based upon its ability to be amplified in vivo. If recombination molecules can compete over their parental molecules, they may survive. For example, the recombinant RNA2 between CMV and *Tomato aspermy virus* (TAV), a cucumovirus, became dominant over its parental RNA 2 molecules of CMV and TAV, although the recombinant emerged under a certain selection pressure.

Replicase-Driven Template-Switching Mechanism (Copy-Choice Mechanism)

Many studies suggest that the template-switching mechanism operates by two steps:

1. viral replicase halts during RNA synthesis and separates from the primary template (donor strand) and subsequently attaches to another molecule (acceptor strand), and
2. the acceptor strand has some physical features which allow the replicase to bind and restart RNA synthesis.

The exact template-switching mechanism is poorly understood. A pausing or terminator signal on a donor strand contains A/U rich sequence (*Brome mosaic virus,* BMV, a bromovirus) or a stem-loop structure (TCV and CMV) (see Figure 15.2). The reinitiation signal of an acceptor strand may contain a subgenomic promoter sequence (CMV), hairpin motif (TCV), or a heteroduplex structure between the acceptor and donor strands (BMV). In addition to these RNA components for recombination, other RNA features can potentially influence recombination frequency and junction site selection.

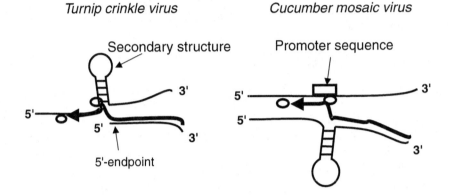

FIGURE 15.2. Two examples of similarity-nonessential recombination observed for *Turnip crinkle virus* and *Cucumber mosaic virus.*

Breakage-Induced Template-Switching Mechanism

The breakage-induced template-switching mechanism was also proposed. This mechanism differs from the template-switching mechanism in that cleavage of the donor strand actually triggers recombination. The 5' ends of the cleaved RNA template are predicted to cause template-switching of the replicase. It has been demonstrated that artificially created 5' ends in a donor strand actually served as recombination hot spots in tombusviruses.

IDENTIFICATION OF RECOMBINATION

In early studies, the isolation of the recombinant molecules with characteristic marker mutations on the parental RNAs was a simple way to identify recombination. More recently, northern blot hybridization with specific probes has been used to detect recombinant molecules in plants that were infected with pseudorecombinants or mixed viruses. As an alternative method, the ribonuclease protection assay is occasionally utilized. Reverse transcription-polymerase chain reaction (RT-PCR) is the most convenient method, however, it carries the possibility that artifact products (due to PCR errors) may arise. As described in cucumovirus RNA recombination, both RT-PCR (Figure 15.3) and Northern blot are effectively used to detect recombinants. In Figure 15.3, CMV RNA3 species containing the 3' end of

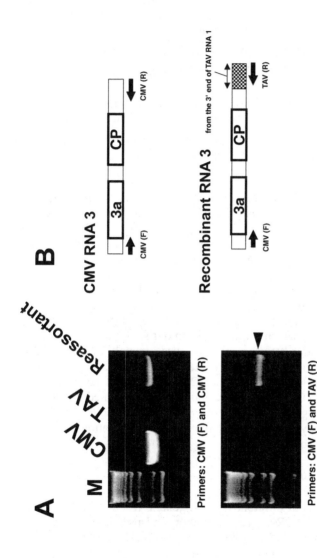

FIGURE 15.3. Detection of recombinant RNAs by RT-PCR. (A) Lane M: a 1-kb ladder marker. Lanes CMV and TAV: RT-PCR products from *Cucumber mosaic virus* and *Tomato aspermy virus* RNAs, respectively. Reassortant lane: RT-PCR products from total RNAs of tobacco plant infected with the reassortant, TAV RNA 1 + CMV RNAs 2 + 3. Primer pairs that were utilized for amplification are indicated below each gel. (B) The location of the primers used for RT-PCR are indicated by arrows. These results revealed that there were recombinant RNA3s with the 3'-termini of TAV RNA1 as well as the authentic RNA3. 3a: movement protein; CP: capsid protein.

193

TAV RNA1 in the reassortant (TAV RNA1 and CMV RNAs 2+3) was identified by RT-PCR using an appropriate primer pair.

In addition to RT-PCR, sequence analysis and phylogenetic studies have been proven to be extremely effective for the detection and characterization of recombination events among RNA viruses.

SIGNIFICANCE IN RNA RECOMBINATION

From an evolutionary point of view, recombination in RNA viruses may result in emergence of a new virus, which may be more virulent than its parental viruses.

Evolution of RNA Viruses

Despite the fact that high levels of RNA recombination can be induced under experimental conditions, natural recombination is a rather rare event in the span of a few years. However, it is clear that RNA recombination and pseudorecombination serve as mechanisms for plant viruses to evolve and survive in their dynamic environment. For example, when viral replicase errors create deleterious effects and disrupt the integrity of viral genomes, RNA recombination is an efficient way to avoid the broken sequences (genome-purging effect). It is possible that a novel virus generated by RNA recombination has a better chance for survival in the given environment than the parental virus from which it originated. In our studies, we made artificial pseudorecombinants between cucumoviruses, namely CMV and TAV. Interestingly, the hybrid CMV RNA2 containing the 3' end of TAV RNA1 was found to prevail over both RNA2 of CMV and that of TAV. Recently, alstroemeria-infected CMV stably maintained the duplicated 3'-UTR sequence through passages in alstroemeria plants but lost the duplication in tobacco plants, thereby suggesting that the RNA recombination of CMV is strongly involved in host adaptation.

Risk of Newly Emerging Virus

In mixed infections of different viruses, genetic resources can be rearranged or exchanged by RNA recombination and pseudorecombination. Because of the high error rate of viral replicase, it is conceivable that a recombinant virus may acquire a new and unexpected virulence potential. However, productive natural recombination events, which result in the evolution of new virus strains, would be expected to be relatively rare over limited periods of time in local populations, considering that it will take substantial time for recombinant molecules to adapt to the environment and eventually survive.

RECOMBINATION IN DNA VIRUSES

Two groups of DNA viruses exist: double-stranded (ds) DNA viruses and single-stranded (ss) DNA viruses.

The dsDNA Reverse-Transcribing Viruses

The dsDNA of reverse-transcribing viruses, sometimes referred to as pararetroviruses, use reverse transcriptase in their replication cycles but lack the phase of integration into the host genome by integrase (see also Chapter 7 in this book). However, integration of their sequences is actually found in plant genomes as episomes and thought to occur by nonhomologous recombination. In some cases, the integrated viral sequences caused an episomal viral infection. Episome is a unit of genetic materials that is sometimes integrated into a chromosome and at other times independently multiplies.

Plant dsDNA reverse-transcribing viruses are included in the families *Caulimoviridae, Pseudoviridae,* and *Metaviridae.*

The family *Caulimoviridae* consists of six genera, namely *Caulimovirus, Soymovirus, Cavemovirus, Petuvirus, Badnavirus,* and *Tungrovirus.*

It has been speculated that recombination between two isolates of the dsDNA reverse-transcribing viruses occurs via template-switching by reverse transcriptase during viral replication. A similar mechanism has also been proposed for RNA viruses, described previously.

Episomal viral infection from the integrated sequences has been reported in three cases: *Petunia vein clearing virus,* a petuvirus, *Banana streak virus* (BSV), a badnavirus, and tobacco vein clearing virus, a caulimovirus. For BSV, it has been suggested that two successive homologous recombination events resulted in the episomal virus.

Recombination between inoculated virus and the viral transgene has been demonstrated in *Cauliflower mosaic virus* (CaMV), a caulimovirus, under moderate selection pressure.

The recombination was explained by the template-switching mechanism at the RNA level.

It has been shown that the CaMV 35S promoter served as a "hot spot" for nonhomologous (or microhomologous) recombination.

The potential risk to create new viruses or other invasive genetic elements will be controversial, because the 35S promoter has been so widely used for the creation of transgenic plants.

The ssDNA Viruses

At present, two families are distinguished, namely *Geminiviridae* and *Nanoviridae*.

The family *Geminiviridae* consists of four genera: *Begomovirus, Mastrevirus, Curtovirus,* and *Topocuvirus*.

Geminiviruses have genomes of one or two circular DNA molecules and replicate via a circular dsDNA intermediate by using virus-encoded replication-associated proteins (*Rep As*). Based on the sequence analysis, it has been speculated that curtoviruses originated from an interspecific recombination between a mastrevirus and a begomovirus.

Intermolecular recombination between two components in bipartite geminiviruses has also been reported. For example, pseudorecombination between the DNA A of *Tomato mottle virus* and the DNA B of *Bean dwarf mosaic virus* (BDMV), both begomoviruses, resulted in generation of a new BDMV DNA B by intermolecular recombination. As the dsDNA viruses were identified in the plant genome, the ssDNA geminivirus-related (GRD) sequences have been also discovered in the tobacco genome. The GRD sequences were cross-hybridized with *Tomato golden mosaic virus,* also a begomovirus, showing similarities to the CP gene, *AR1*. The most likely mechanism for such integration is nonhomologous recombination with the host genome.

BIBLIOGRAPHY

Chen YK, Goldbach R, Prins M (2002). Inter- and intramolecular recombination in the *Cucumber mosaic virus* genome related to adaptation to alstroemeria. *Journal of Virology* 76: 4119-4124.

Frischmuth T (2002). Recombination in plant DNA viruses. In Khan JA, Dijkstra J (eds.), *Plant viruses as molecular pathogens* (pp. 339-363). Binghamton, NY: The Haworth Press, Inc.

Hou YM, Gilbertson RL (1996). Increased pathogenecity in a pseudorecombinant bipartite geminivirus correlates with intermolecular recombination. *Journal of Virology* 70: 5430-5436.

Kohli A (1999). Molecular characterization of transforming plasmid rearrangement in transgenic rice reveals a recombination hotspot in the CaMV promoter and confirms the predominance of microhomology mediated recombination. *The Plant Journal* 17: 591-601.

Lai MMC (1992). RNA recombination in animal and plant viruses. *Microbiological Reviews* 56: 61-79.

Masuta C (2002). Recombination in plant RNA viruses. In Khan J, Dijkstra J (eds.), In *Plant viruses as molecular pathogens* (pp. 203-223). Binghamton, NY: The Haworth Press, Inc.

Masuta C, Ueda S, Suzuki M, Uyeda I (1998). Evolution of a quadripartite hybrid virus by interspecific exchange and recombination between replicase components of two related tripartite RNA viruses. *Proceedings of the National Academy of Sciences USA* 95: 10487-10492.

Nagy PD, Simon AE (1997). New insights into the mechanisms of RNA recombination. *Virology* 235: 1-9.

Suzuki M, Hibi T, Masuta C (2003). RNA recombination between cucumoviruses: Possible role of predicted stem-loop structures and an internal subgenomic promoter-like motif. *Virology* 306: 77-86.

Chapter 16

Virus Variability and Evolution

Fernando García-Arenal
José M. Malpica

INTRODUCTION

Populations of plant viruses are genetically heterogeneous. This phenomenon is known as genetic polymorphism. Population diversity is a way to quantify genetic polymorphism. Heterogeneity allows us to distinguish between individuals or groups of them, for which terms such as strains, isolates, variants, and so on are used.

The frequency distribution of genetic variants in the population of an organism (i.e., the genetic structure of the population) may change with time, and this process is called evolution. This change is the product of evolutionary forces, selection and drift, acting upon the genetic variants produced by mutation and genetic exchange.

MUTATION

Mutation is the process that results in differences between the nucleotides incorporated in the daughter strand during nucleic acid replication and those in the template. Mutations are of two types: base substitutions and insertions and deletions. In base substitutions a base in the template is replaced by a different one. If both the original and replaced bases are purines or pyrimidines, the substitution is a transition. If a purine or a pyrimidine is replaced by a pyrimidine or a purine, respectively, the substitution is a transversion. Insertions are mutations in which bases that are not present in the template are added to the nascent strand. Deletions are mutations in which bases present in the template are not present in the nascent strand.

Mutation is the initial source of variation in populations, hence the interest in estimating the rate at which it occurs. It is important to differentiate here between the mutation rate and the observed mutant frequency in the analyzed population. These two parameters may differ broadly, as an unknown fraction of the generated mutants will be deleterious and will be eliminated from the population by selection. Also, the relationship between mutant frequency and mutation rate may be different according to the replication strategy and life history of the virus. Various estimates of mutation rate have been reported for lytic RNA viruses infecting mammals, with similar values in the range of 10^{-4} to 10^{-5} misincorporations per nucleotide per replication round, which result in the more biologically significant number of about one error per genome per replication cycle. A recent study for the plant RNA virus *Tobacco mosaic virus* (TMV), a tobamovirus, based on the detection of mutants lethal for cell-to-cell movement, gave an estimate of the mutation rate per base of 1.8×10^{-5}, and a mutation rate per genome of 0.11. The nature of the mutations was also analyzed, which is the first report on the mutational spectrum of an RNA virus. Interestingly, a large fraction (69 percent) of the mutations were insertions and deletions—half of them involving from three to many bases. Such a ratio of base substitutions to insertions and deletions had only been reported previously for the retrovirus *Spleen necrosis virus* and for the archeon *Sulfolobulus acidocaldarius*. A second important trait is that a large fraction of the mutants (35 percent) were multiple mutants, a characteristic not shared with any other reported mutational spectrum. These data show that most mutations in TMV, and possibly in other RNA viruses, could hardly become adaptive; they also support the view that the high mutation rates of RNA viruses, rather than an evolutionary strategy, are due to the need to replicate fast their chemically unstable RNA genome. The rate and character of mutations in TMV are also consistent with two classical observations: the characteristic low specific infectivity of RNA viruses and the vulnerability of RNA virus populations to increased mutation rates that rapidly lead to their extinction.

Mutation rates are an order of magnitude smaller for retroviruses than for RNA viruses. These values can perhaps be extrapolated to the plant DNA viruses that replicate by reverse transcription of an RNA intermediate. For viruses with large double-stranded (ds) DNA genomes, mutation rates per base are much smaller than for RNA viruses or for retroviruses (about 10^{-8}), and mutation rates per genome are about 0.003 per replication round. It is not known if these values can be applied to the small single-stranded (ss) DNA plant viruses, for which no estimate of mutation rate is available.

GENETIC EXCHANGE

Genetic exchange is the process by which fragments of nucleic acids are switched between different templates. Genetic exchange may be due to recombination or to reassortment of genomic segments in viruses with a segmented genome.

Recombination is the process by which segments of genetic information are switched between the nucleotide strands of different individual strands during the process of replication. Sequence analyses of populations of various RNA and DNA plant viruses provide evidence that recombination may be a major source of variation for evolution to occur, and it might be particularly important for certain virus groups. At the population level recombination may result in dramatic changes in the biological properties of viruses, with important epidemiological consequences, for example, the appearance of resistance-breaking strains. Also, concern about gene flow from transgenic plants with pathogen-derived resistance to viruses to populations of infecting viruses has resulted in new efforts to analyze the mechanisms of recombination and its role in virus evolution. Nonetheless, if estimates of mutation rates are scant, even less information is available on recombination frequency in the absence of selection. Recombination frequencies in RNA viruses depend on the degree of sequence similarity between the sequences involved and on the presence of recombination hot spots and can be as high as mutation rates.

Reassortment of genomic segments in viruses with a segmented genome, a process often called pseudorecombination by plant virologists, also results in genetic exchange. As for recombination, the effects of reassortment may be dramatic on virus biology, and it may play an important role in virus evolution. Evidence exists that reassortment occurs in natural populations of plant viruses. However, most evidence is for selection against reassortants (i.e., for coadaptation of genomic segments).

STRAINS

Plant virus literature is often inconsistent in the terminology used to name genetic variants, and isolate, variant, mutant, or strain are concepts that have been used with different meanings. According to the most common usage in microbiology, an isolate should be the pure culture of a virus derived from a plant infected in the field or experimentally, by a single isolation event such as a passage to another systemic host, biological cloning through single-lesion passage, or molecular cloning. Isolates from the same virus species that differ in some property are called variants. (See subtopic on population diversity for methods to identify and characterize variants.)

Strains are sets of isolates that share some properties, such as geographic origin, host range, transmission, serology, or nucleotide sequence similarity, that clearly differentiate them from other sets of isolates. The term strain is also used for the isolate that typifies the strain, a practice that may lead to confusion.

Variants are called mutants (of a parental isolate) if it is known that they derive from a particular isolate and, ideally, which mutational event resulted in their generation. Differences among mutants will be smaller than among strains, which in fact arise by genetic divergence through the accumulation of mutations (sensu lato [in the broad sense], i.e., base substitution, insertion, deletion, recombination) with time.

The specific genetic constitution of an organism, in this case a virus, is its genotype. For haploid organisms, haplotype has the same meaning as genotype. Thus, in a virus population different variants can occur, with different genotypes, or haplotypes.

SELECTION

Selection is a directional process by which variants that are fittest in a certain environment will increase their frequency in the population (positive selection), whereas variants less fit will decrease their frequency (negative or purifying selection). A consequence of selection is that in a population large enough (ideally of an infinite size), the frequency of a variant at equilibrium provides an estimate of its fitness. Selection is the process invoked most frequently in the literature to explain virus evolution, but this is not always based on evidence. The effects of selection and genetic drift are often difficult to separate, because selection also results in a decrease of the population diversity and may cause an increased diversity between populations, if under different selection pressures.

Selection Pressures

Selection pressures can be associated with every factor in the virus life cycle. For instance, selection pressures associated with the maintenance of functional structures have been documented for the capsid protein (CP) of the tobamoviruses or for noncoding regions that have a role in the replication of the viral genomes. The maintenance of a functional structure would be a primary factor for noncoding RNAs, as has been shown for satellite RNAs and viroids.

Another obvious group of selection factors will be those associated with the host plant. Evidence for this comes from host adaptation experiments in which consistent selection of different variants in different host plants has been documented. Differentiation of natural populations according to host plant can also be evidence of host-associated selection, as reported for both viruses and viroids. Evidence of host-associated selection also derives from the well-known phenomenon of overcoming of resistance genes. Virulence, defined as the effect of a pathogen in decreasing the fitness of its host (rather than relative to the replicating ability of the virus), is an important property of pathogens that may be selected for and play an important role in their evolution. This is an issue that has received considerable theoretical attention, but on which few experimental results have been reported. Because of its effect on the fitness of the host plant, viruses could determine the size and/or the genetic composition of plant populations, which in its turn could affect virus evolution. It has been shown that virus infection may affect the fitness of wild plants and weeds, as well as of crops, but the effect of viruses on plant evolution is a subject that remains largely unexplored.

A third group of selection factors would be those associated with the interaction of the virus and its vectors. Evidence for vector-associated selection is the loss of vector transmissibility after repeated nonvector passaging of viruses. Evidence for vector-associated selection also derives from reports of the selection of particular genomic combinations in viruses with segmented genomes upon vector transmission. As with their host plants, viruses may also affect the genetic structure and dynamics of the insect vector populations. It has been shown that attraction and preference of aphids may differ between virus-infected or healthy plants, and that virus infection may modify the reproductive potential of aphids positively or negatively.

Selection pressures on plant viruses have been reviewed.

Negative and Positive Selection on Plant Virus Genes

Sequence analyses show that in most instances selection on virus genes is negative selection. The degree of negative selection in genes, or the degree of functional constraint for the maintenance of the encoded protein sequence, can be estimated from the ratio between the nucleotide diversities at nonsynonymous (d_{NS}) versus synonymous (d_S) positions, that is, those nucleotide changes that result (nonsynonymous) or not (synonymous) in a change of the encoded amino acid. As nucleotide diversity is a measure of the probability that the base at a certain position differs between two randomly chosen individuals from the population, the d_{NS}/d_S ratio indicates the amount of variation in the nucleic acid that results in variation in the encoded protein. Analysis of this ratio for structural and nonstructural proteins of a number of RNA and DNA viruses shows that they are similar to those reported for RNA and DNA viruses infecting animals, and all fall within the range reported for DNA-encoded genes of cellular organisms. Thus, variation of virus-encoded proteins is similarly constrained, as in those of their eukaryotic hosts and vectors, which suggests that the need to establish functional interactions with host- and vector-encoded factors is constraining the variability of virus-encoded proteins. Another important source of constraint could be the well-documented multifunctionality of virus-encoded proteins, which would result in different selective constraints corresponding to various functions, and hence the protein would never be optimized for just one of their functions. An extreme case of multiple functional constraints occurs in genomic regions with overlapping reading frames, which is not uncommon in the tightly packaged viral genomes. Negative selection may also reduce variation due to genetic exchange. As for mutants, the fate in the population of a new variant generated by genetic exchange will depend on its fitness. Experiments with viruses with segmented genomes from the genera *Cucumovirus, Nepovirus, Tospovirus,* and *Phytoreovirus* show evidence for selection against some of the possible reassortant types. There are also data that could be evidence for coadaptation among genomic elements, which may result in speciation. On the contrary, recombinants are frequent in natural populations of some virus genera, such as *Begomovirus* or *Potyvirus,* and may be favored in some host species or genotypes. The strength of negative selection may be less in noncoding regions of viral genomes, or in noncoding satellite (sat) RNAs and viroids, in which mutation accumulation, recombination, and genomic rearrangements may be frequent.

Even if the analysis of complete virus genes shows negative selection operating, positive selection can be acting in particular domains of the viral proteins and be evidenced by more detailed analyses of the encoding sequence. Examples of positive selection can be drawn from the documented cases of increased frequency in the virus population of resistance-breaking strains as a consequence of the use of resistance genes.

Complementation

The effects of selection on deleterious mutants may be countered by complementation. Complementation is the process by which the function affected by the mutation is provided in *trans* by fully competent genotypes in multiple-infected cells. This should be particularly important for RNA viruses, which will generate a large number of mutants sharing the cell environment with the competent parental variant. Complementation could result, for instance, in the maintenance of more virulent, less-fit variants, which could have important consequences in pathology at large. Complementation of replication, movement, and transmission has often been described in experiments, but its role in virus evolution has been overlooked.

An obvious case of complementation is the maintenance of satellite viruses and nucleic acids, or of defective RNAs, by helper viruses. The efficiency of complementation has been quantified in a small number of cases.

GENETIC DRIFT

Genetic drift is, with selection, one of two major evolutionary processes that determine the frequency distribution of the genetic variants generated by mutation or genetic exchange in the virus population. Because populations may not be large enough to ensure that all occurring variants will be present in the next generation, random effects would occur during transmission of genetic traits to new generations; this random process is called genetic drift.

Populations of plant viruses can reach very large sizes within one infected plant. For TMV, for instance, it has been estimated that the number of particles in an infected tobacco leaf is in the range of 10^{11} to 10^{12}. This census population size might differ by a large factor from the effective population size, which is the number of individuals that generate (i.e., that pass their genes) to the new generation. Clearly, the effective population size, and not the census population size, is what matters for the evolution of the virus population, a point often overlooked in the virological literature. In a population of an RNA virus such as TMV, the effective population size may be much smaller than the actual population size, because a large fraction of the population will consist of mutants that will not multiply, as suggested by its low specific infectivity (in the range 10^{-3} to 10^{-4}). In addition, infection of a new host may result in a population bottleneck, as it may be started by a very small number of virus particles (one in theory). Also, population bottlenecks will occur in other moments of the life history of the virus (e.g., each time a new plant species becomes a host or a new geographic area is colonized). The new population (in a newly infected plant, area, and so on) is started by a small number of genetic types randomly chosen from the mother population, what is called the founder effect. Founder effects result in a smaller diversity within a population and in bigger diversities between populations. Random changes in the main genotype observed in passage experiments have been mostly interpreted as host selection, but often can be explained by founder effects. It has also been shown that aphid transmission results in founder effects. Founder effects have also been invoked to explain small, within-population diversity and/or high, between-population diversity in field populations of several viruses.

Mutational Meltdown

An important issue regarding the effect of population bottlenecks in the evolution of viruses is that they can result in effective population sizes below the threshold needed to ensure the transmission of the more fit genotypes, as shown experimentally with bacterial and animal RNA viruses. As a result, deleterious mutations will accumulate and the viral population will become progressively dominated by less-fit genotypes, a process known as Muller's ratchet, and will succumb by a mutational meltdown. This has also been described as the error threshold. Mutational meltdown has rarely been shown in field populations of any organism. For plant viruses it has been shown that mutational meltdown can occur in nature as a result of the interaction within a host between two different virus populations. If not uncommon, mutational meltdown should be of importance in populational structure and epidemiology of viruses.

GENETIC POLYMORPHISM
AND POPULATION DIVERSITY

Because of mutation and genetic exchange during virus replication, populations of viruses will be heterogeneous (i.e., will be built of different genetic variants), or will be polymorphic for different genetic traits. The frequency distribution of the different genetic types in the population defines its genetic structure.

Analysis of Genetic Variation

Different approaches may be used to analyze the genetic variation of plant viruses. Initially, variants were characterized by differences in biological properties, such as the symptoms they caused in different host plant species, their host range, or vector transmission properties. The development of techniques that allowed the characterization of properties of the virus other than those related to its interaction with the hosts and vectors allowed more sensitive and reproducible analyses and the typification of a larger number of isolates than bioassays. Moreover, these techniques allow the typification of characters that could be neutral (i.e., not subject to selection) and, hence, would be appropriate to analyze the genetic structure of virus populations. First to be used were techniques based on the characterization of the virus structural protein(s) (including the electrophoretic mobility of CP subunits, peptide mapping, amino acid composition, and sequence analysis of the CP) and its immunology, including the use of monoclonal antibodies and epitope mapping. Later, molecular techniques allowing the analysis of the virus genome became available, and these are most favored presently. The choice of a particular analytical technique should depend on the goal of the analysis, as well as on the sensitivity and cost of the technique. However, it is important to distinguish between techniques that provide only qualitative data (i.e., that can be used to identify variants) and those that provide information that can be used to quantify how different the identified variants are (i.e., to estimate their genetic distance). Estimates of genetic distances can be derived from data on the amino acid composition of the viral proteins and from serological comparisons, using both polyclonal and monoclonal antibodies. Genetic distances can also be estimated from ribonuclease T1 fingerprint analyses and from restriction fragment length polymorphism (RFLP) analyses of nucleic acids. Two procedures often used by plant virologists, ribonuclease protection assay (RPA) of a labeled cRNA probe and single-stranded conformation polymorphisms (SSCPs), yield results that are dependent on sequence context and do not allow the direct estimation of genetic distances. Of course, analyses of the nucleotide sequence of viral genes and the amino acid sequence of the encoded proteins provide the most detailed data, both to identify genetic variants and to estimate the genetic distances between them.

Quasispecies

The heterogeneous structure of virus populations often consists of a major genotype plus a set of minor variants newly generated by mutation or kept at low frequencies by selection. This genetic structure was named a quasispecies and had been associated with the high error rates of RNA-dependent RNA polymerases (RdRp). The quasispecies theory has been capital in making virologists aware of the intrinsic heterogeneity of virus populations; however, its contribution to the understanding of the population genetics of viruses has been seriously questioned. It should be pointed out that the genetic structure of populations of viroids or of plant DNA viruses does not seem to be qualitatively different from that of RNA viruses. In any case, users of the quasispecies term and concept should be aware of its meaning and implications, as it should not be applied to describe just any heterogeneous set of sequences.

Population Diversity

Population diversity can be defined as the probability that two randomly chosen isolates from a population are different. More precise estimates of population diversity are obtained if there is information on how different those randomly chosen isolates are (i.e., what is the genetic distance between them). Thus, population diversity depends on three parameters: number of genetic types present in the population, frequency with which each type occurs, and genetic distances among the existing types. Most work on plant virus populations analyzes only the first of these three parameters (i.e., the number of haplotypes present in the population). Since this number is usually large relative to the number of analyzed isolates, authors often conclude that the analyzed population is highly variable, which perhaps is not the case. It is necessary to stress that the number of genetic types identified and the frequency of the more prevalent one depend on the size of the analyzed genetic target, the sample size, and the analytical method. It should also be kept in mind that the size of the sample needed to estimate the diversity of a population or to compare with that of other populations does depend on the diversity itself (i.e., on the variance of the analyzed trait). It is often considered necessary to analyze a large number of isolates. This approach, however, may not increase the precision of the estimate of the population diversity and it establishes a dangerous trend among researchers and reviewers. Similarly, the size of the genetic target to be analyzed (e.g., the length of the sequenced region) would depend on the hypothesis to be tested, a point also often disregarded.

Data accumulated during the past ten years show that genetic stability is the rule, rather than the exception, in natural populations of plant viruses. Analyses of population diversity from the data available for some virus species indicated in all instances a low genetic diversity. No correlation was found between population diversity and any virus trait, such as mode of transmission or nature of host plant, or nature (DNA or RNA) of the virus genome. The current opinion that RNA viruses are very variable derives mostly from the analysis of viruses such as influenza A, hepatitis C, and foot-and-mouth disease viruses, and these may be exceptions rather than the rule, even for animal viruses. No highly variable virus has yet been reported in plants.

BIBLIOGRAPHY

Domingo E, Holland JJ (1997). RNA virus mutations and fitness for survival. *Annual Review of Microbiology* 51:151-178.

Drake JW, Charlesworth B, Charlesworth D, Crow JF (1998). Rates of spontaneous mutation. *Genetics* 148:1667-1686.

García-Arenal F, Fraile A, Malpica JM (2001). Variability and genetic structure of plant virus populations. *Annual Review of Phytopathology* 39:157-186.

Malpica JM, Fraile A, Moreno I, Obies CI, Drake JW, García-Arenal F (2002). The rate and character of spontaneous mutation in an RNA virus. *Genetics* 162: 1505-1511.

Nei M (1987). *Molecular evolutionary genetics*. New York: Columbia University Press.

Chapter 17

Recombinant DNA Technology in Plant Virology

Huub J. M. Linthorst

INTRODUCTION

With the isolation and characterization of DNA-modifying enzymes such as restriction enzymes and DNA ligases in the late sixties and early seventies of the past century a whole new field of molecular biology was opened, allowing precise modification of DNA (and its encoded proteins) and its amplification and expression in bacteria. Furthermore, using transformation vectors, for animal cells often derived from transforming viruses and for plant cells usually based on *Agrobacterium tumefaciens,* the modified DNA could be transferred to eukaryotic cells, which after propagation permitted researchers to study the effects of the modified DNA.

This chapter provides a concise overview of a number of concepts relating to current molecular plant virology. Via descriptions of the general concepts of reverse genetics, transformation, transgenic plants, and transient gene expression in plants and their relevance to plant virology, the chapter continues with describing the generation of modifiable infectious clones of viral RNAs and their use as vectors for foreign gene expression. It ends with a discussion of how engineered plant viruses may be used to express recombinant proteins, which is illustrated by the expression of antibodies.

REVERSE GENETICS

In order to identify particular genes and understand their function, classical genetics relies on mutants, detectable by their aberrant phenotype and tractable in their offspring. Molecular cloning techniques have made it possible to apply a reverse approach to the study of genetics. It permits the direct introduction of precisely designed mutations in specific genes in vitro, while subsequent introduction of the mutated gene in a suitable organism allows rapid scoring of the mutation's effects.

Studies on plant virus replication and pathogenesis heavily rely on the ability to modify virus-encoded proteins and genomic regulatory sequences at will. The DNA genome of plant viruses, such as *Cauliflower mosaic virus* (CaMV), a caulimovirus, and geminiviruses, is readily accessible to mutagenesis, but the RNA genome of most other plant viruses needs to be reverse transcribed into DNA. The availability of infectious complementary (c)DNA clones allows directed mutagenesis of viral sequences at the DNA level. Numerous approaches are available for mutagenesis of DNA, ranging from deletion of whole genes or parts of genes to very specific changes, in which only one nucleotide of a whole genome is affected. The effects of the mutations can readily be checked by infecting plants or isolated protoplasts with (a transcript of) the modified viral cDNA, as described in the sections discussing construction of infectious cDNA clones of RNA viruses and transient gene expression.

GENETIC TRANSFORMATION

Most plant viruses have genomes consisting of RNA, either single or double stranded, and thus are not directly amenable to DNA technologies. It is possible to synthetically produce stretches of DNA with a specific sequence, but the size limit of approximately 200 nucleotides makes it practically unfeasible to synthesize a complete DNA copy of a viral RNA genome. Fortunately, RNA retroviruses, particularly *Avian myeloblastosis virus* and *Moloney murine leukemia virus,* provide the precise tool that facilitates the application of the powerful techniques of recombinant DNA technology in studies of RNA viruses. The retroviral infection cycle involves a cDNA copy of the genomic RNA and, therefore, these viruses possess an enzyme capable of transcribing the viral RNA genome into a complementary DNA molecule. Since the reverse transcription (RNA→DNA) by an RNA-dependent DNA polymerase (reverse transcriptase, RT) is primed by a short, well-defined DNA oligonucleotide complementary to a specific region of the RNA molecule, this implies that the exact 3' terminus of the RNA will be represented in the DNA copy. RT also harbors nuclease activity (RNaseH) that specifically degrades the RNA moiety of the RNA/DNA intermediate, allowing the synthesis of double-stranded viral cDNA. RT-based techniques have been optimized in recent years such that virtually error-free reverse transcription of RNA with lengths of up to 5,000 to

10,000 bases, a size range spanning the majority of viral RNA genomes, can now be accomplished in one transcription reaction.

TRANSGENIC PLANTS

Plant molecular biology in general was boosted by the development of the *A. tumefaciens* binary vector system. This system is based on the natural transforming capacity of *Agrobacterium,* for which the genetic information in a large part is located onto an extrachromosomal plasmid. This plasmid, referring to its effect on *Agrobacterium*-infected plants, is called *Agrobacterium*-tumor-inducing (Ti) plasmid. In the binary vector system the virulence genes, which encode proteins involved in the transformation process, and enzymes for the synthesis of Ti plant hormones, have been physically separated from the DNA region that is transferred into the plant cell (T-DNA). The main determinants of the T-DNA are left and right border sequences of approximately 40 base pairs and, as there is no further requirement for the DNA to be transferred, the region in between the two borders may be filled in according to the requirements of the researcher. By dividing the virulence and T-DNA regions over separate plasmids, the plasmid harboring the T-DNA could be optimized for DNA manipulation, permitting an effective and efficient transformation methodology.

Although this type of transformation is applicable to most dicot plants, several plant species, notably many monocot species, appeared reluctant to *Agrobacterium*-mediated transformation. To overcome this, ballistic transformation systems (biolistics) have been designed in which the foreign DNA, usually coated onto small gold or tungsten particles, is shot directly into plant tissues. Obviously, the development of these techniques was also of paramount importance to the studies of plant virology.

As opposed to transiently transformed plants, in which the T-DNA has only entered the cell nucleus but not been stably integrated into the chromosomal DNA and, hence, the incorporated transgenes are expressed only as long as the T-DNA remains intact, in transgenic plants the foreign DNA is usually integrated at one or several locations in the plant's chromosomes and is inherited by the plant's offspring. Normally, such stably transformed plants are regenerated from only one cell, which obtained the foreign DNA at the time of the transformation experiment, and in order to select for this one-in-a-million event, the foreign DNA usually contains a selectable marker in addition to the desired transgene(s). Frequently used selectable marker genes encode enzymes that inactivate plant antibiotics or herbicides, such as neomycin phosphotransferase II, which inactivates the antibiotic kanamycin or phosphinothricin acetyltransferase, detoxifying the herbicide glufosinate ammonium (Basta). The rationale behind this approach is that only cells containing the selectable marker gene (and hence also the linked gene of interest) can survive on a medium containing the antibiotic and be regenerated into an intact primary transformant plant. Finally, transgenic offspring may be obtained from such primary-transformed self-fertilizing plants, such as tobacco and *Arabidopsis,* and properly selected and crossed to obtain homozygous transgenic plant lines.

Initial experiments focused on the expression of viral capsid protein (CP), as it was believed that transgenic plants expressing CP would be protected against the homologous virus. In several cases high-level expression of CP indeed resulted in virus resistance, suggesting an effect of the over-expressed transgene product. However, it is now generally assumed that virus resistance of plants transformed with homologous viral sequences is caused by posttranscriptional gene silencing (PTGS), by which gene expression is switched off after transcription. Also, transgenic plants expressing viral proteins involved in movement and replication have been produced. As with CP-transgenic plants, some of the transgenic plants expressing the movement protein (MP) or replicase appeared to have a virus-resistance phenotype, underlining the gene-independency of PTGS. Nevertheless, CP, MP, and replicase transgenic plants have proven (and still prove) to be invaluable tools in studies of processes of virus movement, intracellular transport, and virus replication.

Recent work on *Alfalfa mosaic virus* (AMV), an alfamovirus, may serve as an example for the usefulness of transgenic plants to unravel virus replication. AMV is a tripartite, plus-strand RNA virus. RNAs 1 and 2 encode P1 and P2 proteins, respectively, which are subunits of the viral replicase. RNA3 encodes the viral MP and CP. Tobacco plants transformed with both the P1 and P2 genes (P12 plants) and inoculated with RNA3 support replication of RNA3 and the accumulation of so-called P12-virus, consisting of RNA3-containing virions only. This allowed a comprehensive search for the structural requirements for replication in the RNA3 molecule.

In other studies, transgenic plants were obtained that encoded full-length copies of viral genome segments. For example, transgenic tobacco plants were raised that encoded the RNA1 and RNA2 segments of AMV (R12 plants). R12 plants accumulate low levels of RNAs 1 and 2, which permit translation of the viral replicase proteins P1 and P2. However, these proteins only support transcription into minus-strand RNA1 and -2. Only after inoculation with P12 virus are all viral RNAs are replicated, indicating the indispensability of AMV CP in plus-strand RNA synthesis.

CONSTRUCTION OF INFECTIOUS
cDNA CLONES OF RNA VIRUSES

Modern protocols for the synthesis of full-length viral cDNA usually involve a combination of RT and polymerase chain reaction (PCR) techniques, commonly referred to as RT-PCR, in which the cDNA obtained by reverse transcription is transcribed and amplified in a one-tube series of sequential reactions using heat-stable DNA polymerase from the lower organisms *Thermus aquaticus (Taq)* or *Pyrococcus furiosus (Pfu)*. By choosing the right primers, sets of overlapping DNA products are thus obtained that can be combined in subsequent PCR reactions and/or by linking of restriction-enzyme-digested fragments. With a carefully designed amplification and cloning strategy, full-length cloning of most RNA plant viruses nowadays is a routine job.

For plus-strand RNA viruses of which the phenolized, protein-free RNA genome itself is infectious, an RNA transcript obtained in vitro from an exact, full-length cDNA copy of a viral genome (or genome segment in the case of a multipartite virus) would, in principle, also be infectious. Of course, this is the case only when essential posttranscriptional modifications, such as adding a methylated nucleotide cap to the 5' end of the transcript, can also be provided.

Two different strategies are in use for expression of infectious viral cDNA. One is essentially an in vitro system, in which viral RNA is prepared in the reaction tube, the other approach involves transfection of living plants or plant cells.

The in vitro viral cDNA transcription system typically consists of the viral cDNA (obtained via RT as described previously) cloned in a bacterial plasmid. RNA transcription of the viral sequence requires that the cDNA is coupled to a DNA sequence, providing a strong promoter. A promoter widely used in plant virus studies is one derived from bacteriophage T7. The T7 promoter is attached to the cDNA at a precise location as to allow transcription using T7 RNA polymerase to start at the exact 5' end of the cDNA. Capped RNAs are usually more efficient in establishing viral infection. A 5' cap (a methylated guanine nucleotide, m7GpppG) can be incorporated in the transcript by providing cap donor in the reaction mixture. Usually, transcription then proceeds along the length of the viral cDNA. As long as the entire viral sequence is transcribed, many viruses are not very particular about the 3' end of the transcript. Therefore, a convenient restriction site is incorporated downstream of the 3' end of the viral cDNA. In most cases transcription finishes when the polymerase "falls off" the template at the end of the linearized plasmid DNA. After purification from the reaction mixture, the resulting RNA transcript can then be used for inoculation of plants or protoplasts.

Agrobacterium tumefaciens-mediated transfection offers an alternative method for infection with viral cDNA. This approach is generally known as agro-infection. In this case, the viral cDNA is cloned between a strong plant promoter (usually the CaMV 35S promoter) and a transcription termination signal, and this expression construct is then cloned in an *A. tumefaciens* transformation plasmid between the left and right border repeats that determine the T-DNA region to be transferred to the plant cell. Upon injection into plant tissue of a suspension of *A. tumefaciens* bacteria containing the transformation plasmid, a copy of the T-DNA between the borders is transferred to the nuclei of plant cells. In the nucleus the cell's transcription machinery recognizes the 35S promoter and initiates transcription of the proximal viral cDNA. Upon reaching the termination signal transcription stops and the transcript is polyadenylated. The resulting RNA subsequently translocates to the cytoplasm where it initiates viral infection. An example of agro-infection is the expression of full-length AMV RNA1 and RNA2 from a single T-DNA construct. Modification of the viral sequences in the T-DNA allowed a precise investigation of the role of noncoding regions in virus replication.

TRANSIENT GENE EXPRESSION

As mentioned previously, in transiently transformed cells the foreign DNA is not stably integrated in the plant genome and, hence, its expression is usually limited to only hours or days. Nevertheless, transient expression is frequently used in plant virus expression studies, as it permits much faster results. Furthermore, transgenic proteins could interfere with plant growth and development and, for instance, impede regeneration of transformed cells to "normal" transgenic plants. Transient expression of such proteins would still allow studying their effects in planta.

Several methods may be employed to transiently transform plant cells with foreign DNA. One can use a biolistic approach, in which DNA-coated projectiles are shot into plant tissues. Alternatively, isolated plant protoplasts can be forced by osmotic or electrical treatment to take up DNA supplied in the medium. Finally, in many plant species *A. tumefaciens* is capable of transferring DNA to cells in intact plant tissues. The latter method has the advantage that intact plants can be used and phenotypic changes are more easily detectable. As in stably transformed plants, transiently expressed viral proteins prove to be rewarding tools in the study of virus replication. As an example, we have used *A. tumefaciens*-mediated transient expression of domains of the 126 kDa replicase subunit of *Tobacco mosaic virus* (TMV), a tobamovirus, to map regions involved in triggering the hypersensitive response in tobacco plants possessing the TMV resistance gene *N*. The *A. tumefaciens*-mediated methodology was also used to transiently express AMV RNA1 and RNA2 from full-length viral cDNAs in *Nicotiana benthamiana,* which permitted the isolation of active viral replicase without cross-contaminating CP.

Transient transformation using *A. tumefaciens* also underlies so-called agro-infection, in which infectious viral transcripts derived from full-length viral cDNA enter the cytoplasm, where they become translated and establish a normal infection (see the sections describing plant virus gene vectors and recombinant proteins).

PLANT VIRUS GENE VECTORS

The availability of infectious cDNA clones offers the possibility to express nonviral genes integrated in the viral genome. The reasoning behind viral vectors is that high-level expression may be expected as the nonviral gene is inherited in the viral offspring and, with the virus, moves through the plant to neighboring cells and tissues. For viruses that express their genes via subgenomic mRNAs, the nonviral gene is usually cloned behind an extra viral subgenomic promoter, like that of the CP, of which during the viral life cycle many copies are generated. Alternatively, in the case of polycistronic mRNAs of poty- and closteroviruses the foreign gene may be cloned in frame with a viral gene, possibly via a fusion peptide recognized by the viral processing machinery, to allow release of the intact gene product. Although in principle each virus could serve as a vector, size constraints on the length of the viral genome to be encapsidated favor viruses with rod-shaped or bacilliform particles. For these, packaging of a longer genome is easily accommodated by incorporation of some extra CP subunits in the capsid. Thus, *Potato virus X* (PVX), a potexvirus, *Tobacco rattle virus,* a tobravirus, and TMV have been used to express reporter genes such as *GUS* or *GFP,* but also of proteins such as phytoene desaturase or tumor-specific immunoglobulin fragments that can be used in the treatment of lymphomas.

A special kind of viral vector is involved in a phenomenon called virus-induced gene silencing (VIGS), a special type of PTGS. In VIGS the viral genome contains (part of) a host plant gene. The foreign sequence is not expressed to deliver protein and indeed does not need a promoter. In fact, the double-stranded RNA that is produced during the replication of the viral genome triggers a specific antiviral defense mechanism in the plant that results in degradation of all viral sequences, including the incorporated host sequence. Because the endogenous plant gene shares the same sequence, its mRNA transcript is also degraded and the gene's translation product cannot be formed. As the VIGS signal rapidly moves through the plant, it results in systemic gene silencing, effectively similar to when the gene would have been destroyed by mutation.

RECOMBINANT PROTEINS

In addition to the "classical" stably transformed plants, recently plant viruses have also been employed as transgene vector systems for the production of recombinant proteins for medical and agricultural purposes. In several cases peptides or epitopes from human and animal viral and bacterial pathogens were expressed as plant viral CP fusion products. The fusion proteins were successfully used as vaccines, either in the form of chimeric protein aggregates or exposed at the surface of chimeric CP-encapsidated virus particles such as *Cowpea mosaic virus,* a comovirus. A concise overview of plant virus vector-mediated production of antibodies is given in the following section.

Also in studies of plant virology itself, recombinant plant virus proteins, expressed from modified infectious viral clones, have proved to be useful tools. An example is the use of recombinant MP, in which MP was fused to jellyfish GFP in studies on spread/movement of virus through the plant. As GFP can be easily visualized in UV light in situ without the need to destroy the tissue, this has helped to localize MPs of various plant viruses in the cell and analyze their structural assemblies.

In spite of public reluctance towards transgenic products, recombinant proteins are already widely used in modern medicine and it is expected that more will follow in the coming years. In order to acquire biological activity most recombinant proteins must be synthesized in eukaryotic cells, as these possess the machinery for proper folding, glycosylation, and assembly that is lacking in prokaryotes. Therefore, recombinant protein production is mostly performed in animal and insect cells, yeast, and recently also in transgenic animals expressing the recombinant proteins in their milk. Issues of scale increase, cryptic disease in cell lines and animals, animal health, product toxicity, public opinion, and cost have triggered studies of plant systems as bioreactors for recombinant proteins. Plants are indeed capable of producing biologically active proteins for medical use, as is evidenced by the expanding number of transgenic plant "products" since the first tobacco-produced human growth hormone was obtained in 1986.

ANTIBODY EXPRESSION

The advances in transgenic plant research have led to a rapidly expanding field of research aimed at using transgenic plants for the production of antibodies. Plant scientists as well as scientists with pharmaceutical interests are involved in studies of plantibodies, as plant-produced antibodies are popularly dubbed. Consequently, different strategies underlie plantibody research. On the one hand plants are being exploited as bioreactors for the production of antibodies for medical diagnosis and therapy, while on the other hand plantibodies are used in planta to alter cellular processes or interfere with plant pathogens.

Antibodies are complex, multimeric proteins that play an important role in the animal immune system (see also Chapter 12 in this book). The class of immunoglobulin G (IgG) type antibodies consists of four polypeptide chains, two so-called heavy chains, and two light chains that are linked together by disulphide bonds and are specifically glycosylated. The so-called variable region of each chain carries a number of hypervariable loops and the combined hypervariable loops of covalently linked heavy and light chains are involved in antigen recognition and binding. Successful production of IgG molecules in plants requires a correct folding, disulphide linkage, and glycosylation of the heavy and light chains and their binding by disulphide bridging. As these processes take place in the endoplasmic reticulum (ER), the heavy and light chains need to be translated with signal sequences for ER-targeting.

The approach followed in a number of studies was to cross heavy-chain-transgenic plants with light-chain plants in order to have the complete IgG coding capacity in one plant. While the expression levels of the heavy and light chains in the single-chain transformed plants was usually very low, the level of assembled antibody in the progeny of the crossed plants was much higher, indicating that coexpression of both chains is necessary for assembly and stability of antibody molecules. By choosing particular tissue-specific promoters to control transgene expression, it is possible to direct accumulation of antibodies to an easily harvestable tissue (e.g., the seed). Presently, a number of therapeutic and diagnostic antibodies have thus been produced for which clinical studies are well underway.

It is evident that the size and complexity of plant-produced IgG-like antibodies prohibit their expression through viral vector systems. However, as an alternative to intact antibodies, it is possible to produce small recombinant antibody fragments known as single-chain variable fragments (scFv), in which the variable regions of the light and heavy chain are joined in one polypeptide by a peptide linker. This type of biologically active antibody fragment is compatible with viral vector expression systems and indeed the first biologically active scFv's have been produced using PVX and TMV vector systems.

Both IgG- and scFv-type antibodies have also been successfully used in in planta studies. Plants expressing antibodies against epitopes of viral CPs displayed decreased susceptibility to the corresponding virus, demonstrating that it is possible to use this approach to improve pathogen resistance in plants.

BIBLIOGRAPHY

Abbink TEM, de Vogel J, Bol JF, Linthorst HJM (2001). Induction of a hypersensitive response by chimeric helicase sequences of tobamoviruses U1 and Ob in *N*-carrying tobacco. *Molecular Plant-Microbe Interactions* 14: 1086-1095.

Baulcombe D (1999). Fast forward genetics based on virus-induced gene silencing. *Current Opinion in Plant Biology* 2: 109-113.

Poque GP, Lindbo JA, Garger SJ, Fitzmaurice WP (2002). Making an ally from an enemy: Plant virology and the new agriculture. *Annual Review of Phytopathology* 40: 45-74.

Vlot AC, Neeleman L, Linthorst HJM, Bol JF (2001). Role of the 3'-untranslated regions of alfalfa mosaic virus RNAs in the formation of a transiently expressed replicase in plants and the assembly of virions. *Journal of Virology* 75: 6440-6449.

Chapter 18

Resistance to Viral Infections in Plants

Jennifer L. Miller
Tessa M. Burch-Smith
S. P. Dinesh-Kumar

INTRODUCTION

Viral diseases result in crop losses around the globe that annually cost billions of dollars. For example, during the 1990s, geminiviruses alone destroyed 95 percent of the tomato production of the Dominican Republic and losses to the tomato crop in Florida in 1991-1992 reached U.S. $140 million. However, despite the great losses caused by viruses, plants have evolved complex resistance mechanisms to defend against virus infection. Traditionally plant breeders identified varieties that were resistant to the local pathogens and sought to generate new varieties that showed increased resistance while also producing the desired quality and quantity of yield. Although this method was used to some degree of success, viruses have the ability to evolve around plant defenses. Thus breeders have observed the emergence of a number of resistance-breaking viruses. Globally, scientists are working to understand the interactions between viruses and their hosts, as well as nonhosts. It is now apparent that plants have evolved a battery of mechanisms to defend against viruses. Fascinatingly, in response some viruses have evolved to suppress these defenses. The goal of plant disease researchers now is to understand these phenomena and to ultimately produce crops that are resistant to a wide spectrum of viral pathogens.

Broadly, plant virus resistance can be divided into two categories, natural and engineered. Natural resistance refers to plant systems that inhibit virus replication and spread or attenuate viral symptoms when infection occurs in nature. Examples of natural resistance include cross-protection, resistance conferred by resistance *(R)* genes, and posttranscriptional gene silencing (PTGS)-inhibiting viral replication. Engineered resistance refers to attempts by scientists to introduce into plants genes that cause constitu-

---

Jennifer L. Miller and Tessa M. Burch-Smith contributed equally to this chapter.

tive resistance or that can initiate defense responses upon infection. These attempts have largely made use of virus-derived genes, but have also utilized existing plant antiviral systems and antiviral systems from other organisms, as well as introduced novel mechanisms.

As discussed in previous chapters, a virus host is an organism in which that virus can replicate and, when the host is susceptible, may spread systemically, all the while avoiding or suppressing host defense responses. Effective resistance by a host plant against a virus could act to reduce the infectivity, block the replication and spread, or prevent the transmission of a virus. For example, in a subliminal infection the virus can replicate in the initially infected cell, but cannot move out of that cell. At present there are two well-characterized mechanisms mediating natural plant-host resistance to viruses. One relies on host-encoded resistance (R) proteins eliciting defense responses upon host recognition of a specific pathogen. The other is dependent on homologous RNA degradation that removes transcripts produced from viruses.

RESISTANCE-GENE-DEPENDENT RESPONSES

Host-encoded R proteins confer resistance to a diverse array of pathogens and are hypothesized to function in a gene-for-gene or receptor-elicitor interaction. According to the receptor-elicitor theory, if the plant expresses a specific R gene product, it will recognize a cognate pathogen avirulence (Avr) gene product, triggering a signaling cascade leading to a defense response. This recognition can be either a direct or an indirect interaction between the R gene product and the Avr gene-encoded elicitor. If either the plant or the pathogen lacks its respective gene product, then disease occurs. Recent evidence from a bacterial R-Avr system suggests that these interactions occur within a larger complex, modifying the receptor-elicitor theory. This is referred to as the guard hypothesis. During infection the bacteria modify plant cellular factors to increase their virulence. The relevant R protein, referred to as a guard, monitors one of these host factors. In response to modification the R protein triggers a signaling cascade, resulting in resistance. Since no direct interactions between virus Avr proteins and R proteins have been identified, it is possible that they interact within a larger complex as well.

Resistance-gene-dependent defense has two stages, local and systemic. Local resistance results in the restriction of the virus to the site of infection and the triggering of signaling cascades that cause cell-wall thickening, the production of reactive oxygen species, and the induction of pathogenesis-related (PR) gene expression. Systemic acquired resistance (SAR) is initiated downstream of these cascades and results in the plant developing resistance to not only the initial virus but also to subsequent infections by a few other viruses.

Local and Systemic Defense Responses

Upon recognition of an Avr, a type of programmed cell death referred to as the hypersensitive response (HR) occurs. Programmed cell death is a genetically controlled pathway that leads to the killing of unwanted cells. Virus restriction is associated with the timing of the appearance of HR, but they are separate mechanisms. Some viruses can escape the site of infection and spread systemically, causing necrosis as it moves (systemic HR), leading to the death of the plant. Further, resistance and HR can be uncoupled, as is shown in cases of extreme resistance. Here, defense responses are initiated without the occurrence of HR. The early events of HR include the production of reactive nitric oxide and reactive oxygen species such as superoxide and hydrogen peroxide—key-signaling molecules. The application of hydrogen peroxide alone induces local necrosis.

Additional physiological changes occur concurrent with the rapid cell death associated with HR. These include the production of ethylene, jasmonic acid (JA), salicylic acid (SA), and phytoalexins. It is known that the exogenous application of SA suppresses replication of *Tobacco mosaic virus* (TMV), a tobamovirus, and *Potato virus X* (PVX), a potexvirus, in infected tissue. SA also induces the thickening of cell walls surrounding the necrotic lesions and PR proteins. PR proteins are host proteins whose transcription and translation are induced in response to pathogen stress. Currently the PR proteins are grouped into 14 classes (PR1 to PR14) based on amino acid sequence and function. Several *PR* genes are clustered in cereal genomes, possibly to coordinate regulation. The combinations of PR proteins expressed upon infection differ depending on the host and the pathogen, and also on the developmental stage of the plant.

SAR is a broad-spectrum resistance that results in an organism becoming resistant to subsequent infection by a subset of nonspecific pathogens. Upon a second infection on upper leaves by the same virus or a different virus, small lesions, if any, occur. HR, or even the necrotic lesions that develop as a symptom of some infections, can trigger SAR. A suite of genes called the *SAR* genes are expressed and cause the maintenance of this resistance. Many of these genes are *PR* genes; for example in *Arabidopsis* the *SAR* genes are *PR-1*, *PR-2*, and *PR-5*. The systemic signal that causes SAR is unknown. SA levels are known to increase several hundredfold upon infection and, in fact, exogenous application of SA induces SAR.

Virus Resistance Genes

Several dominant, semidominant, and recessive *R* genes have been identified. Recently, many *R* genes that confer resistance to viral, fungal, bacterial, and nematode pathogens have been cloned and then classified based on their predicted protein motifs. Most *R* genes isolated and characterized to date encode proteins with leucine-rich repeats (LRR) of various lengths and a centrally located nucleotide-binding site (NBS). The NBS region of these *R* genes share sequence homology with the NBS region of cell death genes, such as *CED4* from *Caenorhabditis elegans* and *Apaf-1* from humans. Most of the *R* genes in this NBS-LRR class are subdivided based on N-terminal domain structure: either a TIR (Toll/IL-1 receptor) domain, which has homology to the *Drosophila* Toll protein and the mammalian interleukin 1 receptor, or a putative coiled-coil (CC) domain.

Three of the *R* genes that confer resistance to viruses have been cloned and many more have been mapped. All three cloned *R* genes encode proteins that belong to the NBS-LRR class. The first virus *R* gene cloned was the tobacco *N* gene, which confers resistance to TMV. The N protein is the only cloned virus R protein that belongs to the TIR-NBS-LRR subclass. Deletion analysis showed that all three domains were necessary for function. N transcripts undergo alternative splicing of an alternative exon, giving rise to two products: N and N^{tr}. Both transcripts must be present to confer complete resistance. The longer transcript should encode a protein with a premature stop codon resulting in the translation of only one of the LRRs. The ratio of N and N^{tr} transcripts varies in a pathogen-dependent manner. However, the reason for this is unknown, as the effect on protein levels has not been determined.

The other two cloned virus *R* genes, *Rx1* and *HRT/RCY1,* encode CC-NBS-LRR proteins. *Rx1* confers resistance to PVX in potato. Mutations in the NBS and LRR domains of *Rx1* that result in gain-of-function mutants have been identified. They are hypothesized to cause the disruption of inhibitory domains that would normally prevent signaling when the elicitor is absent. In terms of the guard hypothesis, the gain-of-function mutations could be resulting in disruption of the interaction between Rx1, the guard, and the host factor, therefore constitutively signaling resistance. *HRT* and *RCY1* are alleles of the same gene in *Arabidopsis* ecotypes Dijon-17 and C24, respectively. *HRT* confers resistance to *Turnip crinkle virus* (TCV), a carmovirus, and *RCY1* confers resistance to the Y strain of *Cucumber mosaic virus* (CMV-Y), a cucumovirus. However, *HRT* requires the recessive gene *rrt* to confer complete resistance.

The elicitors for these three virus R proteins differ in structure and their function in pathogenesis. The Avr determinant for the N protein, which is necessary and sufficient to elicit both the HR and SAR defense responses, is the 50 kDa helicase domain of the 126 kDa TMV replicase. The elicitors for Rx1 and HRT/RCY1 are the capsid proteins (CP) of the respective viruses. Rx1 confers extreme resistance to PVX, with no HR response in both potato and in transgenic *Nicotiana* species. However, Rx1 is capable of initiating HR when the PVX-CP is expressed as a transgene in potato and *Nicotiana* species transgenic for Rx1.

Signaling

The protein(s) that directly interacts with the virus R proteins is unknown; however, by studying downstream signaling components it is deduced that more than one signaling pathway leads from *R*-gene-mediated recognition of a pathogen to defense responses. These pathways have been elucidated through the isolation of mutants with disrupted signaling and by degrading known signaling molecules.

As mentioned before, SA plays a role in the induction of *PR* genes during HR and SAR. The bacterial enzyme, salicylate hydroxylase (NahG), degrades SA and can be expressed as a transgene to investigate the role of SA in *R*-gene-mediated resistance. Plants containing *HRT* lost the ability to confer resistance to TCV when NahG was present, indicating the pathway's dependence on SA for defense signaling. Some *RCY1 NahG* plants show an increase in susceptibility to CMV-Y as well. This indicates a partial requirement for SA in RCY1 signaling.

The identification of several defense-signaling mutants has facilitated the dissection of defense signaling pathways. An *Arabidopsis* mutant, *npr1*, does not respond to SA accumulation, is blocked in SAR, and fails to express PR proteins. Signaling pathways that require SA do not always require *NPR1*, providing evidence that SA functions through more than one pathway. *NPR1* encodes at least four ankyrin repeats, domains known to be involved in protein-protein interactions. By yeast two-hybrid analysis it was found to interact with two basic leucine zipper transcription factors, AHBP-1b and TGA6. Interestingly, AHBP-1b directly interacts with the promoter region of *PR-1*. Silencing of *NPR1*-like genes in transgenic *N. benthamiana* plants containing the *N* gene disrupts resistance to TMV, while *RCY1* and *HRT* defense signaling is not disrupted in *NPR1* mutants.

The JA signaling pathway is also involved in mounting a defense response. Two *Arabidopsis* mutants, *coi1* and *jar1*, have been identified that disrupt this pathway. *Coi1* encodes an LRR protein with a degenerate F-box domain, implicating degradation of plant proteins in defense responses. Defense signaling via HRT and RCY1 was not disrupted by coi1. Jar1 belongs to a family of proteins in *Arabidopsis* that, based on fold predictions, may be members of the acyl adenylate–forming firefly luciferase superfamily. Jar1 is not required for all JA pathways, even though it is thought to be directly modifying JA through activating the carboxyl group. Jar1 does not effect RCY1-mediated signaling. The role of jar1 in HRT, N, and Rx1 signaling has yet to be deciphered.

The role of ethylene in defense responses has been examined by isolating mutants and utilizing inhibitors of ethylene action. In *N. tabacum* cv. Samsun NN plants ethylene action inhibitors disrupted the formation of HR lesions, indicating a role for ethylene in defense signaling. The ethylene receptor mutant, *etr1*, had no effect on HRT-mediated resistance to TCV, indicating that the resistance-signaling pathway is independent of ethylene. Contrary to this, the ethylene insensitive mutants *etr1* and *ein2* had a partial effect on the defense response of *RCY1* to CMV-Y. A double mutant *etr1 NahG RCY1* was even more susceptible to CMV-Y, indicating a synergism between SA and ethylene signaling.

Two other interesting proteins in the defense-response signaling cascade have been identified: EDS1 and NDR1. *EDS1* encodes a protein with similarity to a eukaryotic lipase and may be involved in the generation of a second messenger derived from phospholipids or acyl glycerols. Silencing of *EDS1* in transgenic *N. benthamiana* containing the *N* gene disrupted TMV defense signaling. Additional evidence from other R protein pathways indicates that EDS1 typically functions downstream of members of the TIR-NBS-LRR group. *NDR1* encodes a putative basic integral membrane protein whose biochemical activity is unknown. *ndr1* plants are unable to produce SA and they also fail to elicit a defense response for some CC-NBS-LRR-encoding *R* genes. The role of NDR1 in N, HRT/RCY1 and Rx1 signaling has not been determined.

Several other proteins downstream of N have been identified, and the role of protein degradation in defense signaling has found further support. The *N. benthamiana* gene *Rar1 (NbRar1)* is required for the N signaling pathway and interacts directly with a component of the Skp/Cullin/F-box complex, NbSGT1, which also interacts with NbSkp1. Both NbRar1 and NbSGT1 interact with the COP9 signalosome, a multiprotein complex involved in 26S proteasome-mediated protein degradation. The 26S proteasome is a large protein complex required for the degradation of cellular proteins that have been modified by the addition of a small peptide tag. Silencing of *NbSGT1, NbSkp1,* and the COP9 signalosome compromise *N* gene resistance to TMV. Rx1 resistance to PVX was also compromised when *NbSGT1* was silenced. Thus, NbSGT1 is required for both TIR-NBS-LRR and CC-NBS-LRR proteins to mount defense responses, and protein degradation is indicated to be required for defense signaling.

HOMOLOGY-DEPENDENT RESISTANCE RESPONSES

RNA silencing or posttranscriptional gene silencing (PTGS) is an RNA-mediated, homology-dependent gene-silencing phenomenon. Many of the details of the mechanism of PTGS have been determined, although there remain many unanswered questions. The effective initiating molecule in PTGS is double-stranded (ds) RNA. This can originate from a variety of sources, including in vitro transcribed molecules and viral replicative intermediates. Double-stranded RNA can also be produced from highly expressed transgenes, transgenes inserted as inverted repeats, and sense and antisense constructs by RNA synthesis, possibly via RNA-dependent RNA polymerases (RdRp). These dsRNA molecules are cleaved by a dsRNA-specific nuclease, known as DICER in animals and CARPEL FACTORY in plants, to produce fragments called small interfering RNA (siRNA). These are typically about 21 to 23 nucleotides long in animals and about 25 nucleotides long in plants. The siRNAs are part of a nuclease-containing complex that on binding to a homologous mRNA transcript in a sequence-dependent manner lead to the degradation of the transcript by endonucleolytic cleavage. When a foreign RNA sequence of high homology to an endogenous gene is expressed in a cell, the transcript from both the transgene and the endogenous gene is degraded, effectively silencing the gene. PTGS is known to occur in fungi, lower eukaryotes, animals, including *C. elegans, Drosophila* and mammals, as well as in plants.

Some evidence points to PTGS as a natural, ancient defense against cellular parasites, including RNA viruses and transposons. The most convincing evidence is that many plant mutants that are compromised in PTGS also show enhanced susceptibility to viral infection. Viruses have in turn evolved mechanisms to inhibit this PTGS pathway with many encoding suppressors of silencing which act at different steps. The helper-component protease (HC-Pro) of potyviruses is a potent inhibitor of silencing that is also able to reverse silencing in all symptomatic tissue. A complete absence of siRNA exists in plants where silencing has been suppressed by HC-Pro. The proteins 2b of CMV and P19 of *Tomato bushy stunt virus,* a tombusvirus, are weaker suppressors of silencing that are able to inhibit PTGS in growing tissues but not to suppress it in already-silenced tissues. The P25 movement protein of PVX has also been shown to function as a suppressor of PTGS. P25 suppresses the systemic signal of PTGS, preventing the uninfected areas of the plants from initiating silencing until viral infection occurs.

PTGS may also be responsible for the natural antiviral phenomena known as cross-protection and recovery. Recovery was first described in transgenic plants expressing sequence from *Tobacco etch virus* (TEV), a potyvirus. The plants initially developed typical infection symptoms, but after three to five weeks the plants recovered, with the newly emerging tissue being virus and symptom free. This recovered tissue was also resistant to superinfection by TEV, but remained susceptible to the closely related *Potato virus Y* (PVY). Analysis of RNA levels indicated that although transcription rates of the transgene in the recovered and inoculated tissue were the same, the level of accumulated RNA was significantly lower in the recovered tissue. This suggested a homology-dependent RNA degradation mechanism. It was later found that plants use a PTGS mechanism to defend against the replication of *Tobacco rattle virus* (TRV), a tobravirus, and PVX. However, the induction of PTGS did not always lead to recovery. On the basis of its involvement in resistance to diverse viruses, PTGS appears to be a general defense mechanism against viruses.

Cross-protection is defined as the ability of a virus that has infected a plant to prevent or delay the infection by another related virus. It was first demonstrated in tobacco plants infected with the green strain of TMV and subsequently inoculated with the yellow strain of TMV. The yellow strain was then unable to induce disease, or rather, the plant showed resistance to this closely related strain. In another example of cross-protection, tobacco plants infected with a mild strain of PVX became resistant to severe PVX strains. The unrelated viruses PVY and TMV were still able to replicate and spread in these plants. This cross-protection is explained also by RNA-dependent mechanisms involving gene silencing. Cross-protection has found practical application in the control of a number of viruses including *Tomato mosaic virus,* a tobamovirus, and particularly *Citrus tristeza virus,* a closterovirus. Transgenic plants carrying a segment of viral sequence can mimic this cross-protection mechanism. The clones utilized do not produce infectious virus, so they are environmentally safer than previous techniques, which relied on infection with intact mild strains of viruses over large areas.

PATHOGEN-DERIVED RESISTANCE

Apart from the natural resistance to viruses described previously, scientists have also attempted to generate transgenic plants that are resistant or show delayed symptom development on viral infection. Most of these engineering efforts have made use of viral sequence and hence the resulting resistance is termed pathogen-derived resistance (PDR). Although abundant examples of PDR exist in the literature, there have been fewer reports of success in field trials of transgenic plants. Thus, the creation of crop plants with broad resistance to a number of viruses is still an area of ongoing research.

Coat-Protein-Mediated Resistance (CPMR)

The first example of PDR was the generation of transgenic tobacco plants that expressed the coat protein (CP) of TMV strain U1. These plants accumulated CP and showed delayed symptom development on TMV infection. This kind of resistance was thus termed coat-protein-mediated resistance (CPMR). These transgenic plants were also resistant to viruses closely related to TMV-U1 but not to more distantly related viruses. The degree of resistance was later shown to directly correlate with the level of protein accumulation in the plants. However, resistance was overcome when high titers of virus were used for infection and also when TMV RNA was used to inoculate the plants. These data pointed to a resistance mechanism in which the transgenic CP interacts with the TMV virions to interfere with infection. Indeed, further experimentation indicated that it was the disassembly of the TMV particles that was affected by the transgenic CP. In field trials of tomato plants expressing the CP of TMV, the plants proved resistant to the virus and produced high yields. CPMR to a number of viruses has subsequently been engineered, although the mechanism of resistance is not always clear. Viruses to which resistance has been successfully engineered by introducing sequences from the CP include PVX, *Alfalfa mosaic virus* (AMV), an alfamovirus, *Tomato spotted wilt virus,* a tospovirus, and the potyviruses PVY, TEV, and *Soybean mosaic virus.* In some instances the mechanism of resistance resembles that of CPMR to TMV; the involvement of the transgene CP has been demonstrated for CPMR to AlMV. In other instances, as in the cases of TEV, PVY, and TSWV, the mechanism appears to involve the gene transcript and hence may be a homology-dependent gene silencing mechanism such as PTGS.

Replicase-Mediated Resistance (Rep-MR)

Transgenic plants carrying sequences derived from the replicase genes of viruses have also been shown to be resistant to the corresponding viruses. Such resistance is quite strong but is usually limited to the virus strain from which the transgene was obtained. The mechanism by which resistance is achieved is not clear, as there is contradictory evidence pointing to the involvement of both protein and gene transcript. Rep-MR was first described for transgenic tobacco plants carrying a 54 kDa fragment of the TMV-U1 replicase. These plants were resistant to infection by both TMV virions and TMV RNA, with high levels of inocula unable to break the resistance in both instances. In a later study, the expression of a truncated CMV 2a replicase gene in tobacco plants gave high levels of resistance to closely related strains of virus. Interestingly, it appears that this resistance is achieved both through suppression of viral replication and inhibition of systemic viral movement. Although the high levels of resistance to closely related strains and the inhibition of replication point to an RNA-mediated, homology- dependent mechanism of resistance, the suppression of viral spread hints at the involvement of protein. Rep-MR to PVY and AMV has also been engineered and the mechanism of resistance was shown to be protein based since only mutant replicase proteins conferred resistance. In contrast, Rep-MR to PVX; *Cowpea mosaic virus,* a comovirus; and *Pepper mild mottle virus,* a potyvirus, as well as to others, is dependent on the gene transcript and probably involves PTGS.

Movement-Protein-Mediated Resistance (MPMR)

Transgenic plants expressing viral movement protein (MP) are resistant to the virus from which the sequence was derived as well as to other viruses, a phenomenon termed movement-protein-mediated resistance (MPMR). Viral MPs are required for both cell-to-cell and systemic movement of viruses. Cell-to-cell spread is accomplished via plasmodesmata and viral MPs are known to intimately associate with these channels. Evidence suggests that the binding of the MP alters the size exclusion limit of the plasmodesmata allowing the trafficking of virions or viral RNA-protein complexes from cell to cell. Thus it is to be expected that mutant MPs that interfere with the effective interaction between infecting MP and the plasmodesmata will result in resistance to the infecting virus by inhibiting cell-to-cell spread. Indeed, this was found to be true for plants expressing mutant TMV MP. These plants were resistant not only to TMV and other tobamoviruslike *Tobacco mild green mosaic virus* and *Sunn-hemp mosaic virus* but also to unrelated viruses such as TRV, AMV, CMV, *Peanut chlorotic streak,* a soymovirus, and *Tomato ringspot virus* (TRSV), a nepovirus, albeit with varying dynamics and to differing extents. This broad resistance suggests that viral MP interacts with plasmodesmata in similar ways. Thus, transgenic plants carrying mutant MP may be one approach to generating crop plants with broad resistance to many viruses encountered in the field. However, MPMR is not always that broad. Transgenic plants expressing one of the MPs of *White clover mosaic virus,* a potexvirus, were resistant only to other potexviruses.

Satellite RNAs and Defective Interfering RNA/DNA

Small molecules called satellite (sat) RNAs are noninfectious alone and rely on virus-encoded replicases. Satellite RNAs have been used successfully to confer resistance to a variety of viruses. They retain little similarity to the virus from which they originated (helper virus) and when introduced into the plant probably confer resistance by competing with the viral genome for host or viral factors required for successful infection. This has been described as a decoy mechanism. CMV and TRSV have reduced viral replication when sat RNAs are present. In the case of sat RNA of CMV, resistance to *Tomato aspermy virus,* a cucumovirus, was also observed and the mechanism apparently involves the CP of the helper virus. Satellite RNA was also effective in conferring resistance to TCV in *Arabidopsis,* possibly by disrupting virus movement.

Defective interfering (DI) RNA/DNA are subviral molecules produced by recombination of viral genomic DNA or RNA, and they interfere with the replication of the parent virus as they direct replication in favor of themselves. Thus, they can greatly decrease the amount of infectious virus and, hence, disease symptoms. They have been used to confer resistance to *Cymbidium ringspot virus,* a tombusvirus, *Brome mosaic virus,* a bromovirus, and a variety of geminiviruses.

Other Engineered Resistance

Antisense RNA

Plants have been engineered in numerous instances to express a construct-carrying sequence that is the antisense of viral sequence. These studies resulted in varying degrees of resistance. The mechanism behind antisense constructs conferring resistance may be the same as homology-dependent degradation mediated by RNA.

Antibodies to Viral Proteins (Plantibodies)

Transgenic plants are capable of expressing and synthesizing antibodies, so-called plantibodies. Thus, it should be possible to engineer antibodies that are specific for viral proteins and host factors involved in viral replication and spread. Tobacco plants expressing antibodies to intact TMV virions exhibited resistance to the virus with the level of resistance correlating to the level of antibody. Although expressed at very low levels, resistance was also provided by a single-chain antibody (scFv), a fusion (F) product of sequences encoding the variable (v) heavy and light immunoglobin chains of an antibody. The cellular compartment in which the antibody is expressed is a critical factor for achieving resistance. Interestingly, antibodies targeted to the secretory pathway appear to be more effective at mediating resistance than those targeted to the cytoplasm.

Expression of PR Protein

Upon pathogen infection PR proteins are induced (see Local and Systemic Defense Responses, p. 223). Since various pathogen-derived resistance mechanisms appeared to be constrained to families of viruses, the nonspecific resistance that arises as a result of PR induction held much appeal. Transgenic tobacco plants that overexpressed PR genes were generated in an attempt to create extremely resistant plants. This approach was not successful, for the transgenic plants did not show increased resistance to nonspecific pathogens. In fact, the programmed cell death that occurs with HR was not disturbed on TMV infection.

Ribozymes

Two approaches using self-splicing ribozymes that cleave RNA have been developed to introduce resistance in plants. One approach targets viral RNAs and cleaves them. Transgenic plants containing a TMV-targeting ribozyme caused a delay in symptoms upon TMV infection. Ribozyme targeted against *Potato spindle tuber viroid,* a pospiviroid, was introduced into potato and resulted in complete resistance. In the other approach ribozymes have been engineered to recognize an RNA and transsplice the mRNA transcript of a toxic peptide onto it, resulting in cell death after splicing. The idea is to create these ribozymes to recognize viral RNA, therefore killing only cells that are infected. A specific target is recognized by base pairing between the viral RNA and the ribozyme. Yeast cells were transformed with ribozymes that recognize the CMV-CP-encoding RNA, and when they were mated to a strain containing the target the cells died. This approach has not yet been demonstrated in plants. Ribozyme-based resistance requires sequence homology between the ribozyme and the virus, so it will not result in broad-spectrum resistance, and will require the generation of crops carrying many genes to be effective.

Latent Suicide Gene

Another approach to resistance involves the transactivation of cytotoxic genes (latent suicide genes) upon viral infection. These are most often ribosome-inactivating proteins (RIPs). The approach has met with some success. An RIP from pokeweed *(Phytolacca americana)* called pokeweed antiviral protein (PAP) conferred resistance against PVX, PVY, and *Potato leafroll virus,* a polerovirus, in transgenic plants. It appears that the antiviral activity of PAP is distinct from its cytotoxic activity, since constructs carrying a deletion of the domain required for toxicity were still able to confer resistance. Another pokeweed RIP, PAP-II, was also successfully used to confer resistance to TMV and PVX. Other RIPs have also been utilized to generate resistance in plants.

2′,5′-Oligoadenylate System

The 2′,5′-oligoadenylate system in higher vertebrates is activated by interferons in response to viral infection. Upon activation the enzymes of this system degrade dsRNA. Potato was transformed with the cDNA for the human enzyme 2′,5′-oligoadenylate synthetase. The resulting plants were resistant to PVX with the level of virus lower than that in CP-expressing plants. In another study both 2′,5′-oligoadenylate synthetase, and RNase L were introduced into tobacco plants, each under different constitutive promoters. These plants showed increased resistance to TEV.

DISCUSSION

In nature plants are continually attacked by pathogens and in response they have evolved sophisticated defense responses that we are only beginning to understand. In countermeasure, pathogens have evolved mechanisms to circumvent several of these defense mechanisms, and so the dance of evolution continues. Several studies have investigated the evolution of *R* gene specificity and have determined recognition to be a result of positive selection due to pathogen pressure. The signaling pathways downstream of R proteins merge and diverge in a web that results in the increase of PR proteins, whose functions are poorly understood. Plant defenses against viruses also include a non-pathogen-specific resistance mechanism, homology-dependent dsRNA degradation. This response is targeted toward foreign nucleic acids. In response viruses have evolved counterdefensive strategies to modulate host defenses. Attempts to engineer broad-spectrum viral resistance are continuing. Many successful field trials of transgenic crops have been conducted; however, these have not been adopted into widespread cultivation because of a variety of concerns surrounding the presence of transgenes. Some of these include the transcapsidation of an infecting virus, recombination producing new resistance-breaking strains, and the risk of breakdown of resistance under diverse environmental conditions. Despite these concerns, great promise remains in the generation of broad-spectrum virus-resistant plants. With the vast increase in sequence data and gene prediction and analysis methods becoming available to researchers, it should become easier to design effective and enduring resistance.

BIBLIOGRAPHY

Beachy RN (1997). Mechanisms and applications of pathogen-derived resistance in transgenic plants. *Current Opinion in Biotechnology* 8: 215-220.

Cardol E, Van Lent J, Goldbach R, Prins M (2002). Engineering resistance in plants. In Khan JA, Dijkstra J (eds.), *Plant viruses as molecular pathogens* (pp. 399-422). Binghamton, NY: The Haworth Press, Inc.

Dangl JL, Jones JDG (2001). Plant pathogens and integrated defense responses to infection. *Nature* 411: 826-833.

Hull R (2002). *Matthews' plant virology*. New York: Academic Press.

Marathe R, Anandalakshmi R, Liu Y, Dinesh-Kumar SP (2002). The *Tobacco mosaic virus* resistance gene, *N. Molecular Plant Pathology* 3: 167-172.

Tijsterman M, Ketting RF, Plasterk RHA (2002). The genetics of RNA silencing. *Annual Review of Genetics* 36: 489-519.

Valkonen JPT (2002). Natural resistance to viruses. In Khan JA, Dijkstra J (eds.), *Plant viruses as molecular pathogens* (pp. 367-397). Binghamton, NY: The Haworth Press, Inc.

Chapter 19

Virus Diseases: Economic Importance and Control Strategies

A. F. L. M. Derks

DIRECT AND INDIRECT CROP LOSSES

The term loss is used for any detrimental effect of virus infection mostly resulting in economic loss. Direct losses consist of reduction in quality and yield. Indirect losses are of variable nature.

Direct Losses

Often growers first become aware of loss when there is a reduction in quality as a result of virus symptoms such as mosaic, yellowing, necrosis, or malformation of parts of plant. A bad external appearance of marketable products such as fruits, vegetables, or cut flowers leads to degrading of the product or the product will not be fit for marketing at all. For example, fruits of tomato and pepper plants become unattractive due to the presence of brown-necrotic flecks as a result of an infection with *Tomato spotted wilt virus* (TSWV), a tospovirus. However, sometimes virus infections may result in a higher market value as in the case of decorative plants with variegated leaves or flower color symptoms.

Another effect of virus infection is a reduction in the internal quality of the plant product. Sometimes viruses are known to have a negative influence on the chemical composition of fruits, which lowers their shelf life. Plums affected by *Plum pox virus,* a potyvirus, are not only shrunken, but the pulp becomes discolored and loses its taste. The reduced vase-life of virus-infected cut flowers is a loss too, but it does not result in an economic loss, as internal quality is not perceptible at the time of selling.

Besides deterioration of quality, virus infection often leads to yield reduction. When the plants remain small, due to stunting or necrosis, yield reduction is predictable to the grower. Considerable yield loss or even crop failure are reported, for instance, from cassava and tomato affected by

235

begomoviruses and in sugar beet by yellows viruses. Yield reduction in these cases is evident and well documented, but in many instances it is difficult to make a good estimation of yield loss. Yield reduction may even occur when symptoms are either absent or hardly visible (latent infections). For example, total infection of lily stocks with *Lily symptomless virus,* a carlavirus, resulted in a reduction of total bulb weight of up to 20 percent for various cultivars.

Following virus infection, a shift to smaller bulb sizes may lead to more propagation material (smaller sizes), but less saleable, flowering sizes. Over the years this may result in progressive loss of vitality of the stock. Such progressive loss or decline is clearly visible in various virus-infected crops, such as citrus affected by *Citrus tristeza virus,* a closterovirus.

Indirect Losses

Virus infections often cause a weakening in resistance of plants against other organisms, such as fungi. Increased susceptibility to root rots caused by *Pythium* spp. leads to increasing damage as reported from virus-infected plants in crops of pea, sugar cane, and lily.

Although latent virus infections are often considered to be less harmful to the crop, they may be dangerous for other adjacent susceptible crops. Dahlia cultivars, for example, when infected to high percentages with *Tobacco streak virus,* an ilarvirus, are considered a real threat to soybean or tobacco. For quarantine reasons, these cultivars cannot be exported to certain countries unless they are made virus free, which entails substantial indirect loss.

Growers meet indirect losses as a consequence of costs of hygienic and control measures to keep their crops relatively virus free. Several of these costs are paid by the growers indirectly, for instance as levies on field inspection.

Indirect losses are also the result of the (inter)national infrastructure to maintain crops relatively virus free, which includes health inspection and certification, production of virus-tested propagation material, education and research, extension service, and quarantine inspection.

SOURCES OF INFECTION

Cultivated Plants

In general, virus sources have to be searched for in the crop itself, particularly in vegetatively propagated crops and with viruses having a narrow host range. These sources in the crop are most dangerous because of spatial distribution in the stock, presence from the start of growth, and adaptation of the virus to this specific host. The same holds true for seed-propagated crops in case of seedborne viruses, particularly those that are readily spread mechanically, such as tobamoviruses, or those transmitted in a nonpersistent manner by aphids, such as potyviruses. The chance of an epidemic buildup is relatively high in perennial crops because infected plants often act as a virus source for a long time.

Adjacent crops of the same or different plant species may act as a source of infection. Many iridaceous plants, such as gladiolus, are badly infected with *Bean yellow mosaic virus,* a potyvirus, and are a real threat to legumes such as beans. Crops harboring viruses with a wide host range are harmful in this respect.

Plant parts left on or in the soil may act as a virus source for the existing or subsequent crops. Examples are: ground keepers or volunteers; pieces of hairroots from perennial crops infected with *Tobacco rattle virus* (TRV), a tobravirus; or stubbles of sugarcane or rice. With the increase of mechanization, less plant debris is removed, and that acts as an important virus source.

Wild Plants

In Africa *Cacao swollen shoot virus,* a badnavirus, was present in species belonging to different genera, such as *Cola acuminata* (the cola nut) of the same family as the cacao (Sterculiaceae) or even to a genus from a completely different family, such as the silk-cotton-tree *(Ceiba pentandra),* a malvaceous tree. The swollen shoot problem started with the introduction of cacao *(Theobroma cacao)* from Latin America into Africa.

Also, cultivated plants grown at or in the vicinity of plant-breeding farms known to import various wild or cultivated species are running a risk. These species often carry viruses that have become latent as a result of a long period of adaptation and selection.

Weeds in or around a cultivated crop can be harmful, particularly when they are carriers of viruses with a wide host range. Examples are: *Cucumber mosaic virus* (CMV), a cucumovirus, transmitted by aphids in a nonpersistent manner, and nepoviruses spread over larger distances by seeds from weeds. Many nematode- and fungus-transmitted viruses survive, often unnoticed, in the roots of weeds.

Soil and Water

Stable, mechanically transmitted viruses such as tobamoviruses may remain infectious in soil and water for long periods and enter the plants via wounds in the roots. Retention of many soilborne viruses is enhanced by presence of the virus in plant debris, for instance, *Tobacco necrosis virus* (TNV), a necrovirus, or by adsorption of the virus to clay particles (*Tomato bushy stunt virus* [TBSV], a tombusvirus). The growing of plants in circulating nutrient solutions enhances considerably the chance of serious outbreaks of tobamo- and necroviruses in tomato and cucumber crops.

Vectors

Viruses transmitted in a persistent manner may be retained in vectors (insects, nematodes, or fungi) during winter; in eggs of insects (few viruses); in pupae of thrips (tospoviruses); or for many years in resting spores of fungi (varicosa-, ophio-, and necroviruses).

DISEASE FORECASTING

Disease forecasting deals with the prediction of epidemic development of viruses in crops. Risk analysis is one type of forecasting, while the other makes use of models for disease development and/or warning systems.

Risk Analysis

Risk analysis is concerned with the analysis of factors important in the development of (known) virus diseases. Knowledge of the effects of these factors from research or by growers experience is applied in cropping and storage plans and in formulating import restrictions by quarantine agencies.

The decisions taken by a grower before planting may have a big influence on the disease development during the growing season. In areas where a particular virus occurs regularly, virus spread can be reduced by growing more resistant cultivars or using highly certified planting material (reduction of infection sources), by practicing crop rotation, and/or by growing crops in periods having no or low vector population. Before storing planting material, a risk analysis should be worked out carefully, such as combination of stocks or crops in storage rooms, storage conditions, and decontamination of planting material, working places, and storage rooms to reduce virus vectors.

International and interregional trade enhances the chances of an outbreak of (new) virus diseases. Risk analysis in these cases is a problem, because it is difficult to establish the presence of viruses in imported, usually dormant material. Moreover, simple disinfection is not possible for viruses, in contrast to most other pathogens or pests. Sometimes, the virus is present at a low level, but the introduction of a very efficient vector may lead to an outbreak of epidemic proportions. This was seen with the introduction of the thrips species *Frankliniella occidentalis* in the Mediterranean area, as since then the incidence of serious losses caused by tospoviruses in various crops has increased.

Forecasting

During the development of a crop, forecasting implies the registration and estimation of factors determining whether a disease will reach damaging or even epidemic proportions. These factors include (initial) number of infection sources, the time of arrival of the vector and its level, weather conditions determining the development of the vector, as well as the virus in the crop. Early warning systems have been developed for aphid-transmitted viruses in sugar beet and potato. These systems are helpful to efficiently control aphids by the use of chemicals and to plan the date for early lifting or foliage destruction to prevent movement of virus to the tubers.

DIRECT CONTROL

Chemotherapy

Viruses and viroids replicate in close association with the metabolism of the host. Therefore, any chemical affecting a virus would probably cause damage to the host as well. So far, only analogs of uracil and guanine were found to block or delay activity of RNA viruses with limited effect on plant growth. Such products as Virazole reduced the concentration of some viruses and were sometimes successful in the production of virus-free plants by meristem-tip culture. Some growth substances such as cytokinins (added to tissue-culture media) may reduce symptom expression of viruses, but can cause viruslike symptoms by themselves. Some chemicals can inactivate viruses present on the surface of seeds. So far, no chemical is known to cure plants from virus diseases. In contrast, certain organisms causing viruslike symptoms, such as phytoplasmas, can be controlled and their symptoms reduced by repeated application of antibiotics such as tetracycline. A commercially successful control was obtained in this way for pear decline in California.

Heat Treatment

A number of viruses can be eliminated by heat treatment, especially from woody plants. Fruit trees in pots subjected to high temperatures (35 to

40°C) for a number of weeks were freed from viruses. Even when the virus is not completely eliminated, the high temperature may slow down its multiplication and spread so that shoots formed by the plant during heat treatment may be free from virus and their meristem tip may be used to obtain virus-free plants. It has been shown that the effect of high temperature is, most likely, not directly on the virus itself but on the interaction between virus and plant. A number of viruses with high in vitro thermal inactivation points (e.g., TBSV at 80°C) were more readily eliminated from the plant after the plant's exposure to a temperature of 36°C for several weeks than some viruses with low in vitro thermal inactivation points (e.g., TSWV at 45°C).

Potato leafroll virus-infected potato tubers stored for many months at very high temperatures (over 40°C) in parts of India were found to be freed from the virus. However, the heat had also adversely affected the quality of the tubers. As neither chemotherapy nor heat treatment have proven efficient control measures, the only way to control virus diseases is indirectly, through prevention of infection or damage.

INDIRECT CONTROL

Hygiene

Hygiene deals with measures directed toward decontamination of planting material, storage rooms, packing material, and tools, but also rules of conduct to prevent virus spread.

Where possible, it is important to clean the planting material (before storage), storage rooms, packing material, and so on to minimize the presence of virus vectors such as aphids or mites. Seeds of tomato or cucumber can be decontaminated from tobamoviruses by hot-air treatment or soaking seeds in trisodium phosphate. In case of viruses readily transmitted in a mechanical way, it is important to reduce plant handling to a minimum and to decontaminate hands, tools, and so on. Spread of very infectious viruses such as tobamo- and potexviruses can be reduced by dipping hands and tools in skim milk or by washing with green soap in combination with trisodium phosphate.

Important rules of conduct to remember are always handle healthy plants prior to the infected ones inside the greenhouse, destroy possible virus vectors such as thrips or aphids before harvesting a crop to prevent mass flight to adjacent crops, do not move soil infested with virus-carrying nematodes or fungal spores from one field to another, remove plant debris after harvest, and so on.

VECTOR CONTROL

Chemical Control

Transmission of viruses by insects in a persistent manner is most successfully prevented by the use of systemic insecticides. This control measure is not effective for nonpersistently transmitted viruses, because migrating aphids are killed too late to prevent virus transmission over short distances. However, it is relatively effective by periodically spraying the synthetic pyrethroids, which disturb the probing behavior of aphids. Spraying mineral oil, either alone or in combination with a pyrethroid, has given the best results. But these methods are profitable only in high-valued crops, such as lilies and *Zantedeschia,* due to costly frequent treatments at intervals of 7 to 14 days. The cost and yield losses due to plant damage can be reduced by combining the chemical control with disease-forecasting systems, as practiced in the cultivation of potato and sugar beet. In greenhouses, however, it is important to use insecticides to prevent aphids from colonization and thus from formation of winged specimens (alatae).

Nematicides are used on a limited scale to prevent transmission of tobra- and nepoviruses. The nematicides sometimes do not penetrate deep enough, particularly in heavy soils. The same holds true for fungicides to control fungus-transmitted viruses in field conditions. In greenhouses, chemical soil disinfestation and steam sterilization are more effective and applied frequently.

Physical Control

Physical control measures are taken during restricted periods, for instance, during storage or before delivery of planting material or cuttings. With a controlled atmosphere (low O_2 or high CO_2) vectors (e.g., gall mites) can be controlled very efficiently at the beginning of the storage of tulip bulbs. Some of these treatments seem to be very promising to control spread of *Tulip virus X,* a potexvirus.

Vector Avoidance

Various methods can be used to avoid or reduce the attack of a crop by virus-transmitting vectors, for example, cultivation in heavy soils known to be free from trichodorid nematodes transmitting tobraviruses or in windy open or relatively cold areas unfavorable to aphids and thrips *(F. occidentalis).*

Late planting (in the Netherlands in November) is known to be effective in reducing or eliminating transmission of tobra- and necroviruses in bulbous crops; early planting as well as harvesting are favorable to seed potato production.

Another alternative is to avoid cultivation near the plants that act as hosts for vectors during winter. Shrubs and trees are known to be sources of aphids in spring.

Insect repellents, such as reflective aluminum foils, polythene sheets, or mulches, are applied successfully in certain areas or against particular insects.

The cultivation of tomatoes in UV-absorbing plastic tunnels reduces infestation by *Bemisia tabaci,* the vector of *Tomato yellow leaf curl virus,* a begomovirus, and hence the incidence of infection with the latter virus. Cultivation in glass or screen houses is also used to protect crops, particularly nuclear stocks, against virus-transmitting insects.

CULTURAL PRACTICES

Though various cultural practices are helpful in controlling virus diseases, some in fact aggravate virus problems.

Effective Virus Control

Cultural practices to control virus diseases include the removal of sources of infection at the earliest stage. This is achieved by using highly certified planting material, roguing of diseased plants, control of weeds in or around the field, and removal of plant debris. Supplementary measures to control virus spread consist of growing more resistant varieties and avoidance of cultivation near the source of virus (crops with high disease incidence or wild vegetation) or vector (edges of woods). Carefully planned dates for sowing or harvesting can be very useful to prevent virus infection (see also previous sections, Sources of Infection and Vector Control).

Several crops are known to have lower disease incidence at higher plant densities, due to restriction of the insect flight in the crops, or at enlarged field size, in case the vector enters the edges. This edge-effect was observed for various types of vectors, including aphids (transmitting viruses nonpersistently), nematodes, and whiteflies. In case viruses are spread by aphids nonpersistently, it is advisable to plant virus-containing stocks in accordance with the prevailing wind direction, starting with the lowest and ending with the highest infection percentages.

Before harvesting a crop it is recommended to kill possible virus vectors to avoid invasion of these vectors into the neighboring crop. This is an important aspect for greenhouse-grown crops that are planted at intervals and have different stages of development.

Crop rotation and intercropping may restrict the buildup of virus and/or vector sources. In India, so-called companion planting with repellent and trap crops is used to protect the main crop from sucking insects, such as aphids and thrips. Repellent crops such as onion, garlic, and fennel are plants with strong natural aromas that drive away the insect pests. They are raised as intercrops. African marigold, castor, and coriander are grown as trap plants for protection against sucking insects, but the first crop also destroys nematodes in the soil. The cultivation of fodder radish prior to tulips is applied successfully to reduce TRV infections in tulips. Intercropping cowpea with either cassava or plantain reduces the incidence of *Cowpea chlorotic mottle virus,* a bromovirus, and *Cowpea mosaic virus,* a comovirus, compared to cowpea monoculture.

Soil treatments may influence vector life and thus have significant effect on the development of virus diseases. After harvesting rice, flooding or irrigating the fields heavily for a period of three months before planting tulips considerably reduced the incidence of TNV transmitted by zoospores of *Olpidium brassicae* in Japan. Nitrogen gifts to the soil influence the green color of a crop. If their color turns to light green, the plants become more attractive for aphids that may transmit viruses.

In some cases virus transmission can be avoided by removing plant parts. By cutting the top part of dahlia plants (grown for tuber production) periodically, the spread of TSV by thrips is reduced considerably in most cultivars, due to prevention of flowering and thus of virus spread by pollen-carrying thrips.

Enhanced Virus Problems

Irrigation is a common cultural practice to improve yield of a crop, to extend the growing season and to permit production in new areas. In Zimbabwe, irrigated crops of maize are seriously affected by *Maize streak virus,* a mastrevirus, in the dry season due to migration of the leafhopper vector from the deteriorating extensive grassland areas. In other irrigated areas weed growth is intensified or regeneration of crop debris, such as sugar beet roots or rice stubbles, harboring viruses occurs.

Protected cropping in greenhouses sometimes creates a change in the type of viruses infecting the crops. Highly infectious viruses (e.g., tobamoviruses) and viroids are more prominent in greenhouse-grown crops, whereas aphidborne viruses are prevalent in the same crops (pepper, tomato, or cucumber) grown under field conditions. Protective cropping leads under certain circumstances to year-round presence of vectors and/or viruses.

Mechanization of planting has led to more infections by persistent viruses, such as *Bean leafroll virus* and *Pea enation mosaic virus-1,* both enamoviruses, in peas and beans and by *Barley yellow dwarf virus* (BYDV), a luteovirus.

Crop rotations with shorter intervals, less variation in crops, or short distances between related crops are becoming more common due to scarcity of land. This implies more problems with soilborne viruses such as fungusborne viruses in rice and sugar beet and nematode-borne viruses in strawberry and hop, but also with aphid-transmitted viruses such as BYDV.

BREEDING FOR RESISTANCE

Types of Resistance

Different types of resistance to virus diseases can be distinguished based on resistance to virus, tolerance to virus infection, and escape from disease. The oldest form of breeding for resistance is the selection of plants with high vitality and absence or low expression of (virus) symptoms. Often this has resulted in selection for tolerance, meaning the virus is present at low or high level, but with a minimum of damage and virus symptoms. In various vegetatively propagated crops this approach was successful for long periods, but fell out of use with the introduction of sensitive cultivars or species or of material made virus free (by meristem culture or heat treatment). In some situations and crops such as papaya, citrus, and tomato, a variant of selection for tolerance was introduced by making use of "cross-protection": a mild virus strain introduced in the crop gives resistance to severe strains of the same virus.

In case of immunity, plants will not become infected after inoculation with a particular virus. Plants are immune to most viruses and susceptibility is therefore exceptional. The possibility that immunity to a particular virus exists in another cultivar of the crop or in a wild relative is very low. It might be introduced from other species or genera by genetic engineering. The term immunity is used when a virus does not replicate in plant cells or their isolated protoplasts: the plant is a nonhost of the virus. When virus infection is confined to a few cells near the inoculation site, for instance, due to a hypersensitive response of the plant, such a plant is called field resistant. In that case, groups of cells are damaged or killed upon contact with a particular virus, usually visible as chlorotic or necrotic lesions. Hypersensitivity has been exploited in breeding programs for field resistance in many major crops, including bean, potato, and tomato. Important practical advantages are the simple way of testing by mechanical inoculation of the plants, easy assessment of degree of resistance by counting lesions, and the involvement of a small number of major genes. However, hypersensitivity is often specific for a virus strain and, therefore, the risk of breaking this type of resistance is enhanced.

Another type of resistance is based on reduced systemic virus spread, for example, in old plants of potato and tulip, the so-called mature plant resistance. In other cases the virus is not spread to all shoots (sugarcane) or cuttings (dahlia) resulting in virus-free offspring. Probably some barriers exist in or between tissues, through which it is difficult for a particular virus to pass. Differences between selections or cultivars are observed and sometimes applied in breeding programs.

Plants resistant to virus multiplication have a low multiplication rate of the virus. The mechanism involved may be similar to those playing a role in resistance to virus spread. Resistance to virus multiplication has been successfully applied in the breeding of, for instance, sugar beet (*Beet yellows virus,* a closterovirus) and tobacco (*Tobacco mosaic virus* [TMV], a tobamovirus). It is rather easy to determine the degree of this type of resistance by comparing virus concentrations in different selections or cultivars with (semi)quantitative tests, such as ELISA or local-lesion assays, with appropriate test plants.

The last type of resistance of plants to be mentioned is that against vectors. Most of the breeding work has been done to introduce resistance to aphids and to a lesser extent against mites. The effect of vector resistance on (reduction of) virus spread, however, depends strongly on the mode of virus transmission. For instance, when aphids do not like the host plant while probing, they move to other plants in the vicinity. This behavior is favorable to spread of nonpersistently transmitted viruses. In case of persistently transmitted viruses, however, not only the vector population is decreased but virus transmission too, in particular where virus spread is within the crop itself.

Breeding

Breeding for virus resistance is a common practice for a wide range of agricultural and horticultural crops. Crossing (back-crossing), and selection is time-consuming, especially when distant relatives are used for the introduction of suitable resistance genes. For this reason, particularly in ornamental crops with a fast succession of new cultivars, not much attention is paid to breeding for virus resistance. Another reason might be an effective control of virus spread by mineral oils and/or pyrethroids.

It takes several years, ten or more, to select a new cultivar. The time depends on the selected type and number of resistance genes. Resistance genes can be of the recessive (more often) or dominant type, and virus resistance can be monogenic or polygenic (governed by one or several genes, respectively). The chance of breaking down resistance is minimized by combining different types of resistance or by using resistance based on a number of genes. This approach, however, is more time-consuming.

Testing on virus resistance is often done in an open field situation, which has the advantage of including resistance to unknown strains of the virus or to its vectors. Infection pressure may be enhanced by growing the parental (wild) relatives (often harboring viruses) or additional sources of infection. Testing under better-controlled conditions in glass or screen houses, however, has the advantage of higher reproducibility of the tests.

TRANSGENIC PLANTS

Cross-protection of plants with a mild strain of a virus can be seen as a precursor of transgenic plants with coat-protein-mediated virus resistance. Transgenic plants generated with this type of pathogen-derived resistance have shown a reduced rate of infection (e.g., TMV or *Potato virus X* [PVX], a potexvirus) and (very) mild symptoms when challenge-inoculated with the severe strain of the homologous virus. Subsequently, plants were transformed with other viral genes too, for instance, CMV replicase genes and movement protein genes. Resistance based on replicase genes is very strong, but has a small basis in acting against very closely related virus strains. In contrast, movement protein genes induce resistance to a wider range of strains and even to other viruses having similar genome strategies. So far, most field trials have been performed with transgenic plants showing coat-protein-mediated resistance with promising results.

Possible risks associated with the transgenic plants include recombination between transferred genes (transgenes) and genes of another virus, transmission of unrelated viruses by transcapsidation (genomic masking), and escape of transgenes to wild relatives through pollen transfer. The risk of escape of transgenes is not theoretical anymore, since there is a report of gene flow from transgenic maize to wild relatives in Mexico. It has been a subject of public debate followed by delayed or uncleared decisions by several European countries, thus hampering the introduction of transgenic plants into cultural practices. An example of this is the refusal of the use of selection marker genes (such as resistance to certain antibiotics, e.g., hygromycine) at the time when transgenic plants were ready to be planted for field trials.

Besides pathogen-derived genes, other genes have been tested in genetic engineering for virus resistance. Plants transformed with sequences of viral antibodies express proteins, so-called plantibodies, that bind homologous virus particles entering the cell.

Resistance genes from a few plant species have been isolated and manipulated to engineer virus resistance. The *N* gene, from tobacco, responsible for giving a hypersensitive resistance to TMV, has been transferred to tomato, inducing resistance. Another example of successful transfer is the extreme-resistance *Rx* gene against PVX from potato into tobacco.

MERISTEM TIP CULTURE

Meristem tip culture in vitro is applied to regenerate virus-free plants from a fully infected, mostly vegetatively propagated crop (species or cultivar). In the widely used method, apical meristem is aseptically excised and grown on a nutrient medium that is often solidified with agar supplemented with growth-inducing substances. In the past, plants grown from these meristems were transplanted directly into soil, grown isolated from diseased material, and tested thoroughly and repeatedly for the absence of virus(es). With the availability of more reliable and sensitive virus detection assays, specialized laboratories test the plants at different stages of tissue culture and couple this detection technology with a number of multiplication steps. In these cases, clones derived from one meristem or even subclones are tested at each step and discarded if found infected. This method scales up the production of virus-free material considerably if most of the clones remain virus free.

The efficiency of meristem culture depends on many factors, such as the size of the meristem (preferably smaller than 1 mm), the type and number of virus(es) to be removed (difficult for nepoviruses and mixed-virus infections), and the crop itself. Since meristem culture is combined with tissue culture methods, axillary and adventitious meristems are more often used. The efficiency is highest with apical and, in general, also with axillary meristems. Though the ready availability of tissue explants of various crops make adventitious meristems more advantageous, the percentage of virus-free plants remains lower.

To increase the number of virus-free plants, meristem culture is sometimes combined with a preceding prolonged heat (e.g., viruses of *Allium*) or cold treatment (in case of viroids). Another possibility is the addition of chemicals as Virazole to the medium. Although positive results were obtained with various viruses, negative effects have been reported too. The suppression of virus multiplication in the latter cases delayed conclusions on the virus-free status of the material.

Variants of meristem-tip culture make use of somatic embryos or meristem-tip grafting (on rootstock seedlings) in vitro (e.g., citrus).

VIRUS-FREE STOCKS

Virus-free nuclear stocks may be built up in different ways. Selection of the most vigorous (mother) plants and testing these repeatedly on viruses is possible in many vegetatively propagated crops. When a species or cultivar is completely infected, meristem-tip culture is applied, often in combination with rapid multiplication in vitro. For crops propagated by seed, virus-free material is easier to obtain and maintain. Nuclear stocks can be preserved as in vitro cultures for a long period of time in liquid nitrogen (cryopreservation), as seed (gene banks), or as plants in insect-free glass or screen houses.

Propagation of nuclear stock material to basic material normally occurs under controlled conditions to prevent reinfection. In general, insect-free glass or screen houses are used for this purpose. In other cases, basic material is grown in isolated fields, as often occurs with fruit trees. The basic material is regularly inspected and tested and suspected plants are uprooted. Visual selection of virus-infected plants may be difficult in crops generated virus free through meristem culture. Before the advent of sensitive virus detection technology, such crops became completely virus infected because symptoms were hard to observe. However, some examples of reinfections have resulted in severe symptoms.

For economic reasons, propagation following these initial stages will often be in the open. Spatial isolation from old (virus-infected) stocks is as important as other control measures, such as farm hygiene and roguing, to keep stocks virus free (see Vector Control and Cultural Practices).

CERTIFICATION SCHEMES

Certification schemes, applied for the production and propagation of (virus-free) stocks, differ for individual crop and country. In many schemes, for instance those of potatoes, two criteria are used for downgrading (lower grade of certificate). First, each certificate has a maximum number of multiplication cycles in order to reduce chances of (re)infection with an increase of multiplication cycles, particularly for stocks grown in the open. Second, a negative outcome of inspection (based on virus symptoms, varietal purity [trueness to type], other quality aspects) or on virus detection implies a lower certificate. In bulbous crops the first criterion is not used. Stocks are reclassified only in such cases where the standards are not met. In fruit trees, not only are the prospective parent trees thoroughly tested for viruses, other pathogens, and various quality aspects but also the next generation. After growing in isolation for a period of at least five years, with regular inspection and indexing, these trees are registered by specialized nurseries for the production of their budwood. The names of various stages (probasic, basic, foundation, and so on) and certificates vary for individual crop and country.

Certification schemes have in common that the percentage of plants sampled for virus tests decreases with a lower certificate because of labor and other costs involved. In some crops, leaves are used for virus tests, often as composite samples. In other crops, tubers or bulbs are tested after harvesting and have the advantage that primary infections are detected much more reliably than by testing leaf samples. Depending on the crop and virus, testing is done by ELISA, indicator plants (e.g., woody plants), or molecular hybridization tests (in the case of potato viruses). In the later stages (i.e., closer to the stage of commercial production) the field inspection based on symptoms prevails.

The certification of commercial plant propagation material guarantees that the cultivar or variety is true to type, genetically pure, and vital, including relative freedom from viruses, other pathogens, and pests. Certification, labeling of materials, and sealing of bags is done by government officials or special inspection and certification agencies under the supervision of the government. These agencies or state institutions regularly perform the inspections and laboratory tests required for certification programs.

The material, even after testing, is not guaranteed to be virus free. Each certificate specifies the tested viruses, the methods used, and the maximum percentage allowed for each virus separately. Normally, the material should be free of quarantine diseases and pests.

IMPORT AND QUARANTINE

With import and quarantine regulations, countries aim to minimize or exclude alien diseases and pests. Geographically isolated regions (Japan, Australia, East Africa) often have more strict import and quarantine regulations than commercially open regions such as the European Union.

Usually at the port of entry a phytosanitary inspection is performed on plant (propagation) material. Many countries require that plant materials be produced under conditions of "good horticultural practices," either done within the certification scheme of the exporting country or a production system with adapted standards. The material must be accompanied by a phytosanitary certificate issued by the quarantine department of the exporting country certifying the absence of diseases and pests of quarantine status. Within the European Union a plant passport is used for this purpose. This certificate or label describes the identity and origin of the material and provides evidence for the absence of listed pests and diseases in the area of origin within the European Union.

Countries such as Australia and Japan require detailed agreements with exporting countries about import and quarantine regulations based on an import risk analysis, including extensive lists of pests and diseases for specified crops in the countries involved. In some of these agreements appointments are made by the quarantine department of the importing country (pre-inspection) about (field) inspection of selected stocks in the exporting country.

Organizations in most countries or continents (e.g., the European and Mediterranean Plant Protection Organization) have their own lists of quarantine diseases and pests. For quarantine viruses tolerance is zero, which implies that either all plants or plant parts (seeds, tissue culture material, germplasm for breeding, and so on) have to be obtained from plants that reacted negatively in tests suitable for this purpose.

Usually quarantine is performed after entry into the country of import (postentry quarantine), but sometimes it is done in another country (intermediate station) where the risk of pathogens or pests is low. The material is grown or kept isolated for longer periods needed for observation (absence of disease or pest), testing (on absence of harmful organisms), and/or disinfestation (of possible organisms by hot-water treatment, fumigation, and so on). In case of viruses, vector-proof glass or screen houses are required. Many viruses have long incubation periods or are present without visual symptoms for longer periods, or it is otherwise difficult to detect them. Therefore, quarantine is time-consuming, costly, and requires skilled professionals.

BIBLIOGRAPHY

Asjes CJ, Blom-Barnhoorn GJ (2001). Control of aphid-vectored and thrips-borne virus spread in lily, tulip, iris and dahlia by sprays of mineral oil, polydimethylsiloxane and pyrethroid insecticide in the field. *Annals of Applied Biology* 139: 11-19.

Bos L (1999). *Plant viruses: Unique and intriguing pathogens—A textbook of plant virology.* Leiden: Backhuys Publishers.

Hadidi A, Khetarpal RK, Koganezawa H (1998). *Plant virus disease control*. St. Paul, MN: APS Press.

Russell GE (1978). *Plant breeding for pest and disease resistance*. London, Boston: Butterworths.

Thresh JM (1982). Cropping practices and virus spread. *Annual Review of Phytopathology* 20: 193-218.

Van Zaayen A, Van Eyk C, Versluijs JMA (1992). Production of high quality, healthy ornamental crops through meristem culture. *Acta Botanica Neerlandica* 41: 425-433.

Zaitlin M, Palukaitis P (2000). Advances in understanding plant viruses and virus diseases. *Annual Review of Phytopathology* 38: 117-143.

Appendix 1

Description of Positive-Sense, Single-Stranded RNA Viruses

Jeanne Dijkstra
Jawaid A. Khan

FAMILY POTYVIRIDAE

The name has been derived from the genus *Potyvirus.*

Six genera are in the family: *Potyvirus, Ipomovirus, Macluravirus, Rymovirus, Tritimovirus,* and *Bymovirus.*

It is the largest family of plant viruses. Its members are characterized by their flexuous filamentous particles with helical symmetry. The virions have diameters ranging from 11 to 15 nm and a pitch of 3 to 4 nm.

All potyvirids induce cytoplasmic cylindrical inclusions (CI) appearing as pinwheels, laminated aggregates, or scrolls in ultrathin sections of infected leaves. The CI inclusions contain a viral genome-encoded protein which possesses adenosine triphosphatase (ATPase) and helicase activities.

A different type of inclusion bodies, amorphous inclusions (AI), has been found to contain aggregates of another virus-encoded protein (HC-Pro, helper component protease), often in combination with degraded cell organelles and virus particles. The HC-Pro protein possesses vector transmission and proteinase functions.

The capsid protein (CP) of the potyvirids has a highly conserved trypsin-resistant core domain, but the N- and C-terminal residues located on the surface of the virion diverge in length and amino acid sequence.

The potyvirids have a positive-sense, single-stranded (ss)RNA, which comprises about 5 percent of the particle mass. The genome is expressed as a polyprotein.

Genus Potyvirus

The siglum has been derived from "potato Y," part of the name of the type species *(Potato virus Y)*. With nearly 100 species, it is the largest of the six genera.

Type species: *Potato virus Y* (PVY)

Virion Properties

Potyvirus particles range in length from 680 to 900 nm. The virions have sedimentation coefficients ranging from 150 to 160 S and a buoyant density of 1.31 g/cm^3 (in CsCl).

Virions contain an RNA molecule of about 9,700 nucleotides (nt) with a Mr of 3.0-3.5 × 10^6.

Genome Organization, Expression, and Replication

RNA molecules have a VPg (genome-linked viral protein) at their 5' terminus and a poly (A) tract (A_n) with stem-loop structure at their 3' terminus.

The genome has a single, large open reading frame (ORF). Its translation product, a polyprotein, is cleaved to functional proteins by the virus-encoded proteinases, namely the N-terminal 35 kDa P1 protein, the N-terminal part of the 52 kDa HC-Pro (helper component-proteinase) and the 27 kDa C-terminal part of the NIa protein (small nuclear inclusion protein) (see Figure A1.1).

The proteinase P1 cleaves at its C terminus and also has RNA-binding activity.

The N-terminal half of the HC-Pro, the HC, is essential for aphid transmission and efficient RNA replication. The C-terminal half of the HC-Pro is a proteinase (pro) which cleaves at its own C terminus. The HC protein is also involved in long-distance transport and pathogenicity. Single amino acid changes in the HC protein of an isolate of *Plum pox virus* (PPV) drastically altered symptoms in herbaceous hosts. The amorphous inclusions (AI) are made up of HC protein.

The other final proteins are cleaved from the polyprotein by the third proteinase, NIa protein.

The 50 kDa protein (P3) has been found associated with CI protein in *Tobacco vein mottling virus* (TVMV) and with NI protein in *Tobacco etch virus* (TEV).

The region P3+6K1 protein may be associated with membranes.

RNA
9,704 nt

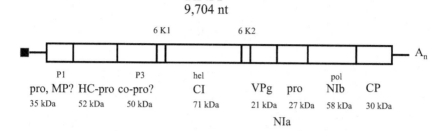

FIGURE A1.1. Organization and expression of the genome of a potyvirus *(Tobacco etch virus)*. Rectangle represents the open reading frame with its final functional proteins produced after processing of the polyprotein indicated. Vertical lines indicate (putative) polyprotein cleavage sites. CI, cylindrical inclusion protein; co-pro?, putative proteinase cofactor; CP, capsid protein; HC-Pro, helper component (HC) for vector transmission and a proteinase (Pro); hel, helicase domain; MP?, putative movement protein; NIa, small nuclear inclusion protein; NIb, large nuclear inclusion protein with polymerase (pol) motif; P1, P3, proteinases; pro, proteinase; VPg, virus protein genome-linked; ■, VPg; A$_n$, poly (A) tract; 6K1, 6K2, small proteins of 6 kDa.

The large CI protein forms the characteristic cylindrical inclusions (pinwheels). The protein has a helicase domain and is most likely responsible for the unwinding of RNA chains. CI protein has also been found in association with the plasma membrane and plasmodesmata and it plays a role in cell-to-cell transport.

The 6K2 protein has membrane-binding activity and is important for virus replication.

Upon further cleaving the NIa protein yields the VPg and a major pro. Besides playing a role in initiation of replication, VPg is also involved in long-distance transport of the virus. Together with the NIb (large nuclear inclusion) protein it forms crystalline inclusions in the nucleus. NIb protein has polymerase domains.

Besides coating, the CP has a number of other functions. Like the HC-Pro protein, the CP is involved in vector transmission. In many potyviruses, a highly conserved motif close to the N terminus of the CP consisting of a triplet of amino acids (Asp-Ala-Gly; DAG) has been found to be essential for successful aphid transmission.

The CP is also required for cell-to-cell and long-distance movement.

Relations with Cells and Tissues

CI are found mainly in the epidermis. Strains of PVY induce crystalline nuclear inclusions containing NIa and NIb proteins.

Host Range and Symptoms

Many potyviruses have a narrow host range, but some species can infect members in a large number of plant families.

Symptoms range from mottling, streaking, yellowing, and mosaic to crinkle, necrosis, puckering, rugosity, and dwarfing.

Transmission and Stability in Sap

Potyviruses are readily transmitted mechanically, but the species differ greatly in stability.

Many potyviruses are inactivated when crude sap from an infected plant is heated at temperatures between 50 and 60°C for 10 min.

Natural transmission is by aphids in a nonpersistent manner.

A number of species are also seed-transmissible.

Antigenic Properties and Relationships

Potyviruses are moderately immunogenic. The species are serologically interrelated, and some are even related to members of the genera *Rymovirus* and *Bymovirus*. By using antibodies to the N-terminal peptide regions of the CPs, virus strains can be distinguished.

Relevant Literature

Adams et al., 2005; López-Moya and Garcia, 1999; Saénz et al., 2001.

Genus **Rymovirus**

The siglum has been derived from the name of the type species, *Ryegrass mosaic virus* (RGMV).

Virion Properties

The virus particles are 690-720 nm long and 11-15 nm in diameter. Rymoviruses have a sedimentation coefficient of about 165 S and a buoyant density in CsCl of 1,325 g/cm^3 (for RGMV).

The CP has a Mr of about 29,000.
Virions contain an RNA molecule of 9,000-10,000 nt.

Genome Organization, Expression, and Replication

RNA molecules have a poly (A) tract at their 3' terminus.

The genome organization and expression are probably comparable to those of the other potyvirids, except the bymoviruses (see Figure A1.1).

Relations with Cells and Tissues

CI (pinwheels), aggregates of granular material and bundles of viruslike particles, have been found in mesophyll and phloem cells of oat plants *(Avena sativa)* infected with *Oat necrotic mottle virus* (ONMV). Similar inclusions have been reported for RGMV in *Lolium* spp. and for *Agropyron mosaic virus* (AgMV) in *Triticum* spp. In addition, ultrathin sections of oat leaves infected with ONMV showed fibrillar wall deposits.

Host Range and Symptoms

Mainly members of the family Gramineae have been found susceptible. Some viruses infect species in a large number of plant genera (e.g., AgMV), whereas others are restricted to members of a few genera only (e.g., ONMV). AgMV inoculated to leaves of *Chenopodium quinoa* induced local lesions in this test plant.

In general, rymoviruses cause chlorosis, mottling, mosaic, and some stunting. However, some species (e.g., ONMV), may also induce necrosis.

Transmission and Stability in Sap

Rymoviruses are readily transmitted mechanically. They are rather stable in crude sap from infected plants. Infectivity is lost in sap heated at temperatures ranging from 50-60°C for 10 min.

Natural transmission by the eriophyid mite *Abacarus hystrix* has been reported.

Antigenic Properties and Relationships

Rymoviruses are moderately to highly immunogenic. Practically no serological relationship exists between the species, and only a weak one between ONMV and *Wheat streak mosaic virus* (WSMV), a species in the genus *Tritimovirus*.

Relevant Literature

Stenger and French, 2004.

Genus **Macluravirus**

The name has been derived from that of the type species *Maclura mosaic virus* (MacMV).

Two species are in the genus, namely the type species and *Narcissus latent virus* (NLV). The latter has previously been considered to be a member of the carlaviruses.

Virion Properties

The virus particles are 650-675 nm long with a diameter of 13-16 nm. The virions have sedimentation coefficients of 155-158 S and buoyant densities of 1.31-1.33 g/cm^3 (in CsCl).

The CP has a Mr of 46,000.

Virions contain an RNA molecule of about 8,000 nt.

Genome Organization, Expression, and Replication

The genome organization and expression are probably comparable to those of potyviruses (see Figure A1.1).

Relations with Cells and Tissues

Pinwheel inclusions are formed in infected cells.

Host Range and Symptoms

Macluraviruses have a narrow host range. Natural hosts of NLV are monocotyledons, but some dicotyledons are used as diagnostic hosts in experimental transmission.

Symptoms in naturally infected plants consist of chlorosis and mosaic. NLV often occurs together with other viruses in *Narcissus* spp., but alone it induces symptomless infections or mild chlorosis in the leaves.

Transmission

The two viruses can be transmitted mechanically. Natural transmission is by aphids in a nonpersistent manner.

Antigenic Properties and Relationships

The viruses are moderately immunogenic. The two species are serologically related. A weak serological reaction has been found between MacMV and the potyvirus *Bean yellow mosaic virus*.

Relevant Literature

Jacob and Usha, 2001; Liou et al., 2003.

Genus **Tritimovirus**

The siglum has been derived from the Latin genus name of wheat *(Triticum)* and mosaic. There are only two species in the genus.
Type species: *Wheat streak mosaic virus* (WSMV).

Virion Properties

The virus particles are 670-700 nm long and about 15 nm in diameter. The sedimentation coefficient is 166 *S* (for WSMV). The CP of WSMV has molecules of 42, 36, and 32 kDa.
Virions contain an RNA molecule of 9,384 nt.

Genome Organization, Expression, and Replication

RNA molecules have a poly (A) tract at their 3' terminus. The genome organization and expression are comparable to those of the other potyvirids, except the bymoviruses (see Figure A1.1). The NI and HC-Pro proteins have been shown to possess proteinase activity, thus resembling those of potyviruses. Like the rymoviruses, tritimoviruses do not have motifs in the HC-Pro and CP for vector transmission.

Relations with Cells and Tissues

Pinwheel inclusions are present in the mesophyll.

Host Range and Symptoms

WSMV has a wide host range which is, however, restricted to members of the Gramineae. *Brome streak virus* (BStV) also infects only gramineous plants, but its host range is narrow.

Symptoms range from mosaic to stunting and necrosis.

Transmission and Stability in Sap

Tritimoviruses are readily transmitted mechanically. They are rather stable in crude sap from infected plants. Its infectivity is lost when the sap is heated at about 55°C for 10 min.

Natural transmission of WSMV is by the eriophyid mite *Aceria tulipae* in a circulative manner. The virus can be acquired only from infected plants by nymphs, but it can be transmitted by adults and nymphs.

Antigenic Properties and Relationships

Tritimoviruses are moderately to highly immunogenic. The two species are serologically interrelated, but no relationship exists between them and other potyvirids.

Relevant Literature

Choi et al., 2000; Stenger and French, 2004.

Genus Ipomovirus

The siglum has been derived from the Latin genus name of sweet potato *(Ipomoea)*.

The type species is the only species in this genus.

Type species: *Sweet potato mild mottle virus* (SPMMV).

Virion Properties

The virus particles are 800-950 nm long and 14 nm in diameter. They have a sedimentation coefficient of 155S.

Virions contain an RNA molecule of about 10,000 nt.

The CP has a Mr of 37,700.

Genome Organization, Expression, and Replication

RNA molecules have a poly (A) tract at their 3' terminus.

The genome organization and expression is comparable to those of the potyviruses (see Figure A1.1).

The only differences are found in the motifs of HC-Pro and CP. As these motifs play a role in vector transmission, the differences may be ascribed to the different vectors of ipomoviruses, namely whiteflies, not aphids.

Relations with Cells and Tissues

Pinwheel inclusions are found in the mesophyll.

Host Range and Symptoms

The virus has a wide host range.

Symptoms consist of mottling, chlorosis, vein banding, blistering, puckering, distortion, and rosetting. The virus induces characteristic epinasty in tobacco.

Transmission and Stability in Sap

The virus can be transmitted mechanically. However, mechanical transmission from sweetpotato plants to other test plants is difficult, because of the presence of strong inhibitors of infection in sap from the source plants.

The virus is rather labile in crude sap from infected plants. It is completely inactivated in sap heated at 60°C for 10 min.

Natural transmission is by the whitefly species *Bemisia tabaci* in a nonpersistent manner.

Antigenic Properties and Relationships

The virus is moderately immunogenic. It is not serologically related to other potyvirids.

Relevant Literature

Mukasa et al., 2003.

Genus **Bymovirus**

The siglum has been derived from the name of the type species, *Barley yellow mosaic virus* (BaYMV).

Virion Properties

The virus particles have two modal lengths, namely 250-300 nm and 500-600 nm, and a diameter of about 13 nm.

The buoyant density is 1.28-1.30 g/cm^3 (in CsCl).

The CP has a Mr of 28,500-33,000.

The longer virus particles contain an RNA molecule of 7,500-8,000 nt (RNA1), the shorter particles one of 3,500-4,000 nt (RNA2).

Genome Organization, Expression, and Replication

Both RNA1 and RNA2 have a poly (A) tract. RNA2 is very unstable. RNA1 and RNA2 encode a polyprotein that is proteolytically processed into functional proteins (Figure A1.2).

The genome of bymoviruses is comparable to that of potyviruses, except for its coding regions being divided between two RNAs. The translation

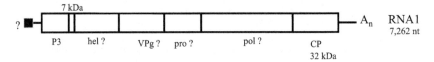

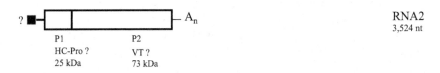

FIGURE A1.2. Organization and expression of the genome of a bymovirus *(Barley mild mosaic virus)*. Rectangles represent the open reading frames with their final functional proteins produced after processing of the polyprotein indicated. Vertical lines indicate putative polyprotein cleavage sites. CP, capsid protein; HC-Pro ?, putative helper component proteinase of potyviruses, hel ?, putative helicase domain; P1, P2, P3, translation products; pol ?, putative polymerase domain; pro ?, putative proteinase; VPg, virus protein genome-linked; VT ?, putative protein involved in transmission by the fungal vector. ? ■, putative VPg; A_n, poly (A) tract.

products P1 and HC-Pro of potyviruses correspond to P1 and P2 expressed by RNA2, whereas the rest of the potyvirus motifs are present in the translation products of RNA1.

Good evidence suggests that P2, a 73 kDa protein, is involved in transmission by the fungal vector.

Relations with Cells and Tissues

In ultrathin sections of infected leaf cells, network structures, pinwheel inclusions, and aggregates of virions are found in the cytoplasm.

Host Range and Symptoms

Bymoviruses have a very narrow host range, restricted to species in the family Gramineae. Barley *(Hordeum vulgare)* is the only host known to be susceptible to BaYMV.

Symptoms consist of yellowing, necrosis, and stunting.

Transmission and Stability in Sap

The viruses are readily transmitted mechanically. BaYMV is rather stable. Natural transmission is by the plasmodiophorid fungus *Polymyxa graminis.*

Antigenic Properties and Relationships

Bymoviruses are moderately immunogenic. The virus species are serologically interrelated, except *Barley mild mosaic virus* (BaMMV).

Relevant Literature

López-Moya and Garcia, 1999.

FAMILY SEQUIVIRIDAE

The family has been named after the genus *Sequivirus* and it comprises two genera, namely *Sequivirus* and *Waikavirus.* The virions are isometric, about 30-40 nm in diameter, and possess a linear, monopartite, positive-sense ssRNA genome of 9,000-12,000 nt in size. The full details on their genome organization and expressions are not available. A VPg is probably present at the 5' end and a poly (A) tail is present at the 3' end of the members of the genus *Waikavirus* but absent in those of *Sequivirus*. The viral genome consists of a single major ORF, which is translated into a polyprotein ranging from 3,000 to 3,500 amino acids. The polyprotein undergoes cleavage by proteinase into functional proteins, as reported in picornalike viruses. In the N-terminal half of polyprotein, three CP species are located, separated from the N-terminus by a polypeptide of Mr $40\text{-}60 \times 10^3$. Downstream of the CP species (i.e., in the C-terminal half of the polyprotein) domains with protease and polymerase functions are present. Two small ORFs are present near the 3' end of a waikavirus genome. Natural transmission is by aphids or leafhopper in a noncirculative, semipersistent manner and requires helper protein. This protein is either self-encoded as in waikaviruses or encoded by a helper virus in the sequiviruses.

Genus Sequivirus

The genus name has been derived from the Latin word *sequor* (to follow or accompany) and refers to the sequiviruses' dependency on helper viruses for transmission by aphids. The helper virus for the type species is *Anthriscus yellows virus* (AYV), a waikavirus.

Type species: *Parsnip yellow fleck virus* (PYFV).

Virion Properties

The virions are isometric, 30 nm in diameter. The capsids consist of three major protein species Mr of about 22.5×10^{-3}, 26×10^{-3}, 31×10^{-3} (parsnip serotype) or 24×10^{-3}, 26×10^{-3}, 31×10^{-3} (*Anthriscus* serotype), which encapsidate the virus RNA genome.

The particles sediment into two fractions: the top component (T) particles appear to be empty shells (60 *S*), while the bottom component (B) contains virus particles with ssRNA (152 *S*); they have buoyant densities of 1.297 (T) and 1.49 (B), respectively.

Genome Organization

The genome consists of monopartite, positive-sense ssRNA of about 10,000 nt with a single ORF. It contains a VPg at its 5' end and a poly (A) tail at its 3' end (see Figure A1.3). The large ORF encodes a polyprotein of about 336 kDa that sequentially undergoes self-proteolytic cleavage (as in picornalike viruses) to yield functional gene products including three distinct CP species (eg., 22.5, 26, and 31 kDa in parsnip serotype) mapped

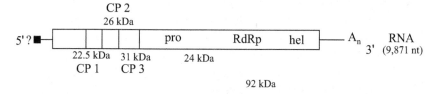

FIGURE A1.3. Organization and expression of the genome of a sequivirus *(Parsnip yellow fleck virus)*. Rectangles represent the open reading frames with their translation products. CP(1-3), capsid protein species; hel, helicase domain; pro, protease; RdRp, RNA-dependent RNA polymerase; ? ■, putative VPg; A_n, poly (A) tract.

within the N-terminal one-third of the polyprotein. The 26 kDa CP species reveals sequence similarity to the VP3 structural protein of picornaviruses. About 400 amino acids are present upstream of the CP species in the polyprotein. The C-terminal one-third of the polyprotein contains sequence motifs resembling the putative NTP-binding motif, protease, and RNA polymerase domains of *Tomato black ring virus* (a nepovirus), *Cowpea mosaic virus* (a comovirus), and other picornaviruses. The 3' UTR (untranslated region) is relatively larger in size and forms a stem-loop structure, which resembles that of mosquito-borne flaviviruses.

Relations with Cells and Tissues

The virions are found in mesophyll cells and epidermal cells. Virus infection induces the formation of inclusion bodies. Infected cells possess plasmodesmatal and cytoplasmic tubules about 45 nm in diameter, containing virus particles. Tubules containing virus particles are also seen in the sieve tubes.

Host Range and Symptoms

Natural host range of sequiviruses is restricted to plants in the family Umbelliferae (PYFV) and to those in the family Compositae (*Dandelion yellow mosaic virus;* DYMV). PYFV causes vein yellowing, yellow flecks, and mosaic symptoms in parsnip *(Pastinaca sativa).*

Transmission and Stability in Sap

Sequiviruses can be transmitted mechanically but are dependent on a helper virus, AYV, for transmission by aphids (*Cavariella* spp. for PYFV and *Aulacorthum solani* and *Myzus* spp. for DYMV) in a noncirculative manner. The virus particles have a thermal inactivation point at 57-65°C (for 10 min) and retain infectivity in crude sap from infected plants stored at room temperature for four to seven days.

Antigenic Properties and Relationships

Sequiviruses are moderately to highly immunogenic. The serotypes of parsnip and *Anthriscus* differ by a serological differentiation index of four to five.

Relevant Literature

Reavy et al., 1993; Turnbull-Ross, 1992, 1993.

Genus Waikavirus

The name has been derived from the Japanese word *waika* describing the symptoms induced by *Rice tungro spherical virus* in rice.

Type species: *Rice tungro spherical virus* (RTSV).

Virion Properties

The virions are isometric, 30 nm in diameter. The capsid consists of three protein species of about (Mr × 10^{-3}) 34, 24, and 18 in *Maize chlorotic dwarf virus* (MCDV), which encapsidate the RNA species.

The virus particles sediment as one component at 175S ($\pm$ 5 S) in RTSV, 183S ($\pm$ 6 S) in MCDV, and have buoyant densities of 1.507 for RTSV and 1.551 for MCDV.

Genome Organization, Expression, and Replication

The genome consists of monopartite, positive-sense ssRNA of about 9,000-12,000 nt. The genome organization and expression is presented in Figure A1.4. It contains VPg at its 5' end, a poly (A) tail at its 3' end, and one large and two small ORFs. The large ORF encodes a polyprotein of about 413 kDa and 390 kDa in MCDV and RTSV, respectively. It sequentially undergoes proteolytic cleavage by a virus-encoded proteinase (as in picornalike viruses) to yield functional products, such as three distinct CP species (18, 24, and 34 kDa in MCDV; 24, 23.7, and 29 kDa in RTSV). The CP species are mapped within the N-terminal one-third of the polyprotein. About

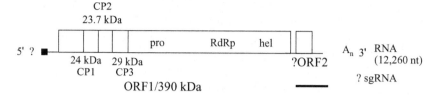

FIGURE A1.4. Organization and expression of the genome of a waikavirus *(Rice tungro spherical virus)*. Rectangles represent the open reading frames (ORF) with their translation products. CP1-3 capsid protein species; hel, helicase; ?ORF2, putative open reading frame2; pro, protease; ?sgRNA, putative subgenomic RNA; RdRp, RNA-dependent RNA polymerase; A_n, poly (A) tract; ? ■, putative VPg.

600 amino acids present upstream of the CP species in the large polyprotein. The C-terminal one-third of the polyprotein contains sequence motifs resembling the putative NTP-binding motif, protease, and RNA polymerase domains of *Cowpea mosaic virus* (a comovirus), poliovirus, and other picornaviruses. The amino acid sequence of the protease domain resembles that of the picornal 3C proteases, the 24 kDa-comoviral protease, and the 49 kDa-potyviral protease. Probably, two small ORFs downstream of the polyprotein are present. The RNA species may correspond to the small ORFs. However, subgenomic (sg) RNAs representatives of these two ORFs have not been identified in MCDV-infected tissues.

Relations with Cells and Tissues

The virions are found in phloem parenchyma and other phloem elements in leaves and roots. Infected cells contain dense, granular, cytoplasmic inclusions with virus particles.

Host Range and Symptoms

Natural host range is restricted to *Sorghum halepense, Zea mays* (MCDV), and *Oryza sativa* (RTSV). The symptoms caused by MCDV in maize consist of leaf reddening or yellowing and chlorosis of the tertiary veins. RTSV causes slight stunting, but when associated with *Rice tungro bacilliform virus,* a tungrovirus, the severity is increased.

Transmission and Stability in Sap

Natural transmission is by leafhoppers (*Graminella* spp. for MCDV and *Nephotettix* spp. for RTSV) in a noncirculative manner. A helper protein, which may be self-encoded, is required for transmission. The RTSV is not mechanically transmissible, but it helps the vector transmission of RTBV. At approximately 4°C infectivity in crude sap from infected plants is lost in a week; thermal inactivation point is approximately 60°C.

Antigenic Properties and Relationships

No serological relationships have been established.

Relevant Literature

Reddick et al., 1997; Shen et al., 1993; Thole and Hull, 1998.

FAMILY COMOVIRIDAE

The family has been named after the genus *Comovirus*. The three genera in the family are *Comovirus, Fabavirus,* and *Nepovirus.*

Virions are isometric with a T = 1 icosahedral symmetry. The comovirids have a bipartite RNA genome consisting of a positive-sense ssRNA molecule.

The viral RNAs are translated in polyproteins, from which functional proteins are processed by proteolytic cleaving.

Purified virus preparations contain three centrifugal components: an empty top component (T) consisting of CP only, and a middle component (M) and bottom component (B), differing in their nucleic acid content.

Genus Comovirus

The siglum has been derived from the name of the type species, *Cowpea mosaic virus* (CPMV).

Virion Properties

The virus particles have a diameter of 28-30 nm. The capsids of the virions consist of two types of polypeptides with Mr values of 37,000 and 23,000. The 60 larger polypeptide species are arranged as 12 pentamers at the 5-fold axis and the 60 smaller ones as trimers at the 3-fold axis.

T, M, and B components have sedimentation coefficients of 58, 95, and 115 S and Mr values ($\times 10^{-6}$) of 3.80, 5.15, and 5.87, respectively.

The buoyant densities in CsCl are 1.29 for T, 1.40 for M, and 1.41 and 1.45 for B.

Purified virus preparations contain two electrophoretic components. Each electrophoretic component consists of the three centrifugal components T, M, and B. The slower migrating form can be found early in infection, the faster one accompanies later infection processes. It has been shown that the faster migrating form originates from the slower migrating form by loss of a polypeptide with a Mr of approximately 2,500 from the 23 kDa protein.

Genome Organization, Expression, and Replication

The two RNA species, RNA1 (B-RNA) and RNA2 (M-RNA), consist of 5,805 and 3,299 nt, respectively. The genome organization and expression is presented in Figure A1.5.

The genomic RNAs have a small protein, VPg, at their 5' terminus and a poly (A) tract at their 3' terminus.

Both RNAs contain a single, large ORF from which the polyproteins are translated. The ORF of RNA1 codes for a 200 kDa protein, that of RNA2 for a protein of 105 kDa. After processing, the polyprotein of RNA1 yields five final products, namely a 32, 58, 2-4, 24, and 87 kDa protein (Figure A1.5).

The 24 kDa protein (pro) is the viral proteinase that is responsible for the cleaving of the polyprotein. For some of the cleavages the 32 kDa protein, a

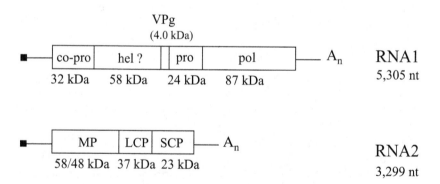

FIGURE A1.5. Organization and expression of the genome of a comovirus *(Cowpea mosaic virus)*. Rectangles represent the open reading frames with their final functional proteins produced after processing of the polyproteins, indicated. Vertical lines indicate polyprotein cleavage sites. Co-pro, proteinase cofactor; hel ?, protein with putative helicase function; LCP, large capsid protein; MP, movement protein; pol, polymerase domain; pro, proteinase; SCP, small capsid protein; VPg, virus protein genome-linked; ■, VPg; A_n, poly (A) tract.

proteinase cofactor (co-pro), is required. This protein is considered to be a regulator of processing. The 87 kDa protein is a replicase (pol) and has an RNA-dependent RNA polymerase (RdRp) motif.

The processed polyprotein of RNA2 yields four final products, namely a 58, 48, 37, and 23 kDa protein. The 37 kDa and 23 kDa products represent the two different CPs (LCP and SCP). The 48 kDa protein (movement protein [MP]) has been shown to be necessary for movement of the virus from cell to cell. The function of the 58 kDa protein is yet unknown, but this protein along with the 48 kDa and CPs, are needed to establish a successful infection.

Infection Process

In protoplasts inoculated with virus suspensions containing particles with RNA1 only, this RNA species is replicated and RNA1-encoded proteins are formed. However, when such virus suspensions are inoculated to leaves of a susceptible plant, replicated RNA1 does not move to adjacent cells, thus leading to subliminal infections. As expected, particles containing RNA2 only are not infectious. When leaves are inoculated with CPMV suspensions, virions accumulate in the cytoplasm and characteristic tubules are formed. The tubules penetrate the plasmodesmata and play a role in cell-to-cell transport of the virions. The 48 kDa protein has been shown to be involved in the formation of the tubules.

Relations with Cells and Tissues

Infected cells show vacuolate inclusions in the cytoplasm often adjacent to the nucleus. The inclusions consist of a kind of reticulum formed by vesicles containing fibrillar elements. In between the vesicles is electron-dense material in which virus particles have been observed embedded. Most likely, these cytopathic structures are the site of virus replication (viroplasm). Crystalloid inclusions, consisting of virus particles have also been found in epidermal cells, mesophyll, and even in phloem cells.

Host Range and Symptoms

The host range is rather limited and most comoviruses are restricted to members of the family Leguminosae. Symptoms range from chlorotic mottling and yellow mosaic to distortion, necrosis, and growth reduction.

Transmission and Stability in Sap

Comoviruses are readily transmitted mechanically. The viruses are rather stable in crude sap from infected plants, in which they retain their infectivity at 20°C for about four days. They are inactivated in sap heated at 55-60°C for 10 min.

Natural transmission is by beetles belonging to different families. A few reports mention transmission by thrips and grasshoppers. Seed transmission has been reported for CPMV, but the transmission percentage was very low.

Antigenic Properties and Relationships

The viruses are strongly immunogenic. Serological relationships have been established between all comoviruses.

Relevant Literature

Lomonossoff and Shanks, 1999.

Genus Fabavirus

The name has been derived from the plant species, *Vicia faba* (broad bean) from which the type species, *Broad bean wilt virus 1* was isolated originally.

Type species: *Broad bean wilt virus 1* (BBWV-1).

Virion Properties

The virus particles have a diameter of about 28 nm.

The two CPs of 44 kDa and 22 kDa are arranged in the same way as the comoviruses.

The T, M, and B components have sedimentation coefficients of 63, 100, and 126 *S*, respectively, and Mr values comparable to those of the comoviruses.

Genome Organization, Expression, and Replication

Both of the genomic RNAs have a poly (A) tract at their 3' terminus and a putative VPg at their 5' terminus.

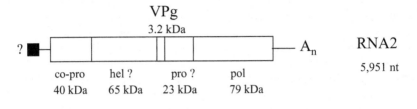

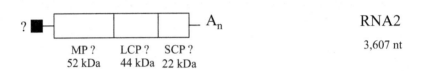

FIGURE A1.6. Organization and expression of the genome of a fabavirus *(Broad bean wilt virus)*. Rectangles represent the open reading frames with their final functional proteins produced after processing of the polyproteins, indicated. Vertical lines indicate (putative) polyprotein cleavage sites. Co-pro, proteinase cofactor; hel ?, protein with putative helicase function; LCP ?, putative large capsid protein; MP ? , movement protein; pol, polymerase domain; pro, proteinase; SCP ? putative small capsid protein; VPg, virus protein genome-linked; ?■, putative VPg; A_n, poly (A) tract.

Expression of the two genomic RNAs, RNA1 (B-RNA) and RNA2 (M-RNA) is comparable to that of the comoviruses (Figure A1.6).

Processing of the RNA1-encoded polyprotein may be similar to that of the comoviruses.

The proteins produced by processing of the RNA2-encoded polyprotein are the LCP and SCP and a 52 kDa protein, a putative MP.

Relations with Cells and Tissues

Depending on the virus strain, amorphous and crystalline inclusions are found in the epidermal cells of BBMV-1-infected broad bean leaves. Virus particles have been observed in tubules in the plasmodesmata.

Host Range and Symptoms

The fabaviruses have wide host ranges infecting both monocotyledons and dicotyledons. Inoculated leaves often show characteristic chlorotic ringspots or concentric ring-shaped necrotic lesions. Systemic symptoms range from chlorosis and oak-leaf pattern to epinasty, (apical) necrosis, and wilting.

Transmission and Stability in Sap

Fabaviruses are readily transmitted mechanically. Their stability in crude sap from infected plants is comparable to that of the comoviruses.

In nature, fabaviruses are transmitted by aphids in a nonpersistent manner.

Antigenic Properties and Relationships

Fabaviruses are strongly immunogenic. BBMV-1 isolates are divided into serotypes I and II. Fabaviruses are more closely related to the comoviruses than to the nepoviruses.

Relevant Literature

Koh et al., 2001.

Genus Nepovirus

The siglum has been derived from "nematode-transmitted viruses with polyhedral particles."

The nepoviruses are divided into three subgroups (a, b, c) based on characteristics of their RNA2 species and serological relationships.

Type species: *Tobacco ringspot virus* (TRSV).

Virion Properties

In contrast to como- and fabaviruses, capsids of TRSV and other officially recognized nepovirus species consist of only one protein with a Mr of 52-60 kDa.

The T, M, and B components of TRSV have sedimentation coefficients of 53, 91, and 126 S and Mr values ($\times 10^{-6}$) of 3.4, 4.7, and 6.1, respectively. The buoyant densities are 1.29 for T, 1.42 for M, and 1.50 and 1.52 g/cm^3 for B.

A number of nepoviruses have linear or circular satellite RNAs associated with them.

Genome Organization, Expression, and Replication

Genome organization and expression resemble those of comoviruses (Figure A1.7).

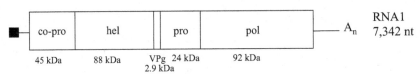

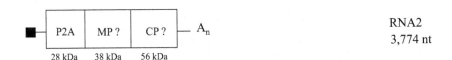

FIGURE A1.7. Organization and expression of the genome of a nepovirus *(Grapevine fanleaf virus).* Rectangles represent the open reading frames with their final functional proteins produced after processing of the polyproteins, indicated. Vertical lines indicate (putative) polyprotein cleavage sites. Co-pro, proteinase cofactor; CP ?, putative capsid protein; hel, helicase domain; MP ?, putative movement protein; P2A, protein required for RNA2 replication; pol, polymerase domain; pro, proteinase; VPg, virus protein genome-linked; VPg; A_n, poly (A) tract.

The genome of nepoviruses consists of two RNA species, RNA1 (B-RNA) and RNA2 (M-RNA). RNA1 is made up of 7,000-8,400 nt, depending on the virus species. The RNA2 species has a much more variable number of nucleotides, varying from 3,400 to 7,200. The latter characteristic, along with the encapsidation strategy of RNA2, has been used to divide the nepoviruses into three subgroups. Subgroup a contains viruses with an RNA2 of 4,300-5,000 nt which is encapsidated in both B and M components. Subgroup b consists of viruses with an RNA2 of 4,600-5,300 nt that is encapsidated in the M component. Subgroup c has viruses with an RNA2 of 6,300-7,300 nt that is encapsidated in the M component which can hardly be separated from B component.

RNA2 codes for three functional proteins, namely a 28 kDa protein (P2A) required for replication of RNA2; a 38 kDa protein, a putative MP that makes up the tubules in the plasmodesmata for cell-to-cell transport; and a 56 kDa protein, the putative CP, for encapsidation and virus spread.

The determinants responsible for transmission of *Grapevine fanleaf virus* by the nematode species *Xiphinema index* are located within the 513 C-terminal residues of the RNA2-encoded polyprotein.

Relations with Cells and Tissues

Virus particles of TRSV have been observed in crystalline arrays or amorphous aggregates in epidermal cells and in meristematic tissues. Particle aggregates have also been detected in tissues of pollen and ovules.

Host Range and Symptoms

Nepoviruses show much variation regarding their host ranges. Some species, such as TRSV, can infect a large number of woody and herbaceous plants, while others are restricted to a single plant species. Symptoms caused by nepoviruses range from ringspots or line pattern to chlorotic mottling, necrosis, distortion, and epinasty.

Transmission and Stability in Sap

In general, nepoviruses are readily transmitted mechanically. The virions are rather stable in crude sap from infected plants, in which they retain their infectivity at about 20°C for one to two weeks. They are inactivated in sap heated to temperatures between 60 and 65°C for 10 min.

In nature, many nepoviruses are transmitted by nematodes belonging to the genera *Longidorus* and *Xiphinema,* and through seed and by pollen. In addition, TRSV has been reported to be transmitted by thrips, flea beetles, grasshoppers, aphids, and mites. Many nepoviruses, however, have no known vector.

Antigenic Properties and Relationships

Nepoviruses are strongly immunogenic. Some serological relationship exists between members of a subgroup.

Relevant Literature

Belin et al., 2001.

FAMILY LUTEOVIRIDAE

The name has been derived from the genus *Luteovirus.* Three genera are in the family, namely *Luteovirus, Polerovirus,* and *Enamovirus.*

The virus particles have diameters of 25-30 nm and possess two proteins. Their genome is a positive-sense ssRNA molecule consisting of approximately 5,700 nt.

Genus Luteovirus

The name has been derived from the Latin word *luteus* (golden) and refers to the symptoms (yellowing) caused by many members of this genus.

Type species: *Barley yellow dwarf virus-PAV* (BYDV-PAV).

Virion Properties

Luteoviruses have sedimentation coefficients of 106-127 *S,* buoyant densities of 1.39-1.40 g/cm³ (in CsCl) and a Mr of about 6×10^{-6}. Virions have a T = 3 icosahedral symmetry.

Genome Organization, Expression, and Replication

The 3' terminal nucleotides are arranged in stem-loop structures, like those in the family *Tombusviridae.*

The genome has six ORFs (ORFs 1-6) (Figure A1.8).

ORF1 and ORF2 code for proteins involved in replication. There is a –1 frameshift (FS) from ORF1 into ORF2. ORF3 encodes the CP and, as a

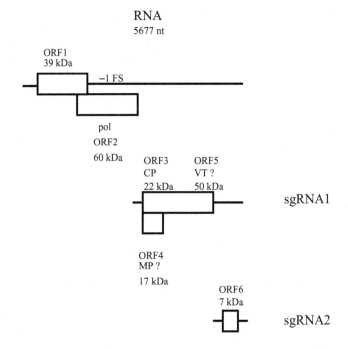

FIGURE A1.8. Organization and expression of the genome of a luteovirus *(Barley yellow dwarf virus-PAV).* Rectangles represent the open reading frames (ORFs) with their translation products. ORFs 1 and 2 are expressed from genomic RNA by –1 frameshift (–1FS). ORFs 3, 4, and 5 are translated from sg RNA1; ORF6 from sg RNA2; CP, capsid protein; MP ?, putative movement protein; pol, polymerase domain; VT ?, putative vector-transmission protein.

result of readthrough of its termination codon, it gives a fusion product with the adjoining ORF5. The product of ORF5 (VT, vector-transmission protein) may play a role in aphid transmission possibly by an interaction with symbionin. The translation product of ORF4 (MP) may be involved in transport from cell to cell. The role of the 39 kDa protein is not clear.

Relations with Cells and Tissues

Luteoviruses are mainly confined to the phloem, where they are found in nuclei and cytoplasm.

Host Range and Symptoms

Luteoviruses have narrow host ranges and are in general restricted to monocotyledons and in particular to species in the family Gramineae. Symptoms consist of chlorosis, yellowing, mottling, red discoloration, and stunting.

Transmission and Stability in Sap

Luteoviruses cannot be transmitted mechanically.

Natural transmission is by aphids in a circulative manner. Aphids fed through parafilm membranes on virus containing sap heated at temperatures between 65 and 70°C are no longer infective.

Antigenic Properties and Relationships

Luteoviruses are strongly immunogenic. A close serological relationship exists between the two species. Besides the similarities in the 3' terminal structures, luteoviruses also display polymerase homologies with the *Tombusviridae*.

Relevant Literature

Koev et al., 2002.

Genus Polerovirus

The siglum has been derived from the name of the type species, *Potato leafroll virus* (PLRV).

Virion Properties

Poleroviruses have sedimentation coefficients of 115-127 *S*, buoyant densities of 1.39-1.42 g/cm^3 (in CsCl) and a Mr of about 6 × 10^{-6}.

Virions have a T = 3 icosahedral symmetry. Virus particles of some species may contain a satellite (sat) RNA.

Genome Organization, Expression, and Replication

The genomic RNA has a VPg and six ORFs (0-5) (Figure A1.9). The genome organization is comparable to that of the luteoviruses, except for one

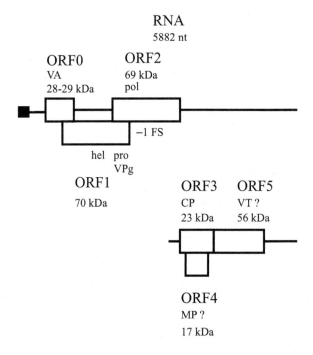

FIGURE A1.9. Organization and expression of the genome of a polerovirus *(Potato leafroll virus)*. Rectangles represent the open reading frames (ORFs) with their translation products. ORFs 0, 1, and 2 are translated from genomic RNA; ORFs 1 and 2 are expressed by −1 frameshift (−1Fs). ORFs 3, 4, and 5 are translated from sg RNA; CP, capsid protein; hel, helicase domain; pol, polymerase domain; pro, proteinase; VA, protein indispensable for virus accumulation; VPg, virus protein genome-linked; VT ?, putative vector-transmission protein; ■, VPg.

extra ORF (ORF0) coding for a protein of about 29 kDa (VA) which is indispensable for virus accumulation. There is a −1 FS from ORF1 into ORF2. ORF1 has a motif for a proteinase. It is possible that a processed product of ORF1 yields the VPg.

Relations with Cells and Tissues

Most virus particles are found in the cytoplasm of phloem parenchyma and companion cells, where they often appear as large aggregates. Crystalline inclusions in the vacuoles of some cells have been shown to consist of virus particles. PLRV-infected plants show vesicles in the cytoplasm and necrosis in phloem tissue. Callose often accumulates in sieve tubes.

Host Range and Symptoms

Some poleroviruses have a wide host range, for instance, *Beet western yellows virus,* whereas others infect mainly members of one plant family, for instance, PLRV in solanaceous plants.

Potatoes grown from PLRV-infected tubers show some chlorosis of the upper leaves, stunting of the shoots and upward curling of leaflets in which carbohydrates have accumulated as a result of impaired transport. Especially the leaflets on lower leaves become brittle and they may develop some marginal necrosis.

Transmission and Stability in Sap

Poleroviruses are not mechanically transmissible.

Natural transmission is by several species of aphids in a circulative manner. Aphids injected with PLRV-containing sap heated at temperatures between 70 and 80°C for 10 min were no longer infective.

Antigenic Properties and Relationships

Poleroviruses are strongly immunogenic. Species within the genus are more or less closely related serologically.

Relevant Literature

Brault et al., 2003; Jaag et al., 2003.

Genus **Enamovirus**

The siglum has been derived from the name of the type species, *Pea enation mosaic virus-1* which is the only species in the genus.

Type species: *Pea enation mosaic virus-1* (PEMV-1).

In PEMV-1-infected plants the virus has always been found accompanied by a second virus, *Pea enation mosaic virus-2* (PEMV-2), a virus belonging to the genus *Umbravirus*. The two viruses have an interdependent relationship: either virus is able to replicate but virions of PEMV-1 cannot move out of the infected cell unless PEMV-2 is also present. PEMV-2, however, cannot be transmitted by aphids in the absence of PEMV-1 and it lacks a gene for the CP.

In purified virus suspensions, two centrifugal components are found, namely B, consisting of particles of PEMV-1 with a diameter of 28 nm and T, containing particles of PEMV-2 with a diameter of 25 nm. Both viruses possess the CP of PEMV-1. Some strains of the PEMV-1/PEMV-2 complex contain a satRNA.

Virion Properties

Virions have a T = 3 icosahedral symmetry. PEMV-1 has a sedimentation coefficient of 107-122 *S*, a buoyant density of 1.42 g/cm^3 (in CsCl) and a Mr of about 5.6×10^{-6}.

Genome Organization, Expression, and Replication

The genomic RNA has a VPg and five ORFs (0,1,2,3,5) (Figure A1.10). The genome organization resembles that of luteoviruses and poleroviruses except for ORF4, which is not found in PEMV-1.

Relations with Cells and Tissues

No inclusion bodies are found. Virus particles occur in the cytoplasm of mesophyll and phloem cells.

Host Range and Symptoms

The PEMV-1/PEMV-2 complex has a narrow host range infecting mainly species in the family Leguminosae.

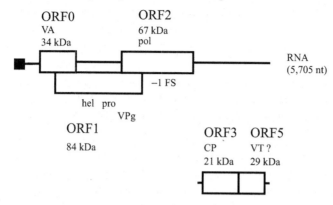

FIGURE A1.10. Organization and expression of the genome of an enamovirus *(Pea enation mosaic virus-1)*. Rectangles represent the open reading frames (ORFs) with their translation products. ORFs 0, 1, and 2 are translated from genomic RNA; ORFs 1 and 2 are expressed by −1 frameshift (−1FS). ORFs 3 and 5 are translated from a subgenomic RNA; CP, capsid protein; hel, helicase domain; pol, polymerase domain; pro, proteinase; VA, protein indispensable for virus accumulation; VPg, virus protein genome-linked; VT ?, putative vector-transmission protein; ■, VPg.

Most leguminous species show chlorotic flecks, mottling, mosaic besides the characteristic outgrowths (enations) on the underside of the midrib, and malformations of leaves and plants.

Transmission and Stability in Sap

PEMV-1 is readily transmissible mechanically.

The virus complex is rather stable in crude sap from infected plants. The viruses are inactivated in sap heated to about 60°C for 10 min.

Natural transmission is by aphids in a circulative manner. Aphid transmissibility is lost after repeated sap inoculations.

Antigenic Properties and Relationships

The PEMV-1/PEMV-2 complex is moderately immunogenic. No serological relationships have been reported between the virus complex and other members of the luteovirids.

Relevant Literature

Skaf et al., 2000; Wobus et al., 1998.

FAMILY TYMOVIRIDAE

The siglum has been derived from the genus *Tymovirus*.

The family comprises three genera: *Tymovirus, Marafivirus,* and *Maculavirus*. The virions are icosahedral in shape with rounded contour and distinct surface structures. They have diameters of about 30 nm. Purified virus preparations reveal two sedimenting components: T represents the non-infectious protein shell and may have small amounts of RNA, whereas B contains infectious virus genome. The genome consists of positive-sense ssRNA with a high content of cytidine. The 5' terminus is capped and the 3' end contains a tRNA-like structure (genus *Tymovirus*) or poly (A) tail (genera *Marafivirus, Maculavirus*). The genome is expressed via polyprotein processing and sgRNA synthesis. Infected plants show cytopathic structures produced as a result of severe modifications of chloroplasts and mitochondria.

Genus Tymovirus

The siglum has been derived from the type species, *Turnip yellow mosaic virus* (TYMV).

Virion Properties

The virions are isometric, about 30 nm in diameter. The capsid consists of 32 capsomeres (20 hexamers and 12 pentamers, total 180 subunits) and has T = 3 icosahedral symmetry. Virions and incomplete virus particles (empty capsids) are present and can be distinguished electron microscopically. Purified particles contain two sedimenting components, T component with protein shells only, and B component with infectious nucleoprotein particles. The virion encapsidates genomic RNA and an sgRNA corresponding to the 3' region of the genome.

The virion Mr is 5.6×10^6 (B component) and 3.6×10^6 (T component). The particles sediment at about 115 S (B components) and 55 S (T component), with buoyant densities of 1.4 and 1.29 g/cm^3, respectively.

Genome Organization, Expression, and Replication

The viral genome consists of one species of linear, positive-sense ssRNA. The genome organization and expression is presented in Figure A1.11. The 5' terminus has a methylated nucleotide cap, the 3' end has a tRNA-like structure accepting valine in TYMV, and RNA has a high

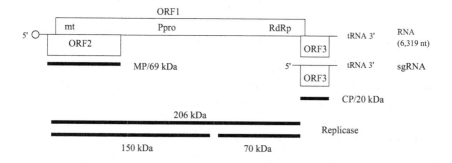

FIGURE A1.11. Organization and expression of the genome of a tymovirus *(Turnip yellow mosaic virus).* Rectangle represents the open reading frame (ORF) with functional proteins. Thick lines indicate corresponding translations. CP, capsid protein; MP, movement protein; mt, methyltransferase; Ppro, papainlike protease; RdRp, RNA-dependent RNA polymerase; O, cap.

content of cytidine. Both RNAs have a tRNA-like region of 36 nt at the 3' end with pseudoknot structure, which is responsible for efficient transcription by RdRp.

The RNA genome of *Tymovirus* is 6,000-6,700 nt in size and it contains three ORFs. ORF1 yields a protein of 206 kDa with conserved motifs of methyltransferase, papainlike protease, helicase, and RdRp. This protein is cleaved by the viral protease, yielding a larger N-terminal 150 kDa product, while the smaller C-terminal product is 70 kDa. The 3' terminal part of the 206 kDa polypeptide is similar in all tymoviruses with two conserved blocks. One contains a 16 nt sequence known as "tymo box," which functions as a sgRNA promoter. The second conserved region overlaps the 5'-terminal part of ORF3 and is the "translation initiation box." ORF2, which nearly overlaps ORF1 and encodes a least-conserved protein, is required for cell-to-cell movement of the virus. ORF3 codes for the viral CP (20 kDa), which is expressed from an sgRNA transcribed from the negative genomic strand using the tymo box sequence as a promoter. The tymo box region of TYMV operates at two levels, namely as a coding region for the expression of RdRp and as a regulatory element for the production of sgRNA. The 206 kDa-replicase polyprotein is cleaved by the viral protease, the cut leading to the separation of helicase and replicase.

Replication takes place probably in vesicles (formed in the margins of chloroplasts of tymovirus-infected plants). These vesicles contain a viral replicase complex and the replicative form of double-stranded (ds) RNA. It is assumed that ORF1 and ORF2 are translated from genomic RNA and CP is translated from sgRNA (694 nt) in which the CP ORF is at 5' proximal end. The transcription initiation site for the sgRNA is mapped to position 5,625 on the genomic RNA, overlapping the end of RdRp. The assembly of virus takes place in vesicles (i.e., where the chloroplast vesicles connect the cytoplasm). Virions and empty particles are also present around vesicles, throughout the cytoplasm. The virus particles accumulate in parenchymat-ous cells, where they are acquired by the vectors.

Relations with Cells and Tissues

Infected plants show characteristic vesicles within their chloroplasts. They are produced by the invaginations of outer chloroplast membranes. Highly characteristic aggregates of swollen and modified chloroplasts are seen.

Host Range and Symptoms

Tymoviruses have host ranges restricted to dicots and they induce bright yellow mosaic, or mottling; all main tissues of host plants are infected.

Transmission and Stability in Sap

Natural transmission of tymoviruses is by beetles of the families Chrys-omelidae and Curculionidae in a noncirculative/circulative manner. Few are transmitted through seed. Mechanical transmission is possible.

Crude sap from an infected plant retains infectivity for about 10 days; the virus particles have a thermal inactivation point at approximately 60-65°C.

Antigenic Properties

Virions are moderately to highly immunogenic and have complex serol-ogy. N-terminal part is a dominant immunogen and located internally. The serological relationships between different species vary from very close, to weak to undetectable.

Relevant Literature

Martelli et al., 2002; Mitchell and Bond, 2005; Shirawski et al., 2000.

Genus **Marafivirus**

The siglum has been derived from the name of the type species *Maize rayado fino virus* (MRFV).

Virion Properties

Virions are icosahedral, 28-32 nm in diameter with high cytidine content, and composed of two serologically related proteins of molecular mass 22 and 28 kDa present in ratio of 3:1. They have one molecule of positive-sense ssRNA with a high cytidine content with a molecular mass of 2.0-2.4×10^6 Da.

The particles sediment as two components, T component, at 47-57 *S* and B component at 118-124 *S*. The T component contains only protein shells, whereas the B component consists of infectious nucleoprotein particles. The two components have buoyant densities (CsCl) of 1.26-1.28 g/cm^3 (T) and 1.37 g/cm^3 (B).

Genome Organization, Expression, and Replication

The ssRNA genome of MRFV is 6,305 nt long.

It is capped at the 5' terminus. A poly (A) tail is present at the 3' end of *Oat blue dwarf virus* (OBDV) and Poinsettia mosaic virus (PnMV), in MRFV there is no evidence of its occurrence. The 3' terminal tRNA structure is absent. Genomic RNA contains the conserved tymo box sg promoter.

The genome organization and expression is presented in Figure A1.12. The RNA genome of OBDV has one ORF, and that of MRFV has a second

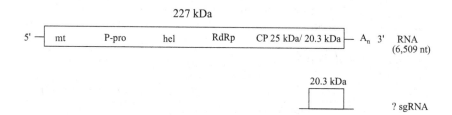

FIGURE A1.12. Organization and expression of the genome of a marafivirus *(Oat blue dwarf virus)*. Rectangles represent the open reading frames with their translation products. CP, capsid protein; hel, helicase domain; mt, methyltransferase domain; P-pro, papain-like protease; RdRp, RNA-dependent RNA polymerase; ? sgRNA, putative subgenomic RNA; A$_n$, poly (A) tract.

smaller putative ORF. The larger ORF encodes a polyprotein of 221-227 kDa. Its N-terminal domain contains conserved motifs characteristic of proteins possessing capping activities, for example, methyltransferase, papainlike protease, NTP-binding helicase motifs, and an RNA-polymerase (RNA motifs). Near the 3' end of the genome the two CPs are encoded. The polyprotein undergoes autocatalytic cleavage by virus-encoded protease to yield functional products. Two strategies could be used for the production of CPs. One would be, as observed in MRFV, posttranslational cleavage of the 224 kDa polyprotein between Gly^{1799} and His^{1800} by papainlike viral encoded protease, yielding a 25 kDa protein and cleavage between Ala^{1802} and Thr^{1843} to yield a protein of 20.3 kDa. A second possible strategy for the production of the latter CP would be a synthesis of 3' coterminal sgRNA, from which the CP of 20.3 kDa is translated.

The second putative overlapping ORF present in MRFV is smaller and encodes a putative protein of 43 kDa. It has limited sequence homology with movement proteins of tymoviruses. It is not known if this protein is expressed in vivo.

Relations with Cells and Tissues

Marafiviruses are found in cytoplasm and vacuoles (in loose aggregates), in mesophyll, epidermis, vascular parenchyma, and phloem (occasionally in sieve tubes) of leaves.

Host Range and Symptoms

The host range of marafiviruses is restricted to the family Gramineae. However, OBDV has a wide host range and infects monocots as well as dicots. This virus causes stunting and leaves become stiffened, bluish, and enations develop along leaf veins.

Transmission and Stability in Sap

Natural transmission is by cicadellid leafhoppers (*Dalbulus* spp.) in a circulative propagative manner.

Marafiviruses are not mechanically transmissible. They are also not transmissible by grafting or through seed. At approximately 20°C infectivity in crude sap from an MRFV-infected plant is lost after four days and ORDV remains infective for 16 to 32 days; the virus particles have a thermal inactivation point at approximately 60-65°C (MRFV) and 60-70°C (OBDV).

Antigenic Properties

Marafiviruses are moderately immunogenic. MRFV and OBDV do not show a serological relationship. However, *Bermuda grass etched line virus* is serologically related to both MRFV and OBDV.

Relevant Literature

Hammond and Ramirez, 2001; Izadpanah et al., 2002.

Genus Maculavirus

The name has been derived from the Latin word *macula* (fleck), and refers to the symptoms it causes in grapevines.

Type species: *Grapevine fleck virus* (GFkV).

Virion Properties

Virions have isometric (icosahedral) particles, about 30 nm in diameter with rounded contours and a distinct surface structure. CP subunits are clustered into pentamers and hexamers like those of tymo- and marafiviruses. Virions and incomplete particles (empty capsids) are present and can be distinguished. Purified particles contain two sedimenting components: the T component represents noninfectious protein shells and may have small amounts of two sgRNAs (approximately 1,000 nt and 1,300 nt, respectively) and the B component with infectious virus genome contains both genomic and sgRNAs. The particles have a buoyant density of 1.43 g/cm^3.

Genome Organization, Expression, and Replication

The viral genome has one species of linear, positive-sense ssRNA encapsidated by CP with a single type of subunits, with molecular mass of 28 kDa. The 5' terminus is capped and the 3' end has a poly (A) tail. The viral RNA has a high content of cytidine.

The RNA genome of GFkV is 7,564 nt in size (excluding the poly [A] tail) and it contains three or four ORFs. ORF1 yields a protein of 215.4 kDa with conserved motifs of replicase-associated proteins consisting of methyltransferase, papainlike protease, and RdRp. Tymo or marafi box present near the viral replicase in tymo- and marafiviruses, respectively, are absent. ORF2 codes for the viral CP (24 kDa). Though the sequence homology with the corresponding proteins of tymoviruses is relatively lower, GFKV

CP contains amino acid triplet "PFQ" (Pro, Phe, Glu), which is conserved in tymo- and marafiviruses. The ORFs 3 and 4 are located at the 3' end of the virus genomes, coding for proline- and serine-rich polypeptides of 31 and 16 kDa, respectively, of unknown function. The 5' UTR of GFkV is larger than those of tymoviruses, rich in pyrimidine, and may facilitate translation of the viral replication-associated protein. The 3' UTR is shorter than that of tymoviruses. The genome organization and expression is presented in Figure A1.13.

Replication takes place in cytoplasm, probably in association with the vesicles in mitochondria. 215 kDa polypeptide undergoes autocleavage by the virus-encoded protease. Replication strategy also follows the expression of the CP via a sgRNA synthesis.

Relations with Cells and Tissues

GFkV-infected plants show characteristic cytopathic structures called multivesiculate bodies produced as a result of severe modification of mitochondria. Maculaviruses are localized to phloem tissues in sieve tubes and companion cells. They often accumulate in the form of crystalline arrays in infected cells.

Host Range and Symptoms

The natural host range of maculaviruses is restricted to *Vitis* species. While Grapevine red globe virus (GRGV, a tentative maculavirus) induces latent infections in all hosts, GFkV causes localized translucent spots (i.e., flecks in *Vitis rupestris*). The leaves may get wrinkled and twisted and curl upward. Severe strains of GFkV induce stunting.

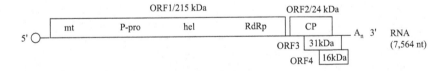

FIGURE A1.13. Organization and expression of the genome of a maculavirus *(Grapevine fleck virus)*. Rectangles represent the open reading frames (ORF) with their translation products. CP, capsid protein; hel, helicase domain; mt, methyltransferase domain; P-pro, papain like protease; RdRp, RNA-dependent RNA polymerase; A_n, poly (A) tract; O, cap.

Transmission and Stability in Sap

Maculaviruses are transmitted neither mechanically nor through seeds. Infected propagative material acts as a virus source for its long-distance dissemination. Natural transmission by a vector is not known.

Antigenic Properties

Virions are efficiently immunogenic. GFkV and GRDV are serologically unrelated.

Relevant Literature

Martelli et al., 2002; Sabanadzovic et al., 2001.

FAMILY TOMBUSVIRIDAE

The family has been named after the genus *Tombusvirus*. It comprises eight genera, namely *Tombusvirus, Carmovirus, Necrovirus, Machlomovirus, Dianthovirus, Avenavirus, Aureusvirus,* and *Panicovirus.*

Its members are characterized by their isometric particles with granular (due to protruding C-terminal domains of the 180 subunits of the CP subunits, a unique feature of tombus-, carmo-, and dianthoviruses) or smooth (machlomo-, necro-, and panicoviruses) appearance that are 25 to 35 nm in diameter. The icosahedral particles have a T = 3 symmetry. Each CP subunit has three structural domains, namely the N-terminal domain, which interacts with RNA; S, the small domain, which constitutes the capsid backbone; and P, the protruding C-terminal domain. Its members have monopartite (except genus *Dianthovirus*), positive-sense ssRNA comprising 17 percent of the particle mass. Total genome is 4,000-5,400 nt long, its 5' terminus probably has a methylated nucleotide cap (demonstrated only in few members) and the 3' end has no poly (A) tail. They mostly have five ORFs, expressed from genomic RNAs. Translation products of the 5'-proximal ORFs1 and 2 are expressed as a result of suppression of an amber codon to produce a readthrough polypeptide. The translation products of 3'-proximal ORFs 3, 4, and 5 are produced through sgRNAs.

Genus Tombusvirus

The siglum has been derived from the name of the type species *Tomato bushy stunt virus* (TBSV).

Virion Properties

The virions are spherical, about 34 nm in diameter, with a 41 kDa CP. Virion Mr is 8.9×10^6, the virus particles sediment at 132-140 *S* and have a buoyant density of 1.35 g/cm^3 (CsCl).

Genome Organization, Expression, and Replication

The ssRNA genome of TBSV is 4,776 nt long.

The genome organization and expression is presented in Figure A1.14. The genome has five ORFs. ORF1-encoded protein (33 kDa) and ORF1 plus ORF2 readthrough (RT) fusion protein (92 kDa) are translated from the genomic RNA and involved in replication of genomic RNA. ORF3 codes for the CP (41 kDa) translated from sgRNA1. The fungal transmission determinants are located on the CP, which is indispensable for virus movement. Subgenomic RNA2 (900 nt) is a bifunctional mRNA. It follows

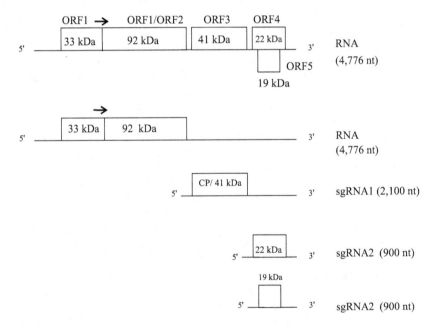

Figure A1.14. Organization and expression of the genome of a tombusvirus *(Tomato bushy stunt virus)*. Rectangles represent the open reading frames (ORF) with their translation products; CP, capsid protein; MP, movement protein; →, translational readthrough.

the leaky scanning translational strategy and expresses two distinct proteins from ORFs 4 and 5. ORF4-encoded MP is of 22-27 kDa and ORF5 (nested within ORF4) encodes a 19 kDa protein that is helpful in long-distance transport (host-specific manner) and responsible for symptom severity. It is not clear if a cap structure is present at the 5' terminus. The 3' end does not contain a poly (A) tail.

The 33 kDa protein is probably involved in the formation of vesicular bodies from the membranes of mitochondria or peroxisomes. Vesicles are the suggested sites for virus RNA replication. Replication does not depend on helper virus, though it acts as helper for the satRNA. During tombusvirus infection, sgRNA accumulates to a high level. The synthesis of sgRNA2 follows that of sgRNA1. Subgenomic RNA2 is a bifunctional mRNA that encodes distinct proteins from ORFs 4 and 5. Mutational studies suggest that the AUG codon of ORF4 is accessed by leaky ribosomal scanning. The longer size of the sgRNA2 leader also increases expression of ORF4 in comparison to ORF5.

Relations with Cells and Tissues

The virions are found in all parts of the host plants. They induce the formation of cytopathic inclusions, that is, multivescular bodies (MVBs) and virus-containing evaginations of the tonoplast. MVBs are sites of the RNA replication. Dense granules having CP accumulations are scattered or loosely aggregated in the cytoplasm of systemically infected cells. Inclusion bodies (crystal or irregular in shape) are present in infected tissues.

Host Range and Symptoms

Tombusviruses have a restricted natural host range, while the experimental range is wide. They systemically infect natural hosts with diffuse mottling and malformation of leaves. The symptoms in most of the experimental hosts are localized with characteristic lesions.

Transmission and Stability in Sap

Tombusviruses can be transmitted by mechanical inoculation, grafting, seed, and pollen. A vector is not involved in transmission, except in the case of *Cucumber necrosis virus,* CuNV, which has a fungal vector. Their infectivity in crude sap from infected plants is lost in four to five weeks at approximately 20°C and their thermal inactivation point is approximately 90°C.

Antigenic Properties and Relationships

Serological relationships have been established among most of the members.

Defective Interfering (DI) RNAs

DI RNAs naturally occur in tombusvirus-infected plants maintained under laboratory conditions. They are generated from genomic RNA and attenuate the virus symptoms. In tombusviruses, the associated DI RNAs range in size from 400-800 nt. A DI RNA consists of four genome segments designated as Regions I to IV. Region I corresponds to the 5' terminus of the virus genome, Regions II and III to noncontiguous internal segments, and Region IV to contiguous or noncontiguous segments at the 3' terminus of the genome. Deletion of regions I, II, or IV abolishes the ability of DI RNA to be replicated by the parent virus. Although region III is missing in the DI RNA of TBSV and *Cucumber necrosis virus,* in *Cymbidium ringspot virus* it is essential for viability.

Satellite RNAs

TBSV contains two species of satRNA, namely satRNA B1 and satRNA B10 (612 nt). Both of them do not contain any functional ORFs and differ in sequence homology except a conserved motif of about 50 nt in *cis* for the replication of tombusvirus RNA. While satRNA B10 attenuates the symptoms caused by TBSV (cherry strain) in *Nicotiana clevelandii,* satRNA B1 does not have any influence on symptom induction.

Relevant Literature

Rochon, 1999; Stuart et al., 2004.

Genus Carmovirus

The siglum has been derived from *Carnation mottle virus* (CarMV), the type species.

Virion Properties

The virions are about 30 nm in diameter, icosahedron, hypothesized to be T = 3, and composed of 180 copies of about 38 kDa CP subunits. See also the general description of family *Tombusviridae.*

The particles sediment at about 118-130 S and have a buoyant density of 1.34 g/cm^3 (CsCl). Virions contain an RNA molecule of about 3,800-4,200 nt in size, about 20 percent nucleic acid and 80 percent protein.

Genome Organization, Expression, and Replication

The genome organization and expression is presented in Figure A1.15. The genome has four ORFs. A 5' leader sequence of about 70 nt occurs before the first AUG. ORF1-encoded protein (26-28 kDa), and ORF1 plus RT fusion proteins (86-89 kDa) are translated from the genomic RNA and involved in its replication. Carmoviruses encode two small ORFs in the middle of genome. ORF2 encodes a protein of 7-8 kDa that is indispensable for cell-to-cell movement, while a 9 kDa protein encoded by ORF3 is involved in systemic movement. ORF4 (3'-proximal gene) codes for CP of 37-42 kDa. CP is followed by a 3' UTR of about 290 nt, lacking a poly (A) tail. Apart from genomic RNA, two sgRNA species of 1,700 and 1,450 nt are

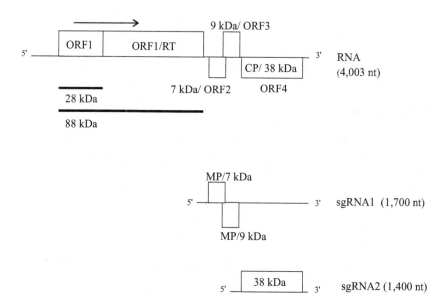

FIGURE A1.15. Organization and expression of the genome of a carmovirus *(Carnation mottle virus)*. Rectangles represent the open reading frames (ORF) with their translation products. Thick lines indicate the corresponding translations. CP, capsid protein; MP, movement protein; →, translational readthrough.

produced in vivo and encapsidated in virus particles. 3'-coterminal with the genomic RNA, they act as mRNAs for ORFs 2 and 3, and for ORF4, respectively. No direct evidence is available for the presence of a cap structure at the 5' terminus.

Most of the carmoviruses exhibit similar genome organizations. However, some unique characteristics are shared by individual carmoviruses. For instance, in CarMV, a second RT extends the translation of the polymerase into the MP gene (9 kDa) and yields a 98 kDa polyprotein. In *Melon necrotic spot virus* (MNSV) two small ORFs of 7 kDa (designated as p7a and p7b) are connected via an amber codon, yielding a 14 kDa-fusion protein. In *Cowpea mottle virus* (CPMoV) a sixth ORF encoding a 28 kDa polypeptide is nested within the CP gene.

After carmovirus infection, the virus transcribes two 3'-coterminal sgRNAs for the expression of MP and CP genes. The smaller sgRNA functions as the mRNA for the CP expression, and the larger sgRNA is mRNA for the production of two MP genes, following leaky scanning mechanism. In *Turnip crinkle virus* (TCV), *cis* sequences responsible for replication, sgRNA synthesis, and encapsidation have been mapped. The sequences necessary for the synthesis of plus and minus strands of RNA are located at the 5' and 3' terminal regions having secondary structure. Further, upstream of the transcription start sites, promoters for the sgRNA1 and sgRNA2 contain secondary structures. The virus assembly origin is located in the 3'-terminal part of TCV CP gene.

Relations with Cells and Tissues

Virions are found in conducting and parenchyma tissues. Viral infection induces the formation of cytopathic inclusions (i.e., MVBs and virus-containing evaginations of the tonoplast). MVBs are sites of the RNA replication. Dense granules having accumulations of CP are seen scattered or loosely aggregated in the cytoplasm of systemically infected cells.

Host Range and Symptoms

Carmoviruses have a narrow natural host range, but their experimental host range is wide. CarMV naturally infects leguminous hosts with serious disease and a number of ornamental plants in the family Caryophyllaceae. It causes a mild mottle symptom in carnation.

Transmission and Stability in Sap

In nature, carmoviruses are transmitted mechanically, through seed and by beetles (*Cowpea mottle virus, Bean mild mosaic virus, Blackgram mottle virus,* TCV). Transmission is also associated with soil/irrigation water or fungus *(Olpidium bornovanus)* zoospores *(Cucumber necrosis virus).*

The crude sap from an infected plant remains infectious for 42 days when sap is stored at 20°C and the virus particles have a thermal inactivation point of approximately 95°C.

Antigenic Properties and Relationships

No serological relationships have been established between virus species.

Defective Interfering RNAs, Satellite RNAs

Carmovirus infections contain three types of small RNAs, such as DI RNA generated from the parental genome (e.g., RNA G 342-346 nt), satRNA of nonviral origin (e.g., RNAs D 194 nt, F 230 nt), and chimeric RNA (RNA C 36, which has a 5' region derived from satellite RNA D and a 3' region from the 3' end of TCV genome). These small RNAs (characterized in TCV) depend on helper virus for their replication and encapsidation. They may or may not affect expression of symptoms or replication of helper virus.

Relevant Literature

Qu and Morris, 1999; Wang and Simon, 1997.

Genus Necrovirus

The name has been derived from *Tobacco necrosis virus.*
Type species: *Tobacco necrosis virus A* (TNV-A).

Virion Properties

The virus particles are about 28 nm in diameter; composed of 180 subunits of a 30 kda CP; see also general description of family *Tombusviridae.*

The virions have Mr 7.6×10^6. The particles sediment at 118 S and have a buoyant density of 1.40 g/cm^3 (CsCl).

Genome Organization, Expression, and Replication

The virus genome contains an RNA molecule of about 3,600-3,780 nt in size. The genome organization and expression is presented in Figure A1.16.

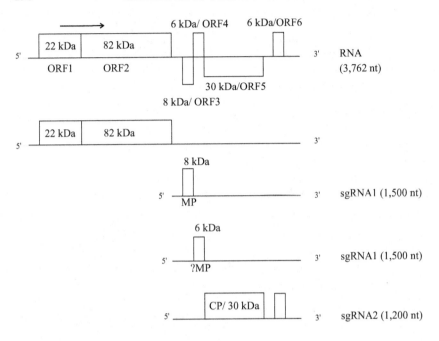

FIGURE A1.16. Organization and expression of the genome of a necrovirus *(Tobacco necrosis virus)*. Rectangles represent the open reading frames (ORF) with their translation products. CP, capsid protein; ?MP, (putative) movement protein; →, translational readthrough.

The genome has five ORFs. The 5'-proximal region of ORF1 encodes a protein of 22-24 kDa and ends with a UAG codon, which is suppressed. Readthrough of the amber termination codon of ORF1 would give an 82 kDa protein (putative polymerase) terminating with a UGA codon (ORF2). Two sgRNAs of about 1,600 and 1,300 nt are synthesized in infected cells. The smaller sgRNA is a translational template for CP and larger one for the ORF3 and possibly ORF4 products. ORF3-encoded protein (7-11 kDa) has a function in the cell-to-cell movement of virus; a 6-7 kDa protein encoded by ORF4 is a putative movement protein. ORF5 is expressed from sgRNA2 (1,300 nt) and its product is a 30 kDa CP. ORF6 is present only in Tobacco necrosis satellite virus of TNV-A and *Beet black scorch virus* (BBSV).

Genomic RNA serves as mRNA for the production of the 22 kDa protein (p22) and its RT (p82) in vivo. In necroviruses, two sgRNAs are coterminal with the genomic RNA. In TNV-A, the shorter sgRNA (1,224 nt) acts as the mRNA for the production of the CP; the longer mRNA (1,501 nt) probably acts as the mRNA for the 8 kDa and 6 kDa proteins from the downstream ORF. The genomic RNA of TNV lacks a 5' cap and 3' poly (A) tail, and the sgRNAs are not capped. Though cap and poly (A) tail are crucial for the translation of most cellular mRNAs, synthesis of the CP of TNV from sgRNA2 has been shown very efficient in infected tobacco protoplasts. The sequences from both the sgRNA leader and trailer are required for an efficient cap-independent translation. The trailer element, consisting of 18 nt sequences conserved in related viruses (e.g. *Barley yellow dwarf virus,* a luteovirus) is required for cap-independent translation.

The full-length minus-strand RNA probably acts as a substrate for synthesis of both the genomic and sgRNA. The 22/23 kDa and 82 kDa proteins are required for the replication of TNV RNA. Further, 3'-*cis*-acting sequences necessary for the replication are located in the last 367 or 342 nt of TNV-D or TNV-A, respectively.

Relations with Cells and Tissues

In roots of infected plants, virions are found in cytoplasm. Virus induces the formation of inclusion bodies, which may be crystals or irregularly shaped bodies in the cytoplasm and stringlike structures in the nucleus.

Host Range and Symptoms

Necroviruses have a wide natural host range that includes monocotyledonous as well as dicotyledonous species. They cause necrotic lesions. Natural infection is restricted to roots and systemic infections are rare. However, some exceptions include Augusta disease of tulip, stipple streak of bean, and necrosis disease of cucumber and soybean. *Leek white stripe virus* causes winter whitening disease in leek, characterized by white striping of the leaf blades.

Transmission and Stability in Sap

Necroviruses can be transmitted mechanically. At approximately 20°C crude sap from infected plants retains infectivity for several days and their thermal inactivation point is approximately 85-90°C.

Natural transmission is by zoospores of chytrid fungus *Olpidium brassicae.*

Antigenic Properties and Relationships

Necroviruses are moderately immunogenic. Two serotypes of TNV are present that can discriminate several strains. TNV is not serologically related to its satellite virus.

Satellite Virus

TNV is commonly associated with satellite viral icosahedral particles, 17 nm in diameter. The Tobacco necrosis satellite virus (TNSV) is dependent for its replication on the presence of specific TNV isolates. Three TNSV serotypes have been described, namely TNSV-1, TNSV-2, and TNSV-C. TNSV consists of a single RNA packaged by 60 CP subunits of 22 kDa (T = 1 symmetry). TNSV interferes with the expression of TNV-A in infected plants, leading to the formation of smaller local lesions.

TNSV is highly immunogenic.

Genome Organization, Expression, and Replication

The satellite virus genome (1,220-1,240 nt long RNA) contains a single ORF encoding for a CP and lacks VPg at the 5' end and poly (A) at the 3' end. The 5' UTR is relatively short and folded into a hairpin structure; the CP-coding region is about 600 nt long followed by a leader sequence of 620 nt. Downstream of the coding region, a stem-loop structure is present, followed by three pseudoknots, one small hairpin, and a long stem-loop structure (that contains the 3' end of the RNA).

Although the deletion of the CP-coding region does not influence replication, leader, and trailer sequences, the 3' terminal hairpin and 5' terminus stem-loop structure are required for replication in vitro.

Relevant Literature

Cao et al., 2002; Meulewater, 1999.

Genus **Machlomovirus**

The siglum has been derived from *Maize chlorotic mottle virus* (MCMV), the type species and only member of the genus.

Virion Properties

The virions are about 30 nm in diameter, icosahedrons, hypothesized to be T = 3, with a CP of 180 copies of a 25 kDa subunit. They possess a single molecule of infectious linear ssRNA. Structure and sequences of the CP are related to those of sobemoviruses.

The virion Mr is 6.1×10^6. The particles sediment at about 109 S and have a buoyant density of 1.365 g/cm^3 (CsCl).

Genome Organization, Expression, and Replication

The virus genome contains an RNA molecule of about 4,400 nt in size. It is capped at the 5' end with an m7G linked to an A residue and a poly (A) tail is absent at the 3' end. The genome organization and expression is presented in Figure A1.17.

The genome has four ORFs. ORF1 encodes a protein of 32 kDa. This polypeptide does not show amino acid sequence similarity to any known viral polymerase. ORF2 encodes a protein of 111 kDa. This ORF is suppressed by an in-frame amber termination codon yielding a prereadthrough ORF encoding a 50 kDa polypeptide. The RT portion of the 111 kDa polypeptide contains a GDD motif present in RdRp. Both the 50 kDa and 111 kDa polypeptides have been expressed in vitro. An internal ORF (ORF3) capable of encoding a 33 kDa polypeptide is interrupted by an in-frame opal termination codon yielding a pre-readthrough 9 kDa polypeptide. These predicted polypeptides have not been expressed in vitro. This polypeptide shows a high degree of sequence similarity to small polypeptides encoded by comoviruses and is involved in cell-to-cell movement. ORF4 encodes

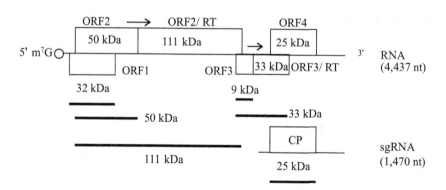

FIGURE A1.17. Organization and expression of the genome of a machlomovirus *(Maize chlorotic mottle virus)*. Rectangles represent the open reading frames (ORF) with their translation products. Thick lines indicate the corresponding translations CP, capsid protein; →, translational readthrough; O, cap.

the 25 kDa CP via sg RNA (1,470 nt). The ORF encoding CP overlaps the 3' half of the preceding ORF in a different translational reading frame.

The genome organization is most similar to that of the genus *Carmovirus*. However, MCMV possesses an additional ORF1 (32 kDa), and ORF3 (33 kDa) is probably not expressed from an sgRNA.

The viral replicase recognizes a structural feature on the viral-sense RNA for the synthesis of the full-length complementary strand. Following complementary strand synthesis, the replicase recognizes its 3' terminus for the synthesis of progeny genomic RNA. The ORF coding for the CP is expressed from an sgRNA (1,470 nt). It is assumed that viral replicase recognizes an internal sequence (including an 11 nt element upstream of the CP initiation codon) identical to the 3' terminus of the full-length complementary strand and synthesizes the sense strand of CP RNA.

Relations with Cells and Tissues

MCMV has been detected in all parts of infected maize plants.

Host Range and Symptoms

Maize is the only known natural host. Experimental host range is restricted to members of the Gramineae.

MCMV causes stunting, shortening of internodes, and chlorotic patches on leaves, followed by epinasty and necrosis, leading to death of plants; inflorescences are affected; panicles and rachis become hard with few spikelets. Infection mixed with potyviruses gives synergetic effects.

Transmission and Stability in Sap

MCMV can be transmitted mechanically. Natural transmission is by six species of chrysomelid beetles belonging to the genera *Oulema, Systena, Chaetonema,* and *Diabrotica.* An isolate of MCMV from Hawaii is transmitted by thrips *(Frankliniella williamsi)* in a noncirculative manner. At approximately 20°C for 12 days infectivity in crude sap from an infected plant is lost, and its thermal inactivation point is approximately 80-85°C.

Antigenic Properties and Relationships

MCMV is moderately to highly immunogenic with three serotypes: Kansas serotype 1 and 2, and the Peru serotype.

Relevant Literature

Nutter et al., 1989; Scheets, 2000.

Genus Dianthovirus

The name has been derived from *Dianthus,* the generic name of carnation.
Type species: *Carnation ringspot virus* (CRSV).

Virion Properties

The virions are icosahedral (polyhedral), 32-35 nm in diameter, with a T = 3 symmetry, and composed of 180 copies of the 37 kDa CP. Genome segments are encapsidated separately into two different types of particles. Two classes of virions are probable, one with RNA-1 and the other with three copies of RNA2. The 5' terminus has a methylated nucleotide cap with m7GpppA and the 3' terminus has no poly (A) tail. Although the 3' terminal 27 nt and 6 nt are identical between RNA1 and RNA2, both segments show limited homology.

The virion Mr is 8.6×10^6. The particles sediment at about 132 S and have a buoyant density of 1.366 g/cm^3 (CsCl).

Genome Organization, Expression, and Replication

The genome organization and expression is presented in Figure A1.18. The virus genome consists of positive-sense, bipartite ssRNAs, namely RNA1 (3,700-4,000 nt) and RNA2 (1,500 nt). RNA1 has three ORFs. The 5'-proximal ORF, interrupted by a −1 ribosomal frameshifting signal, produces a prereadthrough 27 kDa polypeptide (p27) of unknown function and an 88 kDa polypeptide (p88) with an RdRp motif. The p88 polypeptide is a fusion protein serologically related to p27 and p57 polypeptides. An internal ORF (p54-57) encodes a 54-57 kDa and a 3' proximal ORF (p37) encodes the 37-38 kDa CP expressed from a 1,400 nt sgRNA. CP is not necessary for cell-to-cell movement in *Red clover necrotic mosaic virus* and may or may not be required for long-distance movement, but in CRSV it is necessary for systemic infection.

Upstream of the p27 ORF amber termination codon, a heptanucleotide sequence is present that facilitates ribosomal frameshifting. The secondary structure (present downstream of the shifty heptanucleotide) is necessary for the ribosomal shifting as well as infectivity of the virus.

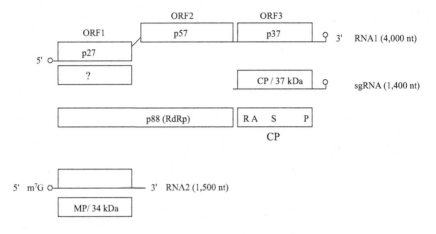

FIGURE A1.18. Organization and expression of the genome of a dianthovirus *(Red clover necrotic mosaic virus)*. Rectangles represent the open reading frames (ORFs) with their translation products. CP, capsid protein; MP, movement protein; R, RNA binding; A, arm; S, shell; P, protruding domain of the RCNV CP; /, translational frameshift; RdRp, RNA-dependent RNA replicase; O, cap; ?, function unknown.

RNA2 contains a single ORF, encoding a putative MP of about 34 kDa.

A dianthovirus replicates in the cytoplasm after producing a negative-sense complementary copy of the viral RNA. The viral replicase identifies the stem-loop structure located at the 3' end of RNA1 and 2 and initiates the synthesis of the complementary strand.

Relations with Cells and Tissues

The virions are found in most of the cells and tissue types in systemic host plants. Infected cells contain large amorphous inclusions composed of large aggregates of virus particles in the cytoplasm. CRSV infection causes crystalline arrays of virus particles in the cytoplasm. Aggregates of virus particles, tubular and spherical inclusions are also seen in nucleoli. *Red clover necrotic mosaic virus* (RCNMV) induces vesiculation in the chloroplast.

Host Range and Symptoms

Natural host range of dianthoviruses is broad; the experimental one is even wider and includes herbaceous species. CRSP systemically infects carnations with ringspots, mottling, and distortions in leaf and flowers. RCNMV and *Sweet clover necrotic mosaic virus* (SCNMV) infections cause necrotic/chlorotic spots. At relatively low temperatures (15-20°C) the symptoms are more pronounced. However, they disappear when the temperature is above 20°C.

Transmission and Stability in Sap

In nature, dianthoviruses are transmitted through the soil, but not with the help of soil-inhabiting organisms, although nematodes and fungi possibly enhance transmission. The virus can be transmitted by mechanical inoculation and grafting but not through seed or by vectors.

At approximately 20°C they retain infectivity in crude sap from infected plants for nearly two months (50 to 60 days) and their thermal inactivation point is approximately 80°C.

Antigenic Properties and Relationships

Dianthoviruses are good immunogens. Strains of CRSV, RCNMV, and SCNMV can be discriminated at the serological level. There is a weak or no cross-reaction at all between serotypes.

Relevant Literature

Mizumoto et al., 2002; Sit et al., 2001.

Genus Avenavirus

The name has been derived from the generic name of oat *(Avena sativa),* the host of type species *Oat chlorotic stunt virus* (OCSV), the only known definitive member.

Virion Properties

The virions are spherical, about 35 nm in diameter with icosahedral symmetry. Virion contains a single CP, encapsidating genomic and a single 3'-terminal subgenomic RNA of 1,772 nt.

Genome Organization, Expression, and Replication

The genome organization and expression strategy of the genus *Avenavirus* appears to be intermediate between the genera *Carmovirus* and *Tombusvirus*.

The genome consists of a positive-sense ssRNA, slightly smaller than tombusviruses. The 3' end has no poly (A) tail. The genome organization and expression is presented in Figure A1.19. The genome of OCSV is 4,114, nt long with four ORFs. ORF1 encodes a protein of 23 kDa and its amber stop codon yields a 84 kDa protein. The latter possesses conserved motifs of RdRp. ORF2 codes for the 48.2 kDa CP, the largest among the members of family *Tombusviridae.* In the CP, the shell domain is most conserved and shows similarities to those of genera *Carmovirus* and *Tombusvirus.*

The genome expression strategy appears to be intermediate between that of the genera *Carmovirus* and *Tombusvirus.* Replication takes place in the cytoplasm.

Relations with Cells and Tissues

The virions are found in the cytoplasm around the organelles in the mesophyll cells of infected plants.

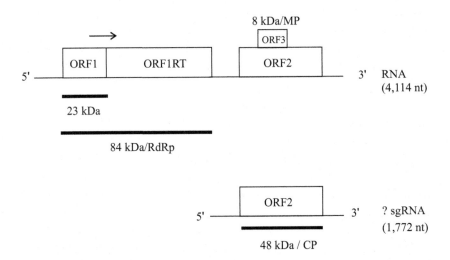

FIGURE A1.19. Organization and expression of the genome of an avenavirus *(Oat chloretic stunt virus).* Rectangles represent the open reading frames (ORF) with their translation products. Thick lines indicate the corresponding transla-tions. CP, capsid protein; MP, movement protein; ORF1RT, readthrough of ORF1; RdRp, RNA-dependent RNA polymerase; ? sgRNA, putative subgenomic RNA; →, translational readthrough.

Host Range and Symptoms

Natural symptoms caused by avenavirus in winter oats consist of bright yellow chlorotic streaking of leaves that turn necrotic with age. Newly emerged leaves are twisted, broader, and darker green than those of healthy oat leaves. Symptomless infection is reported in winter wheat *(Triticum aestivum)*, winter barley *(Hordeum vulgare)*, and annual meadow grass *(Poa annua)*. There are no reports of dicots being infected by avenavirus.

Transmission and Stability in Sap

Avenavirus is poorly sap transmissible. The virus can be propagated by inoculating healthy oat seeds using an embryo-wounding technique. A natural vector is not known.

Antigenic Properties and Relationships

Virions are highly immunogenic. No serological relationships have been established with *Carnation mottle virus, Tomato bushy stunt virus,* or *Maize chlorotic mottle virus,* the type species of genera *Carmovirus, Tombusvirus,* and *Machlomovirus,* respectively.

Relevant Literature

Boonham et al., 1995, 1998.

Genus Aureusvirus

The genus name has been derived from the species name of the host *(Scindapsus aureus)* of type species *Pothos latent virus* (PoLV). The other definitive member is *Cucumber leaf spot virus* (CLSV).

Virion Properties

The virions are spherical, about 30 nm in diameter, with icosahedral symmetry. Each virion contains a single CP. Each CP subunit consists of three domains: the N-terminal domain, the shell, and the C-terminal protruding domain. Besides genomic RNA, virions encapsidate two sg RNAs: sgRNA1 of 2000 nt and sgRNA2 of 800 nt. Satellite or defective interfering RNAs are not present in naturally occurring isolates. See also the general description of family *Tombusviridae.*

The virus particles sediment at 127 *S* (leaf strain of CLSV), 132 *S* (fruit-streak strain of CLSV); and have a buoyant density of 1.36 g/cm³ (CsCl); 1.27 g/cm³ in Cs_2SO_4

Genome Organization, Expression, and Replication

The genome organization and expression strategy of the genus *Aureusvirus* is similar to those of the members of genus *Tombusvirus*. The genome consists of a positive-sense ssRNA, slightly smaller than that of tombusviruses. The 5' and 3' ends are devoid of cap and poly (A) tail, respectively. The genome organization and expression is presented in Figure A1.20. The genomes of PoLV and CLSV are 4,415 nt and 4,432 nt long, respectively. The genome has five ORFs. ORF1 encodes a protein of 25 kDa and its amber stop codon yields an 84 kDa protein. The latter possesses conserved motifs of RdRp. ORFs 1 and 2 are expressed from the genomic RNA. ORFs 3, 4, and 5 are expressed through two sg 3'-coterminal RNAs. ORF3 codes for the CP of 40 to 41 kDa via sgRNA1. In the CP, the shell domain is most

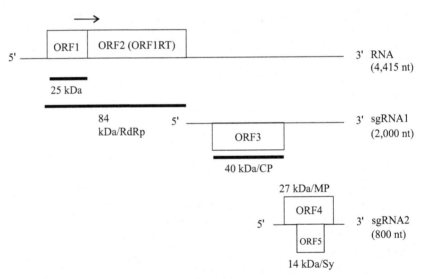

FIGURE A1.20. Organization and expression of the genome of an aureusvirus *(Pothos latent virus)*. Rectangles represent the open reading frames (ORF) with their translation products. Thick lines indicate the corresponding translations. CP, capsid protein; MP, movement protein; ORF1RT, readthrough of ORF1; RdRp, RNA-dependent RNA polymerase; sgRNA1 and 2, subgenomic RNAs; Sy, protein responsible for symptom severity; →, translational readthrough.

conserved, while arm and protruding domains are less conserved. ORF5 is nested in ORF4 but in a different ORF. The latter encodes an MP (27 kDa), while the former codes for a 14-17 kDa protein, suggested to be involved with the pathogenesis. Both the proteins (14-17 and 27 kDa) are expressed from the sgRNA2.

Replication takes place in cytoplasm following the RT and sgRNA synthesis.

Relations with Cells and Tissues

The virions are found in parenchyma and other vascular tissues of systemically infected plants. The prominent cytopathic features are excessive vesiculation of the nuclear envelope and the presence of single-membrane vesiculated bodies protruding into the vacuoles. Virus particles form intracellular crystalline aggregates in the cytoplasm.

Host Range and Symptoms

Aureusviruses have restricted natural host range, while the experimental range is wide. PoLV induces latent infection in pothos, chlorotic or necrotic local symptoms in experimental hosts. Exceptions are *Nicotiana benthamiana* and *N. clevelandii,* where it causes systemic symptoms. CLSV causes systemic mottling and fruit streaking in cucumber.

Transmission and Stability in Sap

Aureusviruses can be transmitted by mechanical inoculation and through seed. While PoLV and CLSV are soilborne, the vector of the former is not known but CLSV is transmitted by the fungus *Olpidium bornovanus.* At approximately 22°C their infectivity in crude sap from infected plants is lost in 22 days and their thermal inactivation point is approximately 80-85°C.

Antigenic Properties and Relationships

Virions are highly immunogenic. Distant serological relationships have been established between PoLV and tombusviruses but no relationship between PoLV and CLSV.

Relevant Literature

Martelli et al., 1998; Rubino and Russo, 1997.

Genus **Panicovirus**

The name has been derived from *Panicum,* the name of the genus of the family Gramineae.

Type species: *Panicum mosaic virus* (PMV).

Virion Properties

The virus particles are about 30 nm in diameter, with icosahedral symmetry hypothesized to be T = 3. They are composed of 180 copies of a 26 kDa CP subunit, and contain an ssRNA genome. See also the general description of family *Tombusviridae.*

The particles sediment at about 109 *S* (St. Augustine decline [SAD] strain) and 102 *S* (type strain).

Genome Organization, Expression, and Replication

The RNA genome is about 4,300 nt in size. The genomic RNA of PMV has five ORFs. By employing several translational strategies such as RT, leaky scanning, noncanonical translational start site, and internal ribosomal entry sites, they yield seven proteins. The relatively larger 5'-terminal part of ORF1 encodes a protein of 48 kDa (p48), and its amber stop codon is suppressed and a 112 kDa RT protein (p112) is generated. Both p48 and p112 are the viral replicase proteins. Though a signature region characteristic of RNA replicases occurs upstream of the GDD motif within the p112 ORF, methyltransferase, and helicase domains, characteristic features of members of the tombusviridae are absent in PMV genome. The genome organization and expression is presented in Figure A1.21.

A single sgRNA of 1,500 nt is present in plants and protoplasts infected with PMV. The 5' terminus caplike structure is absent in the sgRNA. The first seven nucleotides of the sgRNA exactly match the sequences of the 5' end of the genomic RNA.

In vitro translation of the sgRNA transcript yields a putative 8 kDa protein (p8). The start codon of p8 overlaps the p112 replicase ORF by +2 nt. A second ORF, expressed as a –1 FS from the p8 ORF yielding a protein (p8-FS) of 14.6 kDa, has been demonstrated. A p6.6 protein could be expressed in the same reading frame as p8-FS by a leaky scanning mechanism. Internal initiation and frameshift are thought to be the expression strategies for p8, p8-FS, and p6.6 from the sgRNA of PMV. These proteins are involved in cell-to-cell movement.

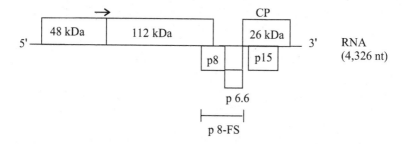

FIGURE A1.21. Organization and expression of the genome of a panicovirus *(Panicum mosaic virus).* CP, capsid protein; →, translational readthrough; FS, frameshift; p, protein.

Two ORFs encoding the 26 kDa CP containing a p15 nested ORF are located in the 3'-proximal region of the sgRNA. CP and p15 may be expressed through internal ribosomal entry sites. Although CP is responsible for PMV spread, p15 does not show any homology with proteins of related viruses. In addition to virus spread, the CP may also influence replication, transcription, and RNA stability of PMV. Inhibition of p8 and p6.6 genes expression in millet protoplast does not affect replication of PMV, but the virus movement is hampered. Expression of p8, p6.6, p15, and CP plays an important role in the systemic infection of PMV in millet.

Downstream of the CP gene, an untranslated sequence of about 325 nt probably contains a small ORF (polypeptide of 34 amino acids). At its 3' end poly (A) tail is absent.

Relations with Cells and Tissues

Viral antigens are detected in infected leaves. Inclusion bodies (crystalline) are present in the cytoplasm, and they cause disruption of mitochondria and tonoplast in infected St. Augustinegrass.

Host Range and Symptoms

Natural host range of PMV is restricted to a few grass species in which the virus causes systemic mosaic. In coinfection with PMSV, it causes a disease named St. Augustine decline (SAD) in St. Augustinegrass *(Stenotaphrum secundatum).* Depending on the host plant, infection either alone (PMV) or coinfected with PMSV induces chlorotic mottle, stunting, and seed reduction in forage grasses. During the summertime, severe symptoms are developed in infected St. Augustinegrass.

Transmission and Stability in Sap

The virus can be transmitted mechanically. The SAD strain is seed transmissible. Infections can spread by virus-contaminated lawn mower blades and from infected St. Augustinegrass plugs or sod.

The virus particles have a thermal inactivation point of approximately 85°C (typestrain) or 60°C (SAD).

Biological vector is not known.

Antigenic Properties and Relationships

Serological relationships have been established between members of the type strain and SAD strain, but they are distinguishable. The type strain reacts with antiserum to *Cocksfoot mild mosaic virus,* a sobemovirus.

Panicum Mosaic Satellite Virus

Panicum mosaic satellite virus (PMSV) consists of a positive-sense ssRNA genome of 824 nt. It depends on the *trans*-acting protein of PMV, its helper virus, for replication and systemic spread. PMSV CP encapsidates PMSV satellite virion. The PMSV shell is composed of 60 protein subunits, and particles are 17 nm in diameter. The symptoms of PMV-infected plants are more severe when its satellite virus is present.

The CP is the sole product of in vitro translation of PMSV RNA, though additional ORFs are present which are not translatable in vitro. Besides symptom-inducing domains, signals for replication and movement are located in the CP.

RNAs of PMV and PMSV do not show any significant sequence homology, except for seven and three identical nucleotides at the 5' and 3' termini, respectively. The entire 3' UTR (263 nt) is required for PMSV RNA replication and the spread of virus. At the 3' end, hairpinlike structures are present that contain essential signals for replication.

Relevant Literature

Qiu and Scholthof, 2000; Turina et al., 1998.

FAMILY BROMOVIRIDAE

The family has been named after the genus *Bromovirus*.

Five genera are in the family: *Bromovirus, Cucumovirus, Alfamovirus, Ilarvirus,* and *Oleavirus*.

Bromovirids have either icosahedral or bacilliform virions.

The tripartite genomes of the members of the family consist of positive-sense, ssRNAs (RNAs 1-3). Their CP is encoded by a sgRNA4.

Genus Bromovirus

The siglum has been derived from the name of the type species, *Brome mosaic virus* (BMV).

Type species: BMV originally isolated from bromegrass *(Bromus inermis)*.

Virion Properties

The virions are isometric, with a diameter of 27 nm. The CP has T = 3 icosahedral symmetry and consists of 180 identical subunits of 20 kDa each, clustered into 12 pentamers and 20 hexamers.

The viruses have a sedimentation coefficient of 88 S (below pH 6.0), a Mr of 4.6×10^6 and a buoyant density of $1.36.g/cm^3$ (in CsCl).

The three genomic RNAs are separately encapsidated. The virion with the smallest genomic RNA (RNA3) also contains sgRNA4.

Genome Organization, Expression, and Replication

Besides the genomic RNAs 1, 2, and 3, BMV has a sgRNA4 transcribed from the 3' terminus of RNA3 for translation of the CP gene (Figure A1.22).

All four RNAs are capped at their 5' terminus and the structure of their highly conserved 3' UTR resembles that of tRNA.

Both RNA1 and RNA2 possess large OFR, each coding for a single protein. These proteins have methyltransferase and helicase motifs (RNA1) and RdRp motifs (RNA2) required for replication of the viral RNA. In ultrathin sections of infected cells they appear as electron-dense inclusions in perinuclear regions in association with endoplasmic reticulum and viroplasms.

The protein translated from the ORF on RNA3 (M protein) is involved in cell-to-cell movement and has been detected in cytoplasmic inclusions and at plasmodesmata. The C-terminal region also plays a role in virus movement.

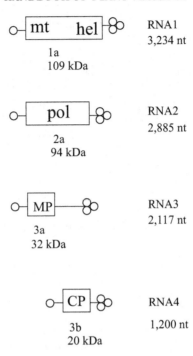

FIGURE A1.22. Organization and expression of the genome of a bromovirus *(Brome mosaic virus)*. Rectangles represent the open reading frames with their translation products 1a, 2a, 3a, and 3b. CP, capsid protein; hel, helicase domain; MP, movement protein; mt, methyltransferase domain; pol, polymerase domain; O, cap; (cloverleaf-like), tRNA-like structure.

Infection Process

After the particles have entered the cell they are uncoated and replication starts on cytoplasmic membranes adjacent to the nucleus. In infected protoplasts, first the presence of RNA1 and RNA2 was detected, followed a couple of hours later by that of RNA3 and RNA4. By that time, the first virions were also observed. An excess of CP is in infected cells. Most virus spread is via the cells around the vascular bundles.

Relations with Cells and Tissues

Early in infection, the endoplasmic reticulum and the nucleus show proliferation. Viral antigen has been demonstrated inside the nucleus, but the virus accumulates in the cytoplasm, forming large inclusions.

Some bromoviruses produce large vacuolate inclusions consisting almost entirely of virus particles. Masses of amorphous inclusions, probably representing viroplasms, are also encountered in infected cells.

Host Range and Symptoms

Bromoviruses have a narrow host range, mainly infecting species in the families of their natural hosts. Symptoms range from yellow mottling and mosaic to necrosis and wilting.

Transmission and Stability in Sap

Bromoviruses are mechanically transmissible. The viruses are rather stable in crude sap from infected plants. They are inactivated in sap heated at temperatures ranging from 70 to 95°C for 10 min.

Natural transmission is by beetles, but BMV is also less efficiently transmitted by aphids in a nonpersistent manner.

Seed transmission has been reported for *Broad bean mottle virus* (BBMV).

Antigenic Properties and Relationships

Bromoviruses are moderately immunogenic. Many species are distantly related serologically.

Relevant Literature

Fusaki et al., 2003; Kao and Sivakumaran, 2000; Okina et al., 2001.

Genus Alfamovirus

The siglum has been derived from the name of the type species, *Alfalfa mosaic virus,* which is the only species in the genus.

Type species: *Alfalfa mosaic virus* (AMV).

Virion Properties

The virions are bacilliform, 18 nm in diameter, and of four different lengths. The three genomic RNAs and a sg mRNA are separately encapsidated.

Depending on their RNA content the virions are 56, 43, 35, or 30 nm long with sedimentation coefficients of 94, 82, 73, and 66 *S* and Mr values ($\times 10^{-6}$) of 6.9, 5.2, 4.3, and 3.8, respectively. The buoyant density (mean for all virions) is 1.278 g/cm^3 (in Cs$_2$SO$_4$).

The CP subunit has a Mr of 24,000.

Genome Organization, Expression, and Replication

The genome organization and expression resembles that of the bromoviruses (Figure A1.23).

Besides the three genomic RNAs (RNAs 1-3), AMV has a sgRNA4 transcribed from the 3'-proximal ORF of RNA3 encoding the CP.

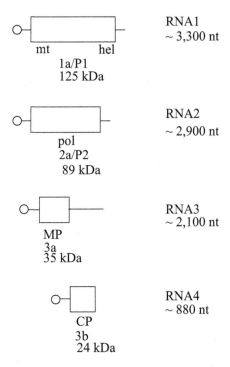

FIGURE A1.23. Organization and expression of the genome of an alfamovirus *(Alfalfa mosaic virus)*. Rectangles represent the open reading frames with their translation products 1a, 2a, 3a, and 3b. CP, capsid protein; hel, helicase domain; MP, movement protein; mt, methyltransferase domain; P1 and P2, replicase proteins; pol, polymerase domain; O, cap.

All RNAs have a cap structure at their 5' terminus and a complex secondary structure at their 3' terminus.

The two largest RNAs, RNA1 and RNA2, encode the proteins, P1 and P2, respectively, that are involved in viral RNA replication. The MP gene on RNA3 codes for a protein which is required for cell-to-cell transport.

Infection Process

The three genomic RNAs are not infective. Infection can start only in the presence of RNA4 or its translation product (CP). The CP binds specifically to a site near the 3' terminus of the three genomic RNAs. Not only the CP of AMV, but also that of ilarviruses can start the replication of AMV.

Virions containing only the RNAs 1 and 2 can replicate in protoplasts in the absence of RNA3.

The virus is synthesized in the cytoplasm, although the presence of CP has been demonstrated both in the cytoplasm and in the nucleus.

Relations with Cells and Tissues

Aggregated and nonaggregated virus particles have been observed in the cytoplasm of infected tobacco leaves. Light microscopy has revealed the presence of vacuolate and dense-staining inclusions in the cytoplasm.

Host Range and Symptoms

AMV has a very wide host range and is often latent in many host plants. Symptoms vary and depend greatly on the virus strain, the host and environmental conditions.

Transmission and Stability in Sap

AMV is mechanically transmissible. The virus is moderately stable in crude sap from infected plants. Its infectivity is lost when the sap is heated at temperatures ranging from 50 to 70°C for 10 min.

Natural transmission is by aphids in a nonpersistent manner.

Transmission through seed and by pollen has also been reported. Many strains can be transmitted by dodder (*Cuscuta* spp.).

Antigenic Properties and Relationships

The virus is moderately immunogenic. Some strains differ sufficiently from one another and can be distinguished serologically.

Genus Cucumovirus

The siglum has been derived from the name of the type species, *Cucumber mosaic virus* (CMV).

Virion Properties

Virions of cucumoviruses resemble those of the bromoviruses. They are about 30 nm in diameter. The virus particles have a sedimentation coefficient of approximately 99 *S,* a buoyant density of about 1.36 g/cm^3 and a Mr of about 6×10^6.

The CP subunit has a Mr of about 24,000.

The three genomic RNAs are separately encapsidated, but RNA3 and sgRNA4 are in the same virion. Some strains have an encapsidated sgRNA4A.

Genome Organization, Expression, and Replication

The genome organization and expression resemble those of bromoviruses (Figure A1.24). All four viral RNAs are capped at their 5' terminus and have a tRNA-like structure at their 3' terminus. Unlike the bromoviruses, cucumoviruses have an extra gene on RNA2 which is expressed from a sgRNA4A (Figure A1.24). Evidence suggests that its protein (2b), which has been found associated with the nucleus, is required for systemic spread of the virus. It also determines the virulence of the virus, possibly by suppressing host posttranscriptional gene silencing. Only genes 1a and 2a are required for replication. The proteins 1a, 2a, 3a, and 3b have functions comparable to those of the bromoviruses. A combination of the proteins 3a (MP) and 3b (CP) has been found to be responsible for spread of CMV and symptom development, but in fact each of the five gene products is involved in virus movement and determines the type of symptoms on an infected plant. Strains of CMV with encapsidated sgRNA4A possess a very small additional RNA5 coterminal with the 3' termini of RNAs 3 and 4, which is also encapsidated.

Besides this viral RNA5, isolates of CMV and *Peanut stunt virus* (PSV) may contain small satellite RNAs with Mr values of about 100,000. The oldest-known satellite RNA of CMV, CARNA5 (*Cucumber mosaic virus-associated RNA5*) has been shown to be responsible for a lethal necrosis of tomato plants.

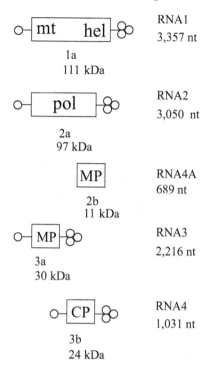

FIGURE A1.24. Organization and expression of the genome of a cucumovirus *(Cucumber mosaic virus)*. Rectangles represent the open reading frames with their translation products 1a, 2a, 2b, 3a, and 3b. CP, capsid protein; hel, helicase domain; MP, movement protein; mt, methyltransferase domain; pol, polymerase domain; O, cap; —8o, tRNA-like structure.

Relations with Cells and Tissues

Light microscopic studies have revealed massive inclusions, mostly in mesophyll cells, but also in the epidermis of infected leaves. These inclusions consist of arrayed virus particles, and they are found most often within the central vacuole, where they appear as large crystalloids.

Virions have also been detected in the nuclei.

Host Range and Symptoms

CMV has a very wide host range, including both monocotyledons and dicotyledons belonging to 85 plant families. The other cucumoviruses, however, are largely restricted to members of two plant families only.

Cucumoviruses cause a variety of symptoms, ranging from chlorosis, mosaic, and yellowing to breaking of flower color, leaf narrowing, necrosis, and stunting.

When satellite RNAs are present, symptoms may be altered drastically.

Transmission and Stability in Sap

Cucumoviruses are readily transmitted mechanically, but they are not very stable in crude sap from infected plants. They are inactivated in sap heated at a temperature of about 70°C for 10 min.

Natural transmission is in a nonpersistent manner by a large number of aphid species, such as those belonging to the genera *Myzus, Macrosiphum,* and *Aphis,* but transmission efficiency varies with the aphid species.

Seed transmission has been reported.

The cucumoviruses can also be transmitted by dodder (*Cuscuta* spp.) and they even multiply in this parasitic plant.

Antigenic Properties and Relationships

The immunogenicity of cucumoviruses is poor, but can be enhanced by treatment with formaldehyde. Two serological subgroups of CMV have been distinguished. A similar division has been made on the basis of nucleotide sequence similarity.

Relevant Literature

Chen et al., 2001; Roossinck, 2001; Takeshita et al., 2001.

Genus Ilarvirus

The siglum has been derived from the original description of this group of viruses, "isometric labile ringspot viruses."

Type species: *Tobacco streak virus* (TSV).

Virion Properties

The shape of the virions is not uniform and varies from spherical to bacilliform.

The sedimentation coefficients of the three different types of virions are approximately 113, 98, and 90 S, their Mr values ($\times 10^{-6}$) 7.45, 5.92, and

4.72, and their sizes approximately 35, 30, and 27 nm in diameter, respectively.

All particle types have a buoyant density of 1.35 g/cm^3 (in CsCl).

The Mr of the capsid subunit is 24,000.

Genome Organization, Expression, and Replication

The three genomic RNAs and the sgRNA are all capped at their 5' terminus and, like AlMV, they have a complex secondary structure at their 3' terminus.

RNA1 codes for one protein (1a) and, like cucumoviruses, RNA2 encodes two proteins, 2a and 2b (Figure A1.25). Proteins 1a and 2a are

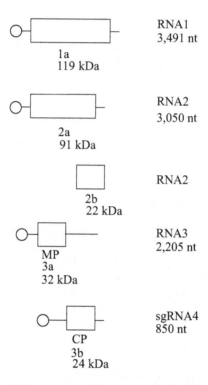

FIGURE A1.25. Organization and expression of the genome of an ilarvirus *(Tobacco streak virus)*. Rectangles represent the open reading frames with their translation products 1a, 2a, 2b, 3a, and 3b. CP, capsid protein; MP, movement protein; sgRNA, subgenomic RNA; O, cap.

involved in replication. The function of 2b may be compared to that in cucumoviruses. The translation products of RNA3 are the movement protein (MP) 3a for transport of the virus from cell to cell and protein 3b, the CP, which is expressed from sgRNA4. An additional small RNA5 which has some homology with RNAs 1, 2, 3, and 4 has recently been found in purified preparations of *Prunus necrotic ringspot virus* (PNRSV) and *Apple mosaic virus* (ApMV). This RNA segment proved to be a copy of the 3' untranslated region of the genomic RNA3.

Infection Process

As with AlMV, CP or its sgRNA4 is required for replication. However, replication also starts when ilarvirus CP is substituted with that of AlMV.

Relations with Cells and Tissues

Aggregates of TSV particles have been found in the cytoplasm and nucleus. In epidermal cells of TSV-infected *Datura stramonium* plants cytoplasmic inclusions of varying size and shape have been detected.

Host Range and Symptoms

TSV has a wide host range, including both monocotyledons and dicotyledons. Many woody plants have been found infected.

Symptoms range from local necrotic ringspots to mottling, mosaic, shoe stringing, and stunting.

Transmission and Stability in Sap

The viruses are readily transmitted mechanically. However, as already indicated by the name, ilarviruses are very unstable. In crude sap from infected plants most of their infectivity is already lost within 10 min at 20°C.

No vector transmission has been established with certainty. Ilarviruses can be transmitted through seed and by pollen.

Antigenic Properties and Relationships

TSV is moderately immunogenic. At least seven subgroups have been distinguished on the basis of serological relationships. Members of each subgroup are serologically interrelated and some cross-react with members of other subgroups.

Relevant Literature

Di Terlizzi et al., 2001.

Genus Oleavirus

The type species is the only species in the genus. The virus has been named after its natural host, *Olea europaea* (olive).
Type species: *Olive latent virus 2* (OLV-2).

Virion Properties

The virions are quasi-isometric or bacilliform. The nonbacilliform particles are 26 nm in diameter, whereas the bacilliform particles have a diameter of 18 nm and four different lengths, namely 37, 43, 48, and 55 nm.

The three genomic RNAs (RNAs 1-3) and an additional sgRNA are separately encapsidated. The sgRNA for the CP is usually not encapsidated.

The virions may also contain additional small RNAs of 200-500 nt.

Genome Organization, Expression, and Replication

The genome organization is similar to that of other bromovirids (Figure A1.26).

The three genomic RNAs 1-3 and the sgRNA4 are capped at their 5' terminus and all RNAs have 3' termini comparable to those of the alfamo- and ilarviruses. An encapsidated sgRNA of 2,073 nt (sgRNA4A) resembles RNA3, but no translation products have been found. CP is not required to start replication.

Host Range and Symptoms

No symptoms are observed in olive.

Transmission

The virus is mechanically transmissible. No vectors are known.

Antigenic Properties and Relationships

The virus is a good immunogen. No serological relationships with other viruses have been found.

Relevant Literature

Grieco et al., 1996; Martelli and Grieco et al., 1997.

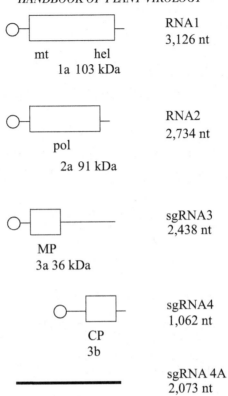

FIGURE A1.26. Organization and expression of the genome of an oleavirus *(Olive latent virus 2)*. Open rectangles represent the open reading frames with their translation products 1a, 2a, 2b, 3a, and 3b. CP, capsid protein; hel, helicase domain; MP, movement protein; mt, methyltransferase domain; pol, polymerase domain; sgRNA, subgenomic RNA; O, cap.

FAMILY CLOSTEROVIRIDAE

The name has been derived from that of the genus *Closterovirus*.

The three genera in the family, are *Closterovirus*, *Crinivirus*, and *Ampelovirus*.

The very long flexuous particles of all members of the family have a diameter of about 12 nm, and they contain a positive-sense ssRNA molecule.

At one end of the virion the capsid has a segmented appearance reminiscent of that of the tail of a rattlesnake. That part of the capsid has been shown to consist of a deviant protein possessing an N-terminal addition to the CP (duplicate capsid protein, CPd).

Virions have sedimentation coefficients ranging from 96 to 140 S and buoyant densities of 1.30-1.34 g/cm^3 (in CsCl) and 1.24-1.27 g/cm^3 (in Cs_2SO_4).

The genome of closterovirids is characterized by genes coding for a homologue of the cellular heat-shock protein (HSP) 70 and for the previously mentioned analogue of the CP.

Genus Closterovirus

The name has been derived from the Greek word *klooster* (spun-out filament) and refers to the morphology of the virions.

Type species: *Beet yellows virus* (BYV).

Virion Properties

The lengths of the virus particles depend on the species and they range from 1,250 to 2,200 nm. *Citrus tristeza virus* (CTV) also possesses shorter-than-full-length particles containing sg or defective RNAs.

Genome Organization, Expression, and Replication

Of all plant viruses with positive-sense ssRNA genomes the monopartite genome of closteroviruses is the longest. The size of the RNA ranges from 15,500 nt for BYV to about 19,300 nt for CTV and is related to particle length.

The 5' termini of the RNAs are thought to be capped. The number of ORFs ranges from 9 for BYV to 10 for *Beet yellow stunt virus* (BYSV) and 12 for CTV. The 5'-untranslated region of CTV varies with the isolate.

The genome organization and expression of BYV is presented in Figure A1.27. Functional proteins are produced by proteolytic processing of the polyprotein coded for by ORF1a, a +1 FS for the translation of the RdRp domain encoded by ORF1b, and by translation of ORFs at the 3' terminus via sgRNAs. The protein translated from ORF1a has proteinase, methyltransferase, and helicase motifs. ORF1b, possibly translated by a +1 ribosomal frameshift from ORF1a, yields a 348 kDa protein. The two types of CP are translated from ORFs 5 and 6. The downstream protein of 22 kDa is the major CP; the deviant minor protein encapsidating the 5' terminus of RNA (CPm, also referred to as CPd) may play a role in transmission by aphids. Mealybug-transmitted closteroviruses have CPs with a high

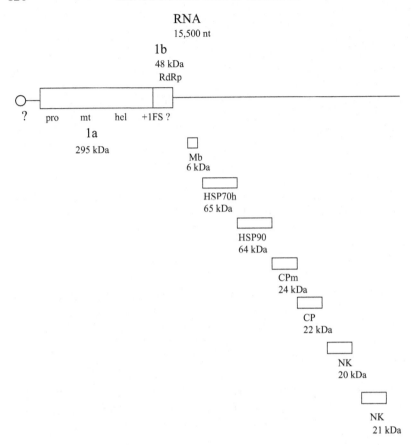

FIGURE A1.27. Organization and expression of the genome of a closterovirus *(Beet yellows virus)*. Rectangles represent the open reading frames (ORFs) with their final functional proteins produced either after processing of the polyprotein translated from ORFs 1a and 1b with a putative ribosomal frameshift (+1FS?) or from a set of seven sgRNAs. CP, capsid protein; CPm, minor capsid protein; hel, helicase domain; HSP70h, a homologue of heat-shock protein 70; HSP90, heat-shock protein-90-like; Mb, membrane-binding protein; mt, methyltransferase; pro, proteinase; NK, function of the protein not known; RdRp, RNA-dependent RNA polymerase; ? O, putative cap.

molecular mass, ranging from 35 to 46 kDa. The ORFs 3-7 produce five proteins involved in movement of the virus from cell to cell, namely a 6 kDa protein, HSP70h, a 64 kDa protein, CPm, and CP. The 6 kDa protein (Mb) is probably a membrane-binding protein. HSP70h has been found associated with virus particles within or near plasmodesmata. The 64 kDa protein is an HSP90, a protein resembling HSP 90.

Replication occurs in the cytoplasm.

Relations with Cells and Tissues

Virus particles are mostly found in the cytoplasm of phloem cells, where they form large aggregates. Infected cells show degeneration of phloem parenchyma and plastids. Light-microscopic studies have revealed the presence of cross-banded or spindle-shaped fibrous inclusions. Ultrathin sections showed that such inclusions consisted mainly of virus particles.

Host Range and Symptoms

Closteroviruses have narrow host ranges.

Symptoms range from color deviations (chlorosis, yellowing, red interveinal discoloration), epinasty and stunting caused by sugar beet-infecting closteroviruses, to yellow flecks on the upper surface of citrus leaves, water-soaked spots on the lower leaf surface, and stempitting in the wood followed by wilting of the trees upon infection with CTV.

Transmission and Stability in Sap

Closteroviruses are either not mechanically transmissible or they are transmitted with great difficulty.

CTV can be transmitted experimentally by slash inoculation of the bark of citron trees. This virus retains its infectivity in sap from infected plants for one day at about 20°C, but it is inactivated in sap heated at about 50°C for 10 min.

Closterovirus species are divided into three groups based on the type of their natural vectors, namely aphids, whiteflies, or mealybugs. Transmission is always in a semipersistent manner.

Antigenic Properties and Relationships

Closteroviruses are moderately immunogenic. There is very little serological relationship between the species.

Relevant Literature

Ayllon et al., 2001; Peremyslov et al., 2004.

Genus Crinivirus

The name has been derived from the Latin word *crinis* (long hair, braid) and refers to the morphology of the virions.

Type species: *Lettuce infectious yellows virus* (LIYV).

Virion Properties

The virions are shorter than those of the closteroviruses. They have two modal lengths, namely 650-850 nm and 700-900 nm.

Genome Organization, Expression, and Replication

Unlike the closteroviruses, criniviruses have a bipartite genome (Figure A1.28). The two RNA molecules, RNA1 and RNA2, are separately encapsidated.

RNA1 is a bicistronic molecule with ORFs 1a,1b, and 2. ORFs 1a and 1b code for the proteins involved in replication, including an RdRp that is translated via a putative +1 FS. ORF2 encodes a 31 kDa protein, the function of which is not known.

RNA2 has seven ORFs and its organization resembles that of the 3' half of the closteroviruses. ORF6 encodes the anomalous CP protein, CPd (duplicate CP = CPm of the closteroviruses) of 52 kDa, but the order of the CP and CPd ORFs is reversed, as compared to that of the closteroviruses. A 62 kDa product encoded by ORF2 is an HSP70h protein and it has been found associated with virus particles.

Host Range and Symptoms

Most criniviruses have a narrow host range.

Commonly encountered symptoms are chlorosis, yellowing, and stunting.

Transmission

Criniviruses are not mechanically transmissible.

Their natural vectors are whiteflies and transmission is in a semipersistent manner.

Antigenic Properties and Relationships

The viruses are moderately immunogenic. Most criniviruses are only distantly related serologically or they are not related at all.

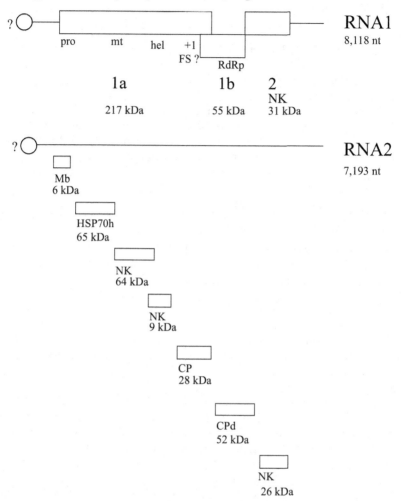

FIGURE A1.28. Organization and expression of the genome of a crinivirus *(Lettuce infectious yellows virus)*. Rectangles represent the open reading frames (ORFs) with their final functional proteins produced either after processing of the polyproteins translated from ORFs 1a, 1b (with a putative ribosomal frameshift [+1FS ?]), and 2 in RNA1 or from a set of seven subgenomic RNAs produced by RNA2. CP, capsid protein; CPd, duplicate CP (= CPm in closteroviruses); hel, helicase domain; HSP70h, a homologue of heat-shock protein 70; HSP90, a protein resembling heat-shock protein 90; Mb, membrane-binding protein; mt, methyltransferase domain; pro, proteinase; NK, function of the protein not known; RdRp, RNA-dependent RNA polymerase; ? O, putative cap.

Relevant Literature

Aguilar et al., 2003; Hartono et al., 2003.

Genus Ampelovirus

The name has been derived from the Greek word *ampelos* (vine).
Type species: *Grapevine leaf roll-associated virus3* (GLRaV-3).
Eight serologically distinct closterovids (designated as GLRaV1-8) are associated with grapevine leafroll diseases of grapevines.

Virion Properties

The virus particles are 950-2,000 nm long and 10-13 nm in diameter.

Genome Organization, Expression, and Replication

The size of RNA ranges from 15,000 to 20,000 nt. The 5' termini of the RNAs are thought to be capped, while a poly (A) tail is absent at the 3' terminus. The genome organization and expression is presented in Figure A1.29. The RNA genome contains up to 12 ORFs. Functional proteins are produced by proteolytic processing of the polyprotein coded by ORF1a, a +1 FS for the translation of RdRp domain encoded by ORF1b, and by translation of ORFs near the 3' terminus via sgRNA. The protein translated from ORF1a has helicase motifs. ORF1b yields a protein of 533 amino acids and shows significant homology with RdRp of closteroviruses. ORF1b overlaps ORF1. ORF2 encodes a small peptide of 6 kDa of unknown function, followed by a long intergenic region of 1,067 nt. A small protein (5 kDa) suggested to be a transmembrane hydrophobic protein is translated from ORF3. ORF4 encodes a 59 kDa protein with similarities to the HSP70 heat-shock proteins. The latter are molecular chaperones and their N-termini may possess adenosine triphosphatase (ATPase) activities, while the C-termini are involved in protein-protein interaction. A protein of 54.8 kDa with unknown function is encoded by ORF5 and ORF6 codes for the CP gene of about 35 kDa. The N-terminal domain of 53 kDa protein encoded by ORF7 contains amino acid residues N, R, G, and D conserved in the CP and identified as the duplicate CP (CPd) products of closteroviruses. Based on the sequences from the CP and their CPds, GLRaV3 was placed in independent lineage, separating from aphid- and whitefly-transmitted closteroviruses. Downstream of CPd, ORFs 8-12 potentially encode proteins of 21 (ORF8), 19.6 (ORF9), 19.7 (ORF10), 4 (ORF11), and 6.7 (ORF12) kDa. Though the genes, with respect to size and position, are organized in similar

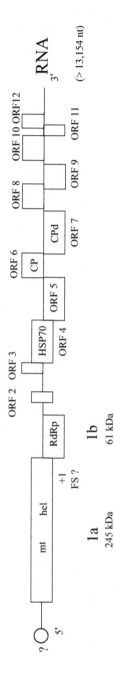

FIGURE A1.29. Organization and expression of the genome of an ampelovirus (*Grapevine leafroll-associated virus3*). Rectangles represent the open reading frames (ORFs) with their final functional proteins produced either after processing of the polyproteins translated from ORFs 1a, 1b (with a putative ribosomal frameshift [+1FS ?]), and 2 in RNA1 or from sgRNAs produced by RNA2. CP, capsid protein; CPd, duplicate CP; hel, helicase domain; HSP70h, a homologue of heat-shock protein 70; mt, methyltransferase domain; RdRp, RNA-dependent RNA polymerase; ? O, putative cap.

fashion as in CTV (a closterovirus), they do not show any significant homology with other known proteins and their functions are not known. At the 3' end of the genome, an untranslated region (277 nt in GLRaV3) is present. Presence of extensive secondary structure suggests that this region is involved in RNA replication. Replication occurs in cytoplasm.

Relations with Cells and Tissues

Characteristic vesicular bodies and fibrillar materials are seen in grapevine plants infected with GLRaV3. The fibrillar content of these vesicles is RNA, the probable replication sites (vesicles) of virus.

Host Range and Symptoms

Ampeloviruses have narrow host ranges.

Symptoms consist of downward rolling, interveinal reddening, or yellowing of leaves in red- and white-fruit grape varieties. An ampelovirus infection causes delay in ripening and reduced level of sugar accumulation.

Transmission and Stability in Sap

GLRaV can be mechanically transmitted to herbaceous plant species via grafting and propagation practices. They are transmitted by five species of mealybugs: *Planococcus ficus, Pseudococcus longispinus, P. viburni, P. calicolarvae,* and by a scale insect *(Pulvinaria vitis).*

Antigenic Properties and Relationships

Ampeloviruses are moderately immunogenic. Very little serological relationship exists between the species.

Relevant Literature

Ling et al., 1998; Fazeli and Rezaian, 2000.

FAMILY FLEXIVIRIDAE

The name of the family refers to the virion morphology of its members. It comprises eight genera: *Carlavirus, Potexvirus, Capillovirus, Trichovirus, Foveavirus, Allexivirus, Vitivirus,* and *Mandarivirus,* previously placed in unassigned genera.

The members are characterized by flexuous filamentous virus particles ranging in size from 470 to 1,000 nm, 12-13 nm in diameter. In some genera nucleocapsids show cross-banded, ropelike features. They possess a mono-partite, positive-sense ssRNA genome, varying in size from 6,400-9,300 nt. The 3' end has a poly (A) tail and the 5' terminus has a methylated cap in some genera. Up to six ORFs are expressed from sgRNAs. At the 5' end an alphalike replicase protein, consisting of conserved methyltransferase, heli-case, and RdRp motifs, is present. The distinction among the genera is bas-ed on the type of movement protein (TGB or a single protein of the 30K superfamily), the number of ORFs, size of the virions, and that of the CP (22 to 44 kDa).

Genus Carlavirus

The siglum has been derived from the name of the type species, *Carna-tion latent virus* (CLV).

Virion Properties

The slightly flexuous filamentous virions with helical symmetry have lengths ranging from 610 to 700 nm, a diameter of 12-15 nm, and a helical pitch of about 3.4 nm.

The virions have sedimentation coefficients ranging from 147 to 176 *S* and a buoyant density of 1.31-1.33 g/cm^3 (in CsCl).

The CP has a Mr ranging from 31,000 to 36,000.

Genome Organization, Expression, and Replication

The genome consists of a single, positive-sense ssRNA molecule which has a putative cap at its 5' terminus and a poly(A) tract at its 3' terminus. The genome organization and expression of *Potato virus M* (PVM) are pre-sented in Figure A1.30. Carlaviruses possess six ORFs (ORFs 1-6).

ORF1 encodes a protein of 223 kDa with methyltransferase, helicase, and polymerase domains. The other ORFs are expressed from two sgRNAs with a length of 3,000 and 1,500 nt, respectively. Both sgRNAs have poly (A) tracts. The ORFs 2, 3, and 4 (triple-gene block; TGB) are most likely trans-lated via the long sgRNA into proteins of 25, 12, and 7 kDa, respectively, known to be involved in the cell-to-cell movement of viral RNA. ORF5 codes for the CP of 34 kDa. It overlaps ORF6, which is translated into a cysteine-rich 11-16 kDa protein. The latter protein has nucleic-acid-bind-ing ability, but its function is not known. ORF5 and ORF6 may be expressed

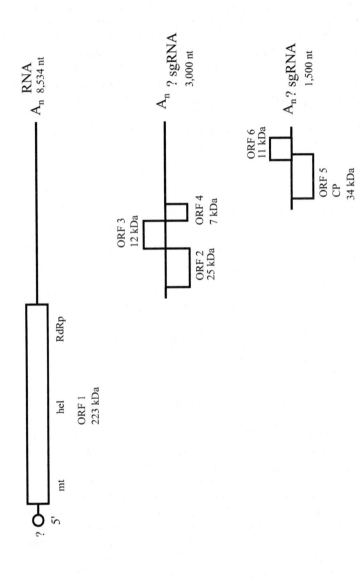

FIGURE A1.30. Organization and expression of the genome of a carlavirus (*Potato virus M*). Rectangles represent the open reading frames (ORFs 1-6) with their translation products. CP, capsid protein; hel, helicase domain; mt, methyltransferase domain; pol, polymerase domain; ? sgRNA, putative subgenomic RNA; ? O, putative cap; RdRp, RNA-dependent RNA polymerase; A$_n$, poly (A) tract.

from the short sgRNA. In some species, sgRNAs have been found encapsidated into small virions.

Relations with Cells and Tissues

Many carlaviruses induce various characteristic inclusions, which can be observed in the light microscope. The paracrystalline, ropey, and fusiform inclusions occurring in the cytoplasm consist of masses of regularly arranged virus particles.

The large vacuolate, irregularly shaped structures and crystalline inclusions, however, are made up mainly of cell material (endoplasmic reticulum, mitochondria, ribosomes) and some aggregated or unaggregated virus particles.

Host Range and Symptoms

Most carlaviruses seem to have restricted natural host ranges. However, many of them are either latent in their host plants or induce such inconspicuous symptoms that their presence may often go undetected. Many carlavirus species have wide experimental host ranges. Some viruses may enhance the symptoms caused by another virus simultaneously present in the infected plant. *Elderberry symptomless virus* (ESLV; syn. Elderberry carlavirus, Elderberry virus A), for instance, has been found to enhance the line-pattern symptoms induced by *Cherry leaf roll virus,* a nepovirus, in elderberry in mixed infections with the latter virus.

Transmission and Stability in Sap

All carlaviruses are readily transmitted mechanically. Many virus species are not very stable in crude sap from infected plants and may become inactivated in less than one hour at 20°C. Their inactivation temperatures range from 55 to 85°C, depending on the virus species.

Natural transmission of most carlaviruses is by aphids in a nonpersistent manner, but also mechanical by contact. Only *Cowpea mild mottle virus* (CPMMV) has been found to be transmitted by whiteflies *(Bemisia tabaci).*

The virus species that naturally infect members of the Leguminosae are also transmitted through seed.

Antigenic Properties and Relationships

Most carlaviruses are good immunogens. Not all virus species are serologically interrelated and some are only distantly related.

Relevant Literature

Choi and Ryu, 2003; Zheng et al., 2003.

Genus **Potexvirus**

The siglum has been derived from the name of the type species, *Potato virus X* (PVX).

Virion Properties

The flexuous filamentous virions with helical symmetry have lengths ranging from 470 to 580 nm, a diameter of 13 nm, and a helical pitch of 3.3-3.7 nm.

The virions have sedimentation coefficients ranging from 115 to 130 *S* and a buoyant density of 1.31 g/cm^3 (in CsCl).

The CP has a Mr ranging from 18,000 to 27,000 (depending on the virus species).

Genome Organization, Expression, and Replication

The genome consists of a single, positive-sense ssRNA molecule. Both the genomic RNA and the sgRNAs are capped at their 5' terminus and have a poly(A) tract at their 3' terminus.

The genome organization and expression of PVX are presented in Figure A1.31. The genome of most potexviruses contains five ORFs (ORFs 1-5). ORF1 encodes a protein of 166 kDa with methyltransferase, helicase, and polymerase domains. The overlapping ORFs 2, 3, and 4 (TGB) are translated via a sgRNA into proteins of 25, 12, and 8 kDa, respectively, known to be involved in cell-to-cell movement of viral RNA. ORF2 also contains a helicase domain. ORF5 codes for the CP of 25 kDa and is expressed from a sgRNA that has been found encapsidated in some virus species.

Relations with Cells and Tissues

Potexviruses induce large cytoplasmic inclusions in many tissues of the host plant. The inclusions are of different types and appear in the light microscope as banded or dense bodies, paracrystals, fibrous structures, spindle-shaped bodies, or laminate inclusion components (LIC). With the exception of LIC, all inclusions are made up almost entirely of aggregated virus particles. The LIC inclusions in PVX-infected plants have been found

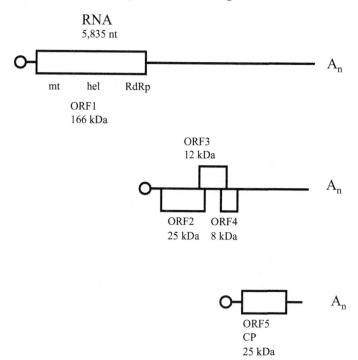

FIGURE A1.31. Organization and expression of the genome of a potexvirus *(Potato virus X)*. Rectangles represent the open reading frames (ORFs 1-5) with their translation products. CP, capsid protein; hel, helicase domain; mt, methyltransferase domain; RdRp, RNA-dependent RNA polymerase; O, cap; A_n, poly (A) tract.

to consist of layers of protein in close association with virus particles and ribosome-like structures. In some PVX infections, nuclear inclusions have been observed.

Host Range and Symptoms

The host range varies with the virus species and is often rather limited. PVX can infect many species in 16 families, although most of the susceptible plants are found in the Solanaceae. Many plant species are infected symptomlessly.

Symptoms range from chlorosis, mottling, mosaic, and ringspots to necrosis and slight stunting.

Transmission and Stability in Sap

Potexviruses are readily transmitted mechanically. They are rather stable. In crude sap from infected plants kept at 20°C PVX retains its infectivity for several weeks. It is inactivated in sap heated at 68-74°C for 10 min.

In nature, potexviruses are transmitted in a mechanical way by contact.

A few cases of transmission of potexviruses by vectors have been reported, but most species have no known vectors.

Potato aucuba mosaic virus (PAMV) has been found transmitted by *Myzus persicae*, but only if the aphids were allowed to feed first on plants infected with *Potato virus Y* (PVY). Other insects, such as grasshoppers and plant bugs, have also been implicated as vectors, but transmission might have resulted from superficial mechanical contact.

Transmission of PVX by zoospores of the phycomycete *Synchytrium endobioticum* released from sporangia produced in PVX-infected potato plants to healthy potato plants has been reported. However, this observation, made about 40 years ago, has never been confirmed.

Antigenic Properties and Relationships

Potexviruses are strongly immunogenic. Not all virus species are serologically interrelated.

Relevant Literature

Chen et al., 2005; Hsu et al., 2004; Lin et al., 2004.

Genus Capillovirus

The name has derived from the Latin word *capillus* (hair) and refers to the morphology of virions.

Type species: *Apple stem grooving virus* (ASGV).

Virion Properties

The virions are flexuous filaments that are threadlike with a distinct cross-banding (ropelike appearance), 600-700 nm in length, and 12 nm wide. They show helical symmetry with a pitch of 3.4 nm. The ssRNA is encapsidated in a single protein species of 27 kDa.

Genome Organization, Expression, and Replication

The capillovirus genome consists of RNA of 6,400-7,400 nt. It is poly-adenylated at the 3' end; it may be capped at the 5' terminus. The genome organization and expression is presented in Figure A1.32. The RNA genome contains two overlapping ORFs. ORF1 encodes a polyprotein of 241-266 kDa in size and possesses all elements of a replication-associated protein with the characteristic motif for (putative) methyltransferase, papainlike protease, NTP-binding helicase, and RdRp. The CP cistron is located in the C-terminal domain and translated from the polyprotein. ORF2 lies in a different frame nested within ORF1. It encodes a polypeptide of 36-52 kDa, possessing protease activity as it contains the sequence Gly-Asp-Ser-Gly (GDSG) present in serine proteases. In its N-terminal part, the amino acid signature of the 30 K "superfamily" of cell-to-cell transport proteins is located. Replication takes place in cytoplasm with accumulation of virus particles in discrete bundles.

Relations with Cells and Tissues

Virus particles are found in bundles in mesophyll and phloem parenchyma cells.

Host Range and Symptoms

Capilloviruses have narrow host ranges of mostly woody species. ASGV induces stem grooves and abnormalities in graft union. Leaves are malformed with interveinal chlorosis and experimentally infected plants mostly show chlorotic, or necrotic local lesions and necrosis. ASGV in *Chenopodium quinoa* induces epinasty, distortion, and mottling in upper leaves.

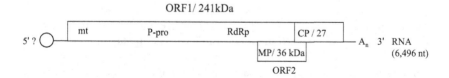

FIGURE A1.32. Organization and expression of the genome of a capillovirus *(Apple stem grooving virus)*. Rectangles represent the open reading frames with their translation products. CP, capsid protein; MP, movement protein; mt, methyltransferase domain; P-pro, papainlike protease; RdRp, RNA-dependent RNA polymerase; A_n, poly (A) tract; O, putative cap.

Symptoms caused by Citrus tatter leaf virus vary seasonally, inducing tatter leaf, blotchy and malformed leaves, and disease that sometimes may be destructive.

Transmission and Stability in Sap

Capilloviruses can be transmitted mechanically and by grafting. Natural spread is mainly through vegetative propagative material. No vectors are known.

The infectivity in crude sap from infected plants is lost in two days. The virus particles have a thermal inactivation point of approximately 60-63°C, dilution end point is 10^{-4}.

Antigenic Properties

Capilloviruses are moderately antigenic.

Relevant Literature

Magone et al., 1997; Yoshikawa et al., 1992.

Genus Trichovirus

The name has been derived from the Greek word *thrix* (hair) and refers to the morphology of the virions.

Type species: *Apple chlorotic leaf spot virus* (ACLSV).

The genus consists of three definitive species: ACLSV, *Potato virus T* (PVT), *Grapevine berry inner necrosis virus* (GINV).

Virion Properties

The virions are very flexuous, show cross-banding (ropelike appearance) depending upon the negative-staining method, are 600-730 nm in length, 12 nm in width with helical arrangement, and have a pitch of about 3.4-3.8 nm.

The virion Mr is 2.2-2.3×10^6. The particles sediment as single or as two very close bands at 92-99 S and have a buoyant density of 1.24 g/cm^3 (CsSO$_4$).

Genome Organization, Expression, and Replication

The trichovirus genome consists of monopartite, linear, positive-sense ssRNA about 7,600 nt long. The 5' terminus has methylated cap and a poly

(A) tail is present at the 3' terminus. The genome organization and expression is presented in Figure A1.33. The genome contains three slightly overlapping ORFs. ORF1 is located at the 5' end and it encodes a polyprotein of about 216 kDa. Its N-terminal region contains a methyltransferase domain, nucleotide-binding site (helicase) in the central region, and the polymerase domain (characterized by GDD triplet) in the C-terminal region. ORF1 is directly expressed from the genomic RNA. ORF2 encodes a polypeptide of about 51 kDa and is translated from sgRNA of 2,200 nt. Due to some homology with movement proteins of other viruses, it is thought to be involved in cell-to-cell movement of virus. ORF3, the CP cistron, is located at the 3' terminus and it encodes a CP of 22-28 kDa; it is translated from the sgRNA (1,100 nt).

Relations with Cells and Tissues

GINV is found as aggregates in the cytoplasm of vascular parenchyma and mesophyll cells. Virus-specific inclusion bodies are not seen in infected cells.

Host Range and Symptoms

ACLSV infects hosts belonging to the Rosaceae and has a narrow host range outside this family. It causes serious diseases in stone fruits: among others, plum (pseudosharka), apricot, (viruela, butteratura), and cherry. GINV causes severe damage to grapevines and spreads naturally in fields. The disease is characterized by inner necrosis in shoots, shortened internodes, and mosaic pattern on leaves. In natural hosts, infections by PVT and ACLSV (in certain plant species), induce little or no symptoms, but sometimes produces mottling, spots, rings, and line pattern. In experimental hosts

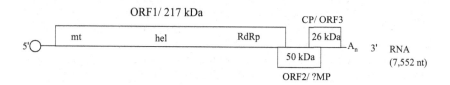

FIGURE A1.33. Organization and expression of the genome of a trichovirus *(Apple chlorotic leaf spot virus)*. Rectangles represent the open reading frames with their translation products. CP, capsid protein; hel, helicase domain; ?MP, putative movement protein; mt, methyltransferase domain; RdRp, RNA-dependent RNA polymerase; A_n, poly (A) tract; O, cap.

(Chenopodium amaranticolor, C. quinoa), chlorotic spots followed by necrotic spots, systemic epinasty, and stunting are induced.

Transmission and Stability in Sap

Trichoviruses are transmitted mechanically, by grafting, and through seeds in PTV. Natural dissemination is mediated by propagative material in clonally propagated hosts. Vectors for ACLSV and PVT are not known.

At approximately 20°C infectivity in crude sap from an infected plant is lost within a day. The virus particles have a thermal inactivation point of approximately 50-60°C (ACLSV, HLV) and 65°C (PVT).

Antigenic Properties

Trichoviruses are moderately immunogenic. Serological relationships between species have not been established.

Relevant Literature

German-Retana et al., 1977; James et al., 2000.

Genus Foveavirus

The name has been derived from the Latin word *fovea* (a pit) referring to the symptoms (wood pitting) caused by *Apple stem pitting virus* (ASPV), the type species.

Virions

The virions have one type of very flexuous filaments, about 800 nm long and 12-13 nm wide. Particles tend to form long threads through end-to-end aggregation. This is often encountered in *Citrus green ring mottle virus* (CGRMV) having 1,000-2,000 nm-long particles which, on the basis of genome sizes, should have a length of 750-800 nm.

ASPV particles sediment as two or three bands in a sucrose density gradient but yield a single band at equilibrium in Omnipaque 350 density gradients with buoyant densities 1.24-1.25 g/cm^3 (CsSO4).

Genome Organization, Expression, and Replication

The foveavirus genome consists of one species of linear, positive-sense ssRNA of 8,300-9,300 nt and is encapsidated by a single species of 25 kDa

CP. Its 5' terminus is capped and 3' end is polyadenylated. The genome organization and expression is presented in Figure A1.34. The RNA genome contains five ORFs; however, an isolate of *Grapevine rupestris stem pitting-associated virus* (GRSPaV) encodes a sixth ORF at the 3' end with a 14 kDa product of an unknown function. ORF1 encodes a polyprotein of 230-250 kDa. It shows similarity to the RdRp, as characterized by a GDD motif and other conserved sequence motifs located in the RdRp, such as methyltransferase domain, a papainlike protease domain, and an NTP-binding motif conserved in helicase proteins: it could undergo a proteolytic processing. ORFs 2-4 encode polypeptides of about 25 kDa, 12 kDa, and 7 kDa, respectively. They constitute a triple-gene block and are thought to be involved in cell-to-cell movement of plant viruses. ORF5 encodes the CP of 28-44 kDa with variable N-terminal and conserved C-terminal regions. CP cistron is followed by an untranslated region of about 135 nt followed by a poly (A) tail at the 3' end.

In CGRMV, two additional ORFs (ORF2a and ORF5a) are located within ORF2 and ORF5, encoding polypeptides of 14 kDa and 18 kDa, respectively. They do not show similarity with sequences of other proteins available in GenBank. It is not known if they are expressed in vivo. The genome organization of GRSPaV is similar to that of ASPV. However, in the former the size of the CP is 28 kDa, while it is 44 kDa in the latter.

Replication is likely to occur in the cytoplasm with a strategy comparable to that of potexviruses, based on direct expression of the 5' proximal ORF, and downstream ORF expression through subgenomic RNAs.

Relations with Cells and Tissues

ASPV seriously affects the cells in wood and bark. Phloem tissue penetrates into developing xylem, leading to the formation of pits in the wood

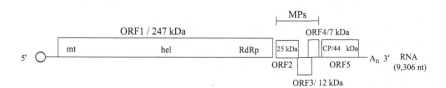

FIGURE A1.34. Organization and expression of the genome of a foveavirus *(Apple stem pitting virus)*. Rectangles represent the open reading frames with their translation products. CP, capsid protein; hel, helicase; MP, movement protein; mt, methyltransferase domain; RdRp, RNA-dependent RNA polymerase; A_n, poly (A) tract; O, cap.

filled with abnormal phloem cells. No specific cytoplasmic structures or inclusion bodies have been observed. Virion accumulates in bundles in the cytoplasm; massive accumulation of CGRMV is also seen in mesophyll cells of infected Kwazan flowering cherry.

Host Range and Symptoms

Foveaviruses have a very narrow natural host range. Symptoms are characterized by wood pitting in natural or indicator hosts, vein clearing, necrotic spots, and apical necrosis. Clear differences in the severity of symptoms induced in *Nicotiana occidentalis* are evident upon mechanical inoculation.

GRSPaV causes rupestris stem pitting disease that appears after graft-inoculation with a chip bud from an infected plant. On the woody cylinder of the indicator *Vitis rupestris,* a narrow strip of small pits extending basipetally from the inoculum bud is induced.

ASPV infects commercial pome fruits (apple and pear), which are symptomlessly infected unless grafted on sensitive rootstocks. Symptoms consist of pitting of the woody cylinder, epinasty, and decline.

CGRMV naturally infects stone fruits (sweet, sour, and flowering cherries; peach; apricot). In Shirofugen flowering cherry it causes epinasty and necrosis of leaves. Necrotic bark tissues, pitting of the woody cylinder and dieback of limes are also seen. However, sweet cherry, peach, and Nanking cherry trees are symptomless.

Transmission and Stability in Sap

Foveaviruses are readily transmitted by grafting and are mechanically transmissible to some hosts (ASPV, ApLV). Natural transmission by vectors is not known.

Their infectivity in crude sap is lost within a day; the virus particles have thermal inactivation point at approximately 55-60°C.

Antigenic Properties and Relationships

No serological relationships have been reported among GRSPaV, CGRMV, and ASPV. However, antisera to ApLV and ASPV cross-react to each other in enzyme-linked immunosorbent assay (ELISA). The chimeric fusion CPs expressed in *Escherichia coli* have been used to raise specific antisera to ASPV and CGRMV.

Relevant Literature

Gentit et al., 2001; Martelli and Jelkmann, 1998.

Genus Allexivirus

The name has been derived from *Allium,* the generic name of shallot and garlic, the main hosts of this genus.
Type species: *Shallot virus-X* (ShV-X).

Virion Properties

The virions are flexuous, filamentous, 800 nm long and 12 nm in diameter, and helically arranged with a pitch of 2.5 nm. They possess one molecule of positive-sense ssRNA and are encapsidated in a single species of 28-36 kDa CP.
The particles sediment at about 170 *S* and have a buoyant density of 1.33 g/cm^3 (CsCl).

Genome Organization, Expression, and Replication

The allexivirus genome consists of RNA of about 900 nt with poly (A) tail at the 3' end. The genome organization and expression is presented in Figure A1.35. It contains six ORFs. ORF1 encodes a polypeptide of 170-195 kDa and displays sequence homology with RNA replicases of potex- and carlaviruses. It contains motifs typical of putative methyltransferase, helicase, and RdRp characteristic of virus-specific RNA replicase. ORF2 encodes a polypeptide of 25-28 kDa, which shows similarity to the first ORF of TGB found in all carla- and potexviruses. ORF3 encodes an 11-12 kDa polypeptide, homologous to the corresponding second ORF of TGB of

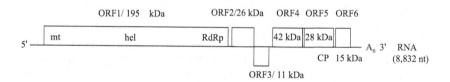

FIGURE A1.35. Organization and expression of the genome of an allexius *(Shallot virus X)*. Rectangles represent the open reading frames with their translation products. CP, capsid protein; hel, helicase domain; mt, methyltransferase domain; RdRp, RNA-dependent RNA polymerase; A$_n$, poly (A) tract.

carla- and potexviruses. TGB has a role in cell-to-cell movement. Between ORF3 and ORF4 is located ORF3-4, which lacks an initiation codon and codes for an amino acid sequence homologous to those of the 7-8 kDa proteins of carla- and potexviruses. It may be a homologue of the third gene in the TGB of carla- and potexviruses. ORF4 encodes a 32-43 kDa polypeptide, rich in serine; it does not show any significant homology with sequences of other proteins. The CP of 26-29 kDa is encoded by the ORF5. It shows a high percentage of homology to the CP of carla- and potexviruses, except for its variable N-terminus. ORF6 encodes a 14-15 kDa protein, which is cysteine rich, a typical characteristic of carlaviruses. It contains highly conserved regions comprising a basic arginine-rich domain and a putative zinc-binding finger motif, suggesting its involvement in regulation of RNA replication.

Relations with Cells and Tissues

ShV-X particles are present in high concentration. Virus particles are found in bundles or in groups and form large amorphous aggregates. During the early infection process, CP is detected in mitochondria; it is detected in the microsomal fractions during the later stages.

Host Range and Symptoms

Allexiviruses have very narrow host ranges. Most of them are latent or induce mild mosaic.

Transmission and Stability in Sap

Allexiviruses can be transmitted mechanically.
Natural transmission is by eriophid mites (*Aceria osschella*, formerly *A. tulipae*) in a noncirculative manner.

Antigenic Properties

Allexiviruses are highly immunogenic. Recombinant ShV-X CP has been used as an immunogen. Though no serological relationships have been established between ShV-X and other allexiviruses, viruses occurring in *Allium* species serologically cross-react.

Relevant Literature

Chen et al., 2001; Song et al., 1998.

Genus Vitivirus

The name has derived from the generic name *Vitis* of the natural host (grapevine) of its type species.

Type species: *Grapevine virus A* (GVA). Other species are *Grapevine virus B* (GVB), *Grapevine virus C* (GVC), *Grapevine virus D* (GVD), *Heracleum latent virus* (HLV).

Virion Properties

The virions have filamentous, flexuous particles 725-825 nm long and 12 nm in diameter. They are cross-banded (ropelike features, depending upon the negative-staining method) with a helical arrangement and a pitch of 3.3 nm.

Virus particles contain a single protein species of Mr of about 22 kDa encapsidating a single species of linear, positive-sense, ssRNA. The CP constitutes 95 percent of particle weight, the RNA 5 percent.

Genome Organization, Expression, and Replication

Genome organization of GVA and GVB are almost identical. The GVA genome is 7,349 nt long, its 5' terminus is capped and its 3' end has a poly (A) tail. The genome organization and expression is presented in Figure A1.36. It contains five ORFs. The 5'- region of the GVA genome possesses UTRs of 86 nt that are rich in AT contents. ORF1 encodes a polypeptide of 194 kDa and shows extensive homology with the replicase-associated proteins of RNA viruses (i.e., methyltransferase domain in the N-terminal region, NTP-binding motifs of the helicases in the central part, and GDD motif of RdRp in the C-terminal region). ORF2 partially overlaps ORF1 and

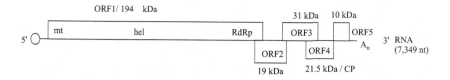

FIGURE A1.36. Organization and expression of the genome of a vitivirus *(Grapevine virus A)*. Rectangles represent the open reading frames with their translation products. CP, capsid protein; hel, helicase domain; mt, methyltransferase domain; RdRp, RNA-dependent RNA polymerase; O, cap; A_n, poly (A) tract.

ORF3 and encodes a polypeptide of about 19 kDa, which shows homology with a similar protein expressed by GVB. ORF3 encodes a polypeptide of 31 kDa (GVA) or 36.5 kDa (GVB), and contains conserved motifs of MP of the 30K superfamily. These proteins are associated with the cell wall and plasmodesmata, and are suggested to be involved in the cell-to-cell movement of virus. The CP of 21.5 kDa is encoded by ORF4, and a polypeptide of 10-14 kDa with cysteine-rich putative nucleotide-binding activity is encoded by ORF5.

Relations with Cells and Tissues

GVA and GVB are found in phloem tissues of their experimental and natural hosts. *(Nicotiana* species, *Gomphrena globosa,* and *Vitis vinifera).* Vascular bundles show extended cell wall thickening originating from deposits of calloselike substances, proliferation of endoplasmic membranes, and vesicular evaginations of the tonoplast containing fibrillar material. Virus particles aggregate in the form of bundles, whirls, banded bodies or staked layers and they sometimes protrude into the vacuole from tonoplast evaginations or fill most of the cell lumens. In *N. benthamiana,* the MP has been transiently expressed in the cell membranes and it accumulates in the cell wall at the level of plasmodesmata and in the cytosol.

Host Range and Symptoms

Natural infection has been detected only in *Vitis vinifera.* Rugose wood is a severe disease of vines (grafted). It is characterized by the presence of pits and grooves on the woody cylinder of the scion. GVA is the agent of Kober stem grooving, a component of the rugose wood complex. It also infects grapevine cultivars without apparent symptoms.

Transmission and Stability in Sap

GVA is transmitted mechanically with great difficulty from field-grown grapevine; it is readily transmitted by grafting from vine to vine and by mechanical inoculation between herbaceous hosts.

Natural transmission is by the pseudococcid mealybugs *Pseudococcus* spp. and *Planococcus* spp. in a noncirculative manner.

At approximately 20°C their infectivity in crude sap from infected plants is lost in six days. The viruses have their thermal inactivation point at approximately 50°C.

Antigenic Properties and Relationships

The virion is moderately immunogenic. Serologically it is distantly related to other species in the same genus. Though the CP of vitivirus shows homology with those of capilloviruses and trichoviruses, serologically they are distantly related.

Relevant Literature

Galiakparov et al., 2003; Minarfa et al., 1997; Saldarelli et al, 1996.

Genus **Mandarivirus**

The name has been derived from mandarin *(Citrus reticulata),* the host of the type species *Indian citrus ringspot virus* (ICRSV).

Virion Properties

The virions are flexuous, filamentous particles with prominent crossbanding, a zipperlike structure, and are about 650 nm long and 13 nm in diameter. Particle morphology is similar to those of allexi- and foveaviruses.

Genome Organization

The genome consists of a monopartite, positive-sense ssRNA of 7,560 nt, excluding the poly (A) tail at the 3' end. The genome organization and expression is presented in Figure A1.37. It contains six ORFs on the

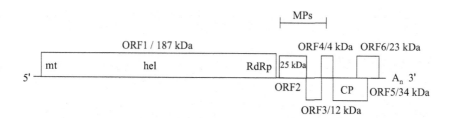

FIGURE A1.37. Organization and expression of the genome of a mandarinvirus *(Indian Citrus ringspot virus).* Rectangles represent the open reading frames with their translation products. CP, capsid protein, MP, movement protein; hel, helicase domain; RdRp, RNA-dependent RNA polymerase; mt, methyltransferase domain; A_n, poly (A) tract.

positive strand and probably follows sg RNA translational strategy. ORF1 encodes an 187.3 kDa protein similar to the RdRp of potex- and allexivirus. It contains a viral helicase 1 and RdRp2 domain. The N-terminal region represents the conserved motifs (I-IV) of the putative methyltransferase (mt) domain of Sindbis-like viruses. The central part has an NTP-binding helicase motif and five other motifs present in nucleic acid helicases. GDD, the highly conserved motif, is located at amino acid positions 1,534 to 1,536. ORFs 2, 4, and partially overlap and encode 25, 12, and 4 kDa proteins, respectively. They constitute a TGB as present in fovea-, carla-, and potexviruses, involved in cell-to-cell movement. The CP (34 kDa) is encoded by ORF5. A 23 kDa protein of unknown function is encoded by ORF6. However, it shows limited homology to nucleic-acid-binding regulatory proteins of allexi- and carlaviruses. At the 3' end of ICRSV genome, ACTTAA, a conserved motif, is present. This motif is also present at the same position in the potex- and carlavirus genomes and is suggested to play a role in the synthesis of negative-strand viral RNAs.

Transmission and Stability in Sap

No data are available.

Symptoms

ICRSV induces ringspot symptoms in infected Kinnow mandarin trees. It causes necrotic local lesions in experimental hosts, such as *Chenopodium* spp., *Glycine max,* and *Vigna unguiculata.* In *Phaseolus vulgaris* necrotic local lesions become systemic.

Antigenic Relationships

Antibodies to allexi- and criniviruses do not react to ICRSV. However, ICRSV reveals faint reaction with PVX.

Relevant Literature

Rustici et al., 2000, 2002.

UNASSIGNED GENERA

Genus Tobamovirus

The siglum has been derived from the name of the type species, *Tobacco mosaic virus* (TMV).

Virion Properties

The rod-shaped virions with helical symmetry have a length of 300-310 nm, a diameter of 18 nm, and a helical pitch of 2.3 nm.

The sedimentation coefficient is 194 S and the buoyant density 1.325 g/cm³ (in CsCl).

The CP has a Mr of 17,000-18,000.

Genome Organization, Expression, and Replication

The genome consists of a single molecule of positive-sense ssRNA, about 6,400 nt in size.

The genomic RNA and the sgRNA of CP are capped at their 5' terminus. All RNAs have a tRNA-like secondary structure at their 3' terminus. The organization and expression of the genome are given in Figure A1.38.

The genome encodes one structural protein, the CP of 17-18 kDa, and at least three nonstructural proteins: a 126-129 kDa protein and a RT product

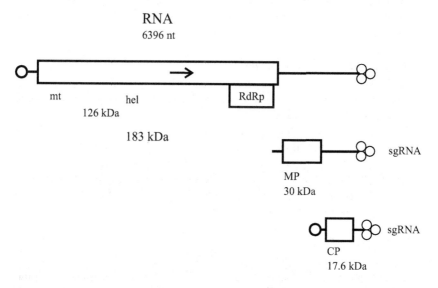

FIGURE A1.38. Organization and expression of the genome of a tobamovirus *(Tobacco mosaic virus)*. Open rectangles represent the open reading frames with their translation products. CP, capsid protein; hel, helicase domain; MP, movement protein; mt, methyltransferase domain; RdRp, RNA-dependent RNA polymerase; O, cap; ⊖, tRNA-like structure; → readthrough.

of 183-187 kDa, both required for RNA replication, and a 28-31 kDa protein (MP) involved in the cell-to-cell movement of the viral RNA. The MP and CP are translated from sgRNAs.

Some strains of TMV and other tobamoviruses produce a third sgRNA (I_1), but nothing is known about its expression. In some species this I_1 RNA is also encapsidated.

Infection Process

After penetration of the cell wall and plasmalemma the virions are uncoated by ribosomes and simultaneously translation starts with the production of replicase, consisting of the 126/183 kDa proteins. The replication proteins, in combination with host-encoded components, are involved in the replication of the viral RNA. First, complementary strands are synthesized, leading to fully dsRNAs (replicative form; RF) and partly dsRNAs (replicative intermediate; RI). New plus-strands are then synthesized from the negative-strand template and sgRNAs, first for MP and later for CP, are transcribed. Translation products are formed and the newly synthesized viral RNA either assembles with CP into virions or it moves from cell to cell in association with MP.

Relations with Cells and Tissues

A wide variety of inclusions can be detected in the light microscope. Characteristic inclusions are the hexagonal crystals abundantly present in hair cells of TMV-infected tobacco leaves. The crystals are made up of layers of virions. Besides these crystalline inclusions, also needle-shaped paracrystals are observed in the epidermis of TMV-infected leaves.

In addition to these inclusions consisting only of virus particles, tobamoviruses induce X-bodies: large, vacuolated structures containing virus particles, proteinaceous granules, and host organelles.

Host Range and Symptoms

The experimental host range of most tobamoviruses is broad, but in nature few plant species have been found susceptible.

Symptoms range from vein clearing and mosaic to distortion and blistering. TMV and some other species induce necrotic local lesions in inoculated leaves of *Nicotiana glutinosa* and other *N*-gene-containing tobacco species or cultivars.

Transmission and Stability in Sap

Tobamoviruses are readily transmitted mechanically. Most tobamoviruses are very stable in crude sap and dried material from infected plants. The infectivity is lost when the sap is heated at about 92°C for 10 min, but *Tobacco mild green mosaic virus* (TMGMV) is already inactivated at 85°C. However, in dried leaf material this virus was still active after 45 years.

No vectors are known. Tobamoviruses are spread mechanically.

Antigenic Properties and Relationships

Tobamoviruses are strongly immunogenic. Serological relationships exist between the species.

Relevant Literature

Kawakami et al., 2004; Kim et al., 2003.

Genus Tobravirus

The siglum has been derived from the name of the type species, *Tobacco rattle virus* (TRV).

Virion Properties

The rod-shaped virions with helical symmetry have two modal lengths of 180-215 nm (long [L] particles) and 46-115 nm (short [S] particles), a diameter of 20-22 nm, and a helical pitch of 2.5 nm.

The sedimentation coefficients are 296-306 S (L particles) and 155-245 S (S particles). The buoyant density is 1.306-1.324 g/cm^3 (in CsCl).

The CP has a Mr of 22,000-24,000.

The L and S particles contain RNA molecules of about 6,800 nt (RNA1) and, depending on the virus isolate, 1,800-4,500 nt (RNA2).

Genome Organization, Expression, and Replication

The bipartite genome consists of positive-sense ssRNA molecules. They are capped at their 5' terminus and have a tRNA-like secondary structure without aminoacylation at their 3' terminus. The organization and expression of the genome are given in Figure A1.39.

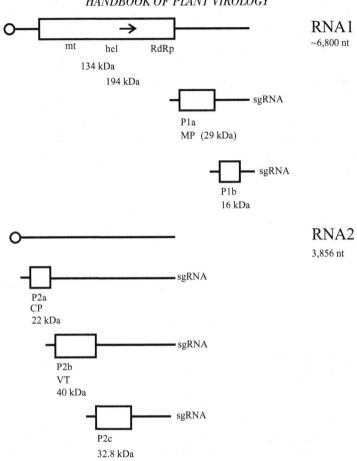

FIGURE A1.39. Organization and expression of the genome of a tobravirus (*Tobacco rattle virus,* isolate PpK20). Rectangles represent the open reading frames with their translation products (P1a, P1b, P2a, P2b, P2c). CP, capsid protein; hel, helicase domain; MP, movement protein; mt, methyltransferase domain; RdRp, RNA-dependent RNA polymerase; VT, protein involved in vector (nematode) transmission; O, cap; →, readthrough.

The RNA1 encodes four nonstructural proteins: a 134 to 141 kDa protein and a RT product of 194 to 201 kDa, both required for RNA replication, a 29-30 kDa protein (P1a) involved in the cell-to-cell movement of the viral RNA, and a protein of 12-16 kDa (P1b) that, at least in the case of *Pea early browning virus* (PEBV), has been shown to play a role in transmission through seed. P1a and P1b proteins are translated from sgRNAs. RNA1 can replicate and move systemically in plants, but no virions are formed and the infections are unstable. This type of infection, which is found in nature and can also be obtained by inoculation of plants with L particles only, was formerly referred to as nonmultiplying (NM) TRV infection, as opposed to the one brought about by normal multiplying virions.

The RNA2, whose length and nucleotide sequence vary with the virus strain, contains the CP gene and genes for nonstructural proteins, their number depending on the virus strain. The RNA2 of TRV isolate PpK20 contains genes for two nonstructural proteins, P2b and P2c, of 40 and 32.8 kDa, respectively, whereas the RNA2 of PEBV isolate TpA56 codes for three nonstructural proteins, P2b, P2c, and P2d, of 29.6, 23, and 9 kDa, respectively.

Besides the CP, the RNA2-encoded P2b protein of isolate PpK20 and, possibly, P2b, P2c, and P2d of PEBV isolate TpA56 are involved in nematode transmission by forming a bridge between the virus particle and the lining of the esophagus. All proteins encoded by RNA2 are translated from sgRNAs.

Natural recombination, both interspecific (between TRV and PEBV) and intraspecific (within TRV), have been reported.

Relations with Cells and Tissues

Some TRV isolates give rise to cytoplasmic inclusions (X-bodies) largely made up of degenerated mitochondria. The inclusions also contain aggregates of mostly L particles, whereas S particles have been found scattered in other parts of the cytoplasm.

Host Range and Symptoms

Tobraviruses have very wide host ranges, including both monocotyledonous and dicotyledonous plants.

Symptoms range from mottling, mosaic, line pattern, ringspotting, and yellowing to necrosis, distortion, stunting, notched leaf (in gladiolus) and abnormal cork formation (in potatoes).

Transmission and Stability in Sap

Tobraviruses are readily transmitted mechanically, except the NM type of TRV, which contains nonencapsidated RNA1 only.

The viruses are stable in crude sap from infected plants. The infectivity is lost when the sap is heated at about 80°C for 10 min.

Natural transmission is by nematode species in the genera *Paratrichodorus* and *Trichodorus*. Specificity exists between virus strains and vector species. Both adults and juveniles can transmit. Nematodes containing tobraviruses remain viruliferous for a long time.

Transmission through seeds of weeds has been reported.

Antigenic Properties and Relationships

Tobraviruses are moderately immunogenic, but some strains have poor immunogenicity. Hardly any serological relationship exists between the species. Within a species a number of serotypes have been distinguished. Sequence similarities are found between the replicase and MPs of tobraviruses and those of tobamoviruses.

Relevant Literature

Masuta, 2002; Visser and Bol, 1999.

Genus Hordeivirus

The name has been derived from the Latin genus name of barley, *Hordeum vulgare,* a natural host of the type species, *Barley stripe mosaic virus* (BSMV).

Virion Properties

The rod-shaped virions with helical symmetry have lengths ranging from 110 to 150 nm (depending on the size of the RNA molecule they contain), a diameter of about 20 nm, and a helical pitch of 2.5 nm.

Depending on the species, the sedimentation coefficients range from 182-193 S (BSMV) to 165-200 S (other species).

The CP has a Mr of 22,000.

Genome Organization, Expression, and Replication

The genome consists of three molecules of positive-sense ssRNA, designated α, β, and γ (comparable to the terms RNA1, RNA2, and RNA3 commonly used for other viruses with a tripartite genome).

The genomic RNAs and the sgRNAs are capped at their 5' terminus. A tRNA-like secondary structure is at their 3' terminus, preceded by a poly(A) tract of varying length.

The genome organization and expression are presented in Figure A1.40. The ORF on RNAα encodes a protein of 130 kDa (αa), with methyltransferase and helicase domains representing a part of a replicase.

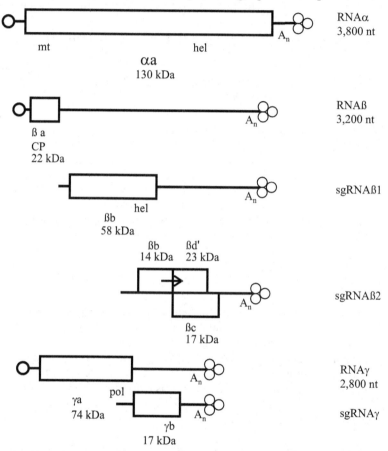

FIGURE A1.40. Organization and expression of the genome of a hordeivirus *(Barley stripe mosaic virus)*. Rectangles represent the open reading frames with their translation products (αa, βa, βb, βc, βd, βd',γa, γb). CP, capsid protein; hel, helicase domain; mt, methyltransferase domain; pol, polymerase domain; sgRNA, subgenomic RNA. O, cap; A$_n$, poly (A) tract; tRNA-like secondary structure; →, readthrough.

The ORFs on RNAβ encode the CP and three nonstructural proteins (βb, βc, βd), the TGB, required for cell-to-cell movement of the virus. Protein βd' is a readthrough product of βd. Its function is not known. In BSM, protein βb of 58 kDa has a helicase motif and is expressed from sgRNAß1. Proteins βc, βd, and βd' of 17, 14, and 23 kDa, respectively, are expressed from sgRNAβ2.

The ORFs on RNAγ, whose length varies with the virus strain, encode two proteins, one of 74 kDa (its size is strain-dependent) with a polymerase domain (γa) and a 17 kDa protein (γb). The latter protein regulates expression of the βRNA genes and plays a role in pathogenicity.

Barley protoplasts can become infected upon inoculation with a mixture of genomic RNAs α and γ, but all three genomic RNAs are required for systemic infection of inoculated leaves.

Some evidence suggests a natural recombination between RNAs α and γ, and between RNAs α and β of some strains.

Relations with Cells and Tissues

Virus particles are present in cells of all tissues, even in root tips. They are found mostly in the cytoplasm, but also in the nucleus. Particles of seed-transmitted strains of BSMV occur in ovules, pollen, and embryos. Vesicles are observed in plastids of infected leaves and virus particles have been found in association with microtubules.

Host Range and Symptoms

Natural host ranges of three of the four species are restricted to members of the Gramineae, but the experimental host range also includes members of the Amaranthaceae, Chenopodiaceae, and Solanaceae.

Symptoms induced by the species infecting gramineous plants are rather aspecific, consisting mainly of mosaic and chlorotic and necrotic stripes. They can easily be confused with symptoms caused by other viruses.

Transmission and Stability in Sap

Hordeiviruses are readily transmissible mechanically. They are rather stable in crude sap from infected plants. The infectivity is lost when the sap is heated at 63-70°C for 10 min, but in dehydrated leaf material and infected seeds BSMV retains its infectivity for many years.

Natural transmission is mainly through seed and to a lesser extent by pollen, and also by contact.

No vectors are known.

Antigenic Properties and Relationships

Hordeiviruses are good immunogens. The species are serologically interrelated, but more of a relationship exists between the species infecting grasses than between hordeiviruses and *Lychnis ringspot virus* (LRSV).

Relevant Literature

Bragg et al., 2004; Goshkova et al., 2003; Harsanyi et al., 2002.

Genus Furovirus

The siglum has been derived from the name of the group, "fungus-transmitted viruses with rod-shaped particles."
Type species: *Soil-borne wheat mosaic virus* (SBWMV).

Virion Properties

The rod-shaped virions with helical symmetry have two modal lengths of 260-300 nm and 140-160 nm and a diameter of 20 nm.
A large number of particles shorter than 140 nm may contain deletion mutants.
The sedimentation coefficients are 220-230 S (L particles), 170-225 S (S particles), and 126-177 S (deletion mutants).
The CP has a Mr of 19,000-20,500.

Genome Organization, Expression, and Replication

The genome consists of two molecules of positive-sense ssRNA (RNA1, RNA2). The two genomic RNAs are capped at their 5' terminus and have a tRNA-like secondary structure at their 3' terminus. The genome organization and expression are presented in Figure A1.41.
RNA1 encodes three proteins: a 150 kDa protein with methyltransferase and helicase motifs, an RT product of 209 kDa, and a putative MP of 37 kDa which is translated from a sgRNA.
RNA2 encodes four proteins: a protein of 25 kDa initiated from a CUG codon, the CP of 19 kDa initiated from its AUG codon, an RT product of the CP of 84 kDa, and a protein of 19 kDa translated from a sgRNA. Strong evidence suggests that the RT portion of RNA2 is involved in transmission by the fungal vector.

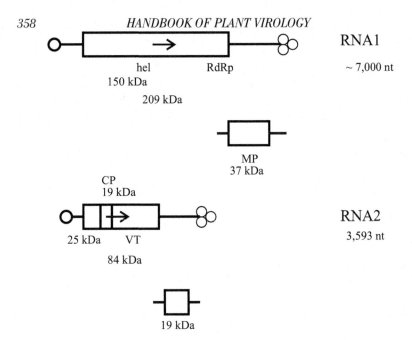

FIGURE A1.41. Organization and expression of the genome of a furovirus *(Soil-borne wheat mosaic virus)*. Rectangles represent the open reading frames with their translation products. CP, capsid protein; hel, helicase domain; MP, movement protein; RdRp, RNA-dependent RNA polymerase; VT, readthrough portion of the 84 kDa protein, involved in vector transmission. O, cap; ⌾, tRNA-like secondary structure; →, readthrough.

Relations with Cells and Tissues

Furoviruses induce different types of inclusions: vesicular, irregularly shaped inclusions with loose virus particles associated with microtubules; crystalline inclusions, often in the vacuole, composed of virions; crystalline inclusions enclosed in a membrane with virus particles, often adjacent to the latter; and aggregated virus particles in regular arrays in the cytoplasm.

Virions are also found scattered in the cytoplasm and vacuoles.

Host Range and Symptoms

Furoviruses have narrow natural host ranges. SBWMV is restricted to members of the Gramineae and some *Chenopodium* spp. The other three species also have few natural hosts with *Hypochoeris mosaic virus* (HyMV) infecting only *Hypochoeris radiata* and *Leontodon autumnalis*.

Their experimental host ranges, however, are rather wide and include species in ten or more families.

Transmission and Stability in Sap

Furoviruses are mechanically transmissible, but SBWMV cannot easily be transmitted in this way.

The stability of furoviruses in crude sap from infected plants varies with the species. SBWMV is inactivated in sap heated at 60-65°C for 10 min, but HyMV already loses its infectivity at 45-50°C. Some furoviruses (e.g., SBWMV) retain their infectivity in dried leaf material for more than ten years.

Furoviruses are soilborne and SBWMV is transmitted by the plasmodiophorid fungus *Polymyxa graminis*. Virus particles are acquired by zoospores and resting spores of the fungus inside the host plant (internally borne virus). Viruliferous resting spores in the soil remain infective for many years.

Antigenic Properties and Relationships

Many furoviruses are good immunogens. Some species are related serologically.

Relevant Literature

An et al., 2003; Koenig et al., 2002.

Genus Pomovirus

The siglum has been derived from the name of the type species, *Potato mop-top virus* (PMTV).

Virion Properties

The labile rod-shaped particles with helical symmetry have either three modal lengths of 290-310, 150-160, and 65-80 nm (*Beet soil-borne virus;* BSBV) or two modal lengths of 250-300 and 100-150 nm (PMTV).

The sedimentation coefficients of the former particles are about 230, 170, and 125 S, respectively.

The major CP has an Mr of 20,000. An RT protein of the CP of 104 kDa has been detected near one extremity of PMTV particles. This RT protein may be responsible for the loosely coiled helix structure observed at the ends of some PMTV virions.

Genome Organization, Expression, and Replication

The genome consists of three molecules of positive-sense ssRNA (RNAs 1-3). The three genomic RNAs are capped at their 5' terminus and have a tRNA-like secondary structure at their 3' terminus. The genome organization and expression of PMTV isolate Sw are presented in Figure A1.42.

RNA1 encodes a protein of 148 kDa with methyltransferase and helicase motifs, and an RT protein of 206 kDa with polymerase motifs in its RT portion.

RNA2, the second largest of the three RNA molecules, encodes the CP and an RT protein of 91 kDa. The RT domain of the CP is involved in transmission by the fungal vectors.

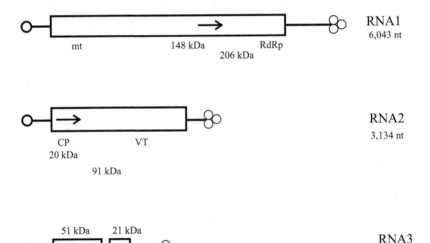

FIGURE A1.42. Organization and expression of the genome of a pomovirus (*Potato mop-top virus* isolate Sw). Rectangles represent the open reading frames with their translation products. CP, capsid protein; hel, helicase domain; mt, methyltransferase domain; RdRp, RNA-dependent RNA polymerase; VT, readthrough protein of the 111 kDa protein involved in vector transmission, O, cap; →, readthrough; ‑⊗, tRNA-like secondary structure.

RNA3 encodes four polypeptides of 51, 21, 13, and 8 kDa, respectively. The first three polypeptides share sequence similarity to the proteins of the TGB in other viruses, known to be involved in movement from cell to cell. The function of the fourth (cysteine-rich) polypeptide is not known.

Relations with Cells and Tissues

Particles of PMTV are found in bundles in the cytoplasm. Other pomoviruses, such as *Broad bean necrosis virus* (BBNV), induce irregularly shaped inclusions in the cytoplasm of epidermal cells, consisting of enlarged endoplasmic reticulum, membranous network structures, and bundles of virions.

Host Range and Symptoms

Pomoviruses have very narrow natural host ranges, but in experimental transmission some species can infect a number of plant species in two or more families of dicotyledonous plants.

Symptoms consist of chlorosis, mottling, yellow blotching, and extreme stunting of the shoots in potato affected by PMTV, which has led to the name "mop-top." PMTV induces brown arcs ("spraing") in the tubers of some potato cultivars. BBNV causes chlorosis, (veinal) necrosis, stunting, and defoliation in broad bean plants.

Transmission and Stability in Sap

Pomoviruses can be readily transmitted mechanically. They are not very stable in crude sap from infected plants. PMTV is inactivated in sap heated at 75-80°C for 10 min, but the inactivation temperature of BBNV is between 55 and 60°C. In sap kept at 20°C most pomoviruses lose their infectivity in less than one day.

Pomoviruses are soilborne and are transmitted by the plasmodiophorid fungi *Polymyxa betae* (BSBV) and *Spongospora subterranea* (PMTV). The viruses are acquired by zoospores and resting spores inside the host plant (internally borne viruses).

Antigenic Properties and Relationships

Pomoviruses are moderately immunogenic. The virus species are distantly related serologically.

Relevant Literature

Sandgren et al., 2001.

Genus Pecluvirus

The siglum has been derived from the name of the type species *Peanut clump virus*. There are two species in the genus.

Type species: *Peanut clump virus* (PCV).

Virion Properties

The rod-shaped particles with helical symmetry have two modal lengths of 245 and 190 nm, a diameter of 21 nm, and a helical pitch of 2.6 nm. In some preparations of PCV a third class of short particles with a modal length of 160 nm has been distinguished.

The sedimentation coefficients are 224 and 183 S for the two types of larger particles, respectively. The buoyant density is about 1.32 g/cm^3 (in CsCl).

The CP has an Mr of 23,000.

Genome Organization, Expression, and Replication

The genome consists of two molecules of positive-sense ssRNA (RNA1, RNA2). The genomic RNAs are capped at their 5' terminus and have a tRNA-like secondary structure at their 3' terminus. The genome organization and expression are presented in Figure A1.43.

RNA1 encodes three proteins of 131, 191, and 15 kDa. The 131 kDa protein encoded by the first ORF has methyltransferase and helicase motifs. The 191 kDa protein is an RT product of this ORF and it has a polymerase motif in its RT moiety. The cysteine-rich 15 kDa protein is expressed from a sgRNA and is involved in the regulation of replication.

RNA2 encodes five polypeptides of 23, 39, 51, 14, and 17 kDa. The CP is expressed from the first ORF and the 39 kDa protein is a product of the second ORF expressed by leaky ribosome scanning. The latter protein is probably involved in transmission by the fungal vector. The 51, 14, and 17 kDa proteins are produced by a TGB essential for movement of the virus from cell to cell. These proteins are most likely expressed from sgRNAs.

Relations with Cells and Tissues

Pecluviruses are detected in most cells of inoculated gramineous plants. Virus particles, singly or in aggregates, occur in the cytoplasm, near the nucleus, and along the plasmalemma.

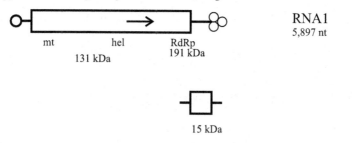

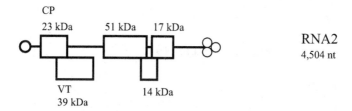

FIGURE A1.43. Organization and expression of the genome of a pecluvirus *(Peanut clump virus)*. Open rectangles represent the open reading frames with their translation products. CP, capsid protein; hel, helicase domain; mt, methyltransferase domain; RdRp, RNA-dependent RNA polymerase; VT, protein involved in vector transmission; O, cap; →, readthrough; —⟨⟩⟨⟩, tRNA-like secondary structure.

Host Range and Symptoms

The natural host ranges are narrow. Besides groundnut *(Arachis hypogaea)*, *Sorghum arundinaceum* has been found infected symptomlessly by PCV. Many species in the Gramineae have been reported to be susceptible to *Indian peanut clump virus* (IPCV). The experimental host range of PCV is wide and includes several species in many families.

Symptoms in groundnut range from mottle, chlorotic ringspots and line pattern to severe stunting (clump) of the infected plants in later stages of the infection.

Transmission and Stability in Sap

Pecluviruses can be transmitted mechanically, but transmission is hampered by the presence of inhibitors of infectivity in the sap of groundnut plants. PCV is rather stable. In crude sap from infected plants of *Chenopodium amaranticolor* kept at 27°C the virus retains its infectivity for four weeks. It is inactivated in sap heated at 64°C for 10 min.

Natural transmission of the soilborne viruses is by viruliferous zoospores and resting spores of the plasmodiophorid fungus *Polymyxa graminis.*

Seed transmission in groundnuts has been reported.

Antigenic Properties and Relationships

Pecluviruses are good immunogens. Three serotypes of IPCV are distinguished. No serological relationship exists between PCV and IPCV.

Relevant Literature

Dunoyer et al., 2002; Erhardt et al., 1999.

Genus Benyvirus

The siglum has been derived from the name of the type species, *Beet necrotic yellow vein virus.* Two species are in the genus.

Type species: *Beet necrotic yellow vein virus* (BNYVV).

Virion Properties

The labile rod-shaped particles with helical symmetry have four modal lengths of about 390, 265, 100, and 85 nm; a diameter of 20 nm; and a helical pitch of 2.6 nm.

The major CP has an Mr of 21,000-23,000. An RT protein of the CP gene, possibly required for assembly of the virus particles, has been detected serologically near one extremity of the BNYVV particles.

Genome Organization, Expression, and Replication

The genome consists of four or five (depending on the virus isolate) molecules of positive-sense ssRNA (RNAs 1-5) whose size corresponds to the length of the virions.

The genomic RNAs and some sgRNAs are capped at their 5' terminus and have a poly (A) tract at their 3' terminus. The genome organization and expression are presented in Figure A1.44.

RNA1 contains one ORF that codes for a polyprotein of about 237 kDa with methyltransferase, helicase, and polymerase domains. The polyprotein

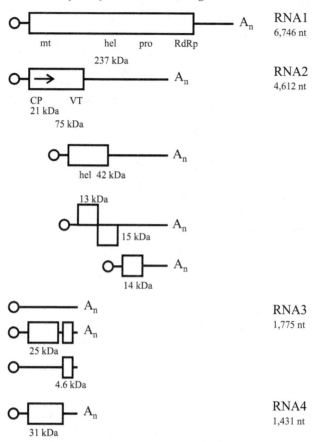

FIGURE A1.44. Organization and expression of the genome of a benyvirus *(Beet necrotic yellow vein virus)*. Rectangles represent the open reading frames with their translation products. CP, capsid protein; hel, helicase domain; mt, methyltransferase domain; RdRp, RNA-dependent RNA polymerase; pro, proteinase; VT, protein involved in vector transmission; O, cap; →, readthrough; A_n, poly (A) tract.

is processed by a proteinase. Protoplasts inoculated with RNA1 alone become infected.

RNA2 has six ORFs. The 5'-proximal ORF encodes the CP of 21 kDa, and together with the second ORF an RT protein of 75 kDa is produced. This RT protein is involved in virus assembly and its C-terminal region (KTER domain) plays an essential role in transmission by the fungal vector.

The following three ORFs (TGB) encode polypeptides of 42, 13, and 15 kDa. Like the TGB proteins in other viruses, these proteins are involved in cell-to-cell movement of the virus. The 42 kDa protein has helicase motifs. This protein, the 13/15 kDa proteins, and the 14 kDa protein are all expressed from sgRNAs. Some circumstantial evidence suggests that the 14 kDa protein has a nucleic-acid-binding activity. It may have a regulatory function.

Chenopodium quinoa inoculated with a mixture of RNA1 and RNA2 develops local lesions.

RNA3 is expressed from sgRNAs. Its major translation product, a 25 kDa protein, enhances vascular movement of the virus and determines the type of leaf symptoms. It is also responsible for the development of rhizomania symptoms in sugar beet infected with BNYVV. A 4.5 kDa protein is expressed from a second sgRNA.

RNA4 encodes a protein of 31 kDa. It is essential for transmission by the fungal vector, but transmission is enhanced in combination with either RNA3 or RNA5. RNA3 and RNA4 are always present in plants naturally infected with BNYVV.

A fifth RNA, present in some isolates of BNYVV, encodes a 19 kDa protein. In infected sugar beet plants this protein has been found to affect severity of symptoms in the roots, probably by stimulating systemic infection.

Relations with Cells and Tissues

Particles of BNYVV are found scattered or in aggregates in the cytoplasm of root, stem, and leaf cells of systemically infected sugar beet plants.

Host Range and Symptoms

Benyviruses have a very narrow host range, but their experimental host range includes many species in the Chenopodiaceae, and one each in the Aizoaceae and Amaranthaceae.

The most characteristic symptom caused by BNYVV in sugar beet is rhizomania (root madness), consisting of abnormally proliferated rootlets from the tap root and lateral roots. Some rootlets become necrotic and in the often-stunted tap root there is also browning of vascular bundles.

In field-infected sugar beet plants BNYVV seldom induces the symptoms in the leaves consisting of vein yellowing and veinal necrosis, after which the type species has been named.

Transmission and Stability in Sap

Natural transmission of the soilborne viruses is by viruliferous zoo-spores and resting spores of the plasmodiophorid fungus *Polymyxa betae.* Benyviruses can be transmitted mechanically, but after repeated mechanical transmission deletions may occur in RNA3 and RNA4, and these RNAs may even be eliminated.

The viruses are not very stable. In crude sap from infected plants kept at 20°C BNYVV retains its infectivity for five days. It is inactivated in sap heated at 65-70°C for 10 min.

Antigenic Properties and Relationships

The viruses are good immunogens. BNYVV and *Beet soil-borne mosaic virus* (BSBMV) are only distantly related serologically.

Relevant Literature

Lee et al.; Rush, 2003.

Genus Sobemovirus

The siglum has been derived from the name of the type species *Southern bean mosaic virus* (SBMV).

Virion Properties

The virions are isometric, about 30 nm in diameter with icosahedral symmetry (T = 3), composed of 180 subunits of CP of about 26-30 kDa, each of which has two domains: the S domain constitutes the parts of the icosahedron and the R domain forms an N-terminal "arm" into the interior part of the virus. Sobemovirus particles are highly stable in vivo and in vitro.

The virion Mr is about 6.6×10^6. The particles sediment at about 109-120 S and have a buoyant density of 1.36 g/cm^3 (CsCl).

Genome Organization

Virus genome consists of a single molecule of positive-sense, ssRNA of 4,100-4,600 nt. The 5' terminus has a VPg essential for infectivity and the 3' terminus lacks a poly (A) tail or t-RNA-like structure. The virion contains 21 percent RNA and 79 percent protein. An sgRNA is present in SBMV.

The genome organization and expression is presented in Figure A1.45. The viral genome contains four ORFs. ORF1 encodes an 11-24 kDa protein that is involved in the cell-to-cell movement of the virus. ORF2 encodes a polyprotein of about 100 kDa that is processed by a virus-encoded protease to yield the VPg, the protease, the helicase, and the RNA polymerase. This is a continuous ORF in most of the sobemoviruses. But in *Cocksfoot mottle virus* (CfMV), this polyprotein is encoded by two overlapping ORFs (2a, 2b) through a −1 ribosomal FS mechanism. The ORF2a codes for the putative VPg and the same protease, and ORF2b codes for the putative RdRp. ORF3 is nested within ORF2 in a −1 reading frame and encodes the putative protein P3 (10-18 kDa) of undefined function. This may be expressed as an FS fusion with the 5'-proximal part of the ORF2. ORF4 overlaps at the 3' end of ORF2, and encodes the CP (21-30 kDa), which is translated from sgRNA. CP of *Rice yellow mottle virus* (RYMV) is not required for RNA replication, but is necessary for RNA spread.

Relations with Cells and Tissues

Virions are found in the cytoplasm, vacuoles, and nuclei. Some members invade vascular tissues. The cytoplasm of infected cells may contain fibrillar structures associated with vesicles or arranged as crystalline aggregates in the cytoplasm and vacuole.

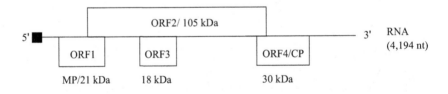

FIGURE A1.45. Organization and expression of the genome of a sobemovirus *(Southern bean mosaic virus)*. Open rectangles represent the open reading frames (ORFs) with their translation products; CP, capsid protein; MP, movement protein; sgRNA, subgenomic RNA; ■, VPg.

Transmission and Stability in Sap

In nature, sobemoviruses are transmitted through seed and pollen and by beetles and mirids (Hemiptera, Miridae) in the noncirculative/circulative manner. Some species have leaf miner flies as vectors.

Host Range and Symptoms

Sobemoviruses have narrow host ranges and symptoms mainly consisting of vein clearing, mosaic, mottling, and leaf distortion.

Antigenic Properties

Sobemoviruses are efficient immunogens. Serological relationships have been established among strains and members of the genus.

Satellite RNAs

In addition to genomic and sgRNA, virions of some sobemoviruses encapsidate a low-molecular-weight RNA with satellite properties. Rice yellow mottle virus satRNA of 220 nt is the smallest known viroidlike RNA. Satellite RNAs associated with members of four plant virus groups share a common sequence GAUUUU, which is also in the same position and loop structure for all four RNAs. In biological properties, these RNAs are undoubtedly satellites. In some properties such as their small size, circularity, high degree of base pairing, and lack of mRNA activity they resemble to viroids. They share structural similarities with viroids and other circular satellite RNAs. All sobemovirus-associated satRNAs contain hammerhead ribozyme motifs believed to be involved in autocatalytic processing during a circle mode of replication. However, some differences exist. Although the helper virus can replicate independently, the small RNAs do not show any sequence similarity with the helper virus and they cannot replicate independently. Further, their thermodynamic properties and ability to self-cleave are quite different from those of viroids, except possibly *Avocado sunblotch viroid.*

Relevant Literature

Lokesh et al., 2001; Makinen et al., 2000; Zhou et al., 2005.

Genus Idaeovirus

The name has been derived from the species name of raspberry, *Rubus idaeus,* the host of type species *Raspberry bushy dwarf virus* (RBDV), and the only known definitive member.

RBDV shares some characteristic features with those of members of the family *Bromoviridae,* such as their flattened isometric particles and the multipartite genome having the CP expression strategy and pollen-mediated transmission as occur in ilarviruses.

Virion Properties

The virions are quasi-spherical, appear flattened in electron microscopy, unless fixed, and are about 33 nm in diameter. The nucleocapsids appear to be angular.

The virion Mr is 7.5×10^6. The particles sediment at about 115 *S,* and have a buoyant density of 1.37 g/cm^3 (CsCl).

Genome Organization, Expression, and Replication

The genome organization and expression is presented in Figure A1.46. The genome consists of a positive-sense ssRNA with three components

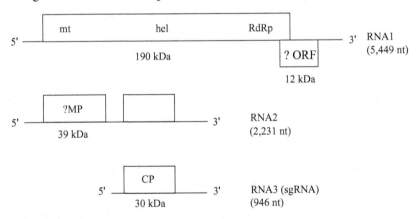

FIGURE A1.46. Organization and expression of the genome of an idaeovirus *(Raspberry bushy dwarf virus).* CP, capsid protein; hel, helicase; ?MP, putative movement protein; mt, methyltransferase; hel, helicase domain; ? ORF, putative open reading frame; RdRp, RNA-dependent RNA polymerase; sgRNA, subgenomic RNA.

(i.e., two genomic and one sg with RNA1 of 5,449 nt, RNA2 of 2,231 nt, and sgRNA3 of 946 nt). The 3'-terminal regions of RNA form a stem-loop structure. The three RNA species appear to be encapsidated in the same particle. RNA1 has a single ORF and encodes a 190 kDa polypeptide. This protein is suggested to be involved in virus replication, as it contains conserved motifs of methyltransferase, helicase, and RNA polymerase domain. Near the 3' end a small putative ORF (12 kDa) is present, though nothing is known about its expression. RNA2 has two ORFs and encodes two polypeptides of 39 and 30 kDa at the 5' and 3' ends, respectively. Based on the limited sequence homology of 39 kDa protein to the genera *Dianthovirus, Furovirus,* and that of the family *Bromoviridae,* it is suggested to be a putative MP. RNA3 is derived from the 3'-terminal region of RNA2 and expresses CP following the sgRNA synthesis.

Relations with Cells and Tissues

RBDV infects all tissues of an infected plant, erratically distributed. In raspberry leaves with yellow disease symptoms, chloroplasts are excessively enlarged, cell vacuole volume decreases and vesicle production increases.

Host Range and Symptoms

Idaeovirus has a restricted natural host range confined to raspberry, while the experimental range is wide. It infects raspberry across the globe where it is grown. Most often infection causes latent infection, yet a few cultivars show symptoms. RBDV is the causal agent of yellows disease characterized by formation of chlorotic line-patterns or yellowing of leaves. It also causes "crumbly fruit" in *Rubus* species, characterized by a high proportion of aborted drupelets.

Transmission and Stability in Sap

Idaeovirus can be transmitted by mechanical inoculation, grafting, and through pollen, both vertically to the seed and horizontally to the pollinated plant. Vegetative propagation of plants seems to be the main cause of long-distance dissemination of the virus. Natural vector is not known. At approximately 22°C their infectivity in crude sap from infected plants is lost after four days, or after heating for 10 min at 65°C.

Antigenic Properties and Relationships

Virions are moderately immunogenic. Most isolates are indistinguishable serologically, either among themselves or to several other ilarviruses.

Relevant Literature

Jones et al, 2000; Ziegler et al., 1993.

Genus Ourmiavirus

The name has been derived from that of the district of Ourmia in northern Iran, where the type species was first found in infected melon.

Type species: *Ourmia melon virus* (OuMV).

Virion Properties

Ourmiaviruses have bacilliform particles, 62, 46, 37, and 30 nm long and 18 nm in diameter. The particles have conical ends and possess hemi-icosahedral symmetry. The buoyant density of all particles is $1.375 g/cm^3$ (in CsCl).

Genome Organization, Expression, and Replication

The genome consists of a positive-sense ssRNA with an Mr of approximately 1.58×10^6. It is divided into three segments of approximately 2,900, 1,100, and 1,000 nt with an estimated Mr of 0.91, 0.35, and 0.32×10^6, respectively.

Presence of a cap or Vpg at the 5' end, or a poly (A) tract at the 3' end has not been established.

The CP of approximately 21 or 25 kDa (depending on the species) is encoded by RNA3. A fibrous nonstructural protein is expressed from RNA1 or RNA2.

Relations with Cells and Tissues

Large numbers of virus particles are found in the cytoplasm of parenchyma cells in association with the fibrous protein.

Host Range and Symptoms

The natural host range of the three recognized ourmiaviruses is restricted to cucurbits, cherry, and cassava. However, their experimental host range is wide and includes species from 14 families.

Symptoms consist of systemic ringspots and mosaic, as well as local lesions on some experimental hosts.

Transmission and Stability in Sap

Ourmiaviruses are readily transmissible mechanically. In some plants, seed transmission occurs to a low degree (one to two percent).

No natural vector is known.

The viruses are rather stable and are inactivated in sap heated at a temperature of 80°C for 10 min.

Antigenic Properties and Relationships

The viruses and the nonstructural protein are good immunogens. No cross-reaction occurs between the virus species.

Relevant Literature

Accotto et al., 1997.

Genus Umbravirus

The siglum has been derived from the Latin word *umbra* (shadow) and refers to the occurrence of these viruses as shadow (constant companion) of other viruses (members of the family *Luteoviridae*) on which they depend for survival in nature.

Type species: *Carrot mottle virus* (CMoV).

Virion Properties

Umbraviruses do not form conventional virus particles, as they lack the gene for CP. However, another virus-encoded protein binds to umbraviral RNA, thus leading to the formation of filamentous ribonucleoprotein (RNP) particles. This protein provides stability to the viral RNA and facilitates its long-distance movement. For encapsidation and vector transmission, umbraviruses depend on the CP of their helper (assistor) viruses (members of the family *Luteoviridae*) and in case of Groundnut rosette virus (GRV) even on the latter's satRNA (Groundnut rosette virus satRNA; GRVsatRNA).

Genome Organization, Expression, and Replication

The genome consists of a single positive-sense ssRNA molecule. No information has been gathered about structures at the 5' end, and its 3' end is most likely not polyadenylated.

The genome organization and expression are presented in Figure A1.47. ORF1 encodes a putative 31-37 kDa protein. ORF2, which slightly overlaps the 3' end of ORF1, might code for a protein of 63-65 kDa, but it lacks an AUG initiation codon near its 5' end. However, before the stop codon of ORF1 there is a sequence of 7 nt often associated with frameshifting in

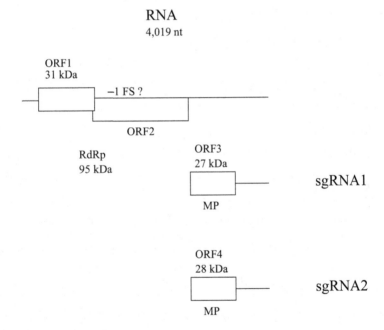

FIGURE A1.47. Organization and expression of the genome of an umbravirus *(Groundnut rosette virus)*. Rectangles represent the open reading frames (ORFs) with their translation products. ORFs 1 and 2 are expressed from genomic RNA by a putative −1 frameshift (−1FS?). ORF3 and ORF4 are most likely translated from subgenomic (sg) RNAs (sgRNA1 and sgRNA2, respectively). MP, movement protein, RdRp, RNA-dependent RNA polymerase.

plant and animal viruses. Thus it is likely that ORF1 and ORF2 are together encoding a 94-98 kDa protein by a −1 FS mechanism. On the basis of predicted amino acid sequence the putative ORF1/ORF2 fusion protein is supposed to be an RdRp.

The ORFs 3 and 4 show a great overlap and each of them yields a 26-29 kDa polypeptide, most likely translated from sgRNAs. The ORF3 protein protects viral RNA and enables transport through the phloem, whereas the ORF4 product plays a role in cell-to-cell movement.

Some umbraviruses possess satRNAs. The GRVsatRNA, consisting of approximately 900 nt, accompanies all naturally occurring isolates and depends for its replication on GRV. Remarkably, GRV also depends on its satRNA for encapsidation in the CP of a member of the *Luteoviridae*, namely *Groundnut rosette assistor virus* (GRAV), an enamovirus, and hence for its transmission by aphids. Moreover, this satRNA is mainly responsible for the symptoms of groundnut rosette disease.

Relations with Cells and Tissues

Ultrathin sections of the cells of many plants infected with umbraviruses showed membrane-bound, vesiclelike structures, approximately 50 nm in diameter, in the vacuole and tonoplast. Such structures were also present in partially purified preparations of umbraviruses.

Cytoplasmic granular inclusions consisting of filamentous RNP particles were found embedded in electron-dense matrix material. The filaments, which partly showed a helical structure, had a diameter of 13-14 nm and an electron-lucent central hole, approximately 4 nm in diameter. Immunogold-labeling and in situ hybridization assays showed that the RNP particles consisted of viral RNA and the ORF3 protein. Long-distance movement of umbraviruses through the phloem may take place in the form of these RNP particles. ORF3 protein was also found in nuclei and nucleoli of infected cells. ORF4 protein localizes to plasmodesmata and forms tubular structures. Its involvement in cell-to-cell movement has been shown.

Host Range and Symptoms

The host range of umbraviruses is rather limited. Symptoms consist mainly of mottling and mosaic, but plants affected by pea enation mosaic disease show typical enations associated with the veins on the lower side of the upper leaves. In this complex disease, *Pea enation mosaic virus-2* (PEMV-2) is responsible for the systemic spread of its assistor virus, *Pea enation mosaic virus-1* (PEMV-1), an enamovirus. PEMV-2 even enables

PEMV-1 to move out of the phloem into the mesophyll, a feature highly unusual for members of the *Luteoviridae*.

Transmission and Stability in Sap

Umbraviruses are not readily transmitted mechanically. In nature, umbraviruses are transmitted by aphids in a circulative-nonpropagative manner with the help of their respective assistor viruses. For successful aphid transmission of GRV, its satRNA also must be present.

In crude sap from infected plants umbraviruses retain their infectivity for several hours at room temperature and for several days at 5°C. Infectivity is lost when the leaf extract is treated with organic solvents, possibly because of loss of the lipid-containing vesicles with their presumed viral-RNA-protecting role.

Relationships

Members of the genus *Umbravirus* have much in common with those of the family *Tombusviridae*.

Relevant Literature

Talinsky and Robinson, 2003; Talinsky et al., 2000.

Genus Sadwavirus

The siglum has been derived from *Satsuma dwarf virus* (SDV), the type species.

This genus contains viruses that were earlier placed as tentative members of the genus *Nepovirus* in the family *Comoviridae*. The genome organization of SDV has some features in common with those of como-, faba-, and nepoviruses. Like nepoviruses, its members have a bipartite genome encapsidated in polyhedral particles and each genome segment is translated into a polyprotein. While nepoviruses have a single CP, SDV has two subunits of different sizes.

The purified particles sediment as three components. The T component has empty protein shells. The M and B components contain nucleoproteins, sediment at 96 and 128 *S*, and have buoyant densities of 1.43 and 1.46 g/cm^3 (CsCl), respectively.

This description pertains to the characteristic features of SDV, in particular.

Virion Properties

Virions are polyhedral, 26 to 30 nm in diameter with hexagonal outlines.

Genome Organization, Expression, and Replication

SDV has a positive-sense ssRNA genome with two RNA species, desig-nated as RNA1 with 6,795 nt and RNA2 with 5,346 nt. A poly (A) tail is present at the 3' end, while the 5' end has a VPg. The genome organization and expression is presented in Figure A1.48. Each RNA species is trans-lated into a polyprotein, which is sequentially processed into functional gene products.

RNA1 contains a single ORF encoding a 230 kDa protein. This protein has conserved motifs, such as NTP-binding motifs, cysteine proteinase, and RdRp. Phylogenetic studies suggest that SDV is distinct from those of the genera como-, nepo-, and fabaviruses.

RNA2 also contains a single potential ORF encoding a 174 kDa protein. By analogy with como- and nepoviruses, two components of CP species (20-23 kDa) are present. They show low amino acid sequence similarity with como-, faba- and nepoviruses. RNA2 contains a region with similarity to the conserved regions of the putative MPs. The genome organization of SDV and that of como-, nepo-, fabaviruses have some features in common.

RNA1 and RNA2 share significant sequence similarity in the N-terminal re-gions of their respective polyproteins, such as their duplication. Duplication in

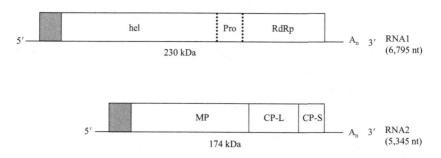

FIGURE A1.48. Organization and expression of the genome of a sadwavirus *(Satsuma dwarf virus)*. Open rectangles represent the open reading frames with their translation products. Gray rectangles represent the sequence similarity in the N termini of polyproteins. Vertical lines indicate the known and dashed lines indicate the putative cleavage sites. CP-L, capsid protein (larger); CP-S, capsid protein (smaller); MP, movement protein; Pro, proteinase domain; hel, helicase domain; RdRp, RNA-dependent RNA polymerase; A_n, poly (A) tract.

N-terminal regions of polyprotein encoded by RNA1/RNA2 has been reported in *Tomato ringspot virus,* a nepovirus.

The 3' and 5' UTRs are equal in size in the corresponding regions of RNA1 and RNA2, and have significant homology. The sequences (3', 5' UTRs), however, do not show any homology with other known sequences.

Relations with Cells and Tissues

Parenchymatous cells of leaves from infected herbaceous plants show tubular structures containing rows of virus particles. Virus particles are in a lattice array and characteristic inclusion bodies are also observed in infected tissues.

Host Range and Symptoms

Natural host range is restricted to citrus. Experimental host range is wider and includes herbaceous species in eight families of dicotyledonous plants.

SDV causes leaf malformation, rosetting of Satsuma orange *(Citrus unshui),* and stunting. Early in the season, infected leaves may have the shape of an inverted boat or spoon.

Transmission and Stability in Sap

Transmitted through sap by mechanical inoculation and through seeds. Natural vectors are not known.

SDV loses infectivity in sap from infected *Physallis floridana* when heated at 50-55°C for 10 min or stored for 8 to 12 days at room temperature.

Relevant Literature

Iwanami et al., 1999; Kasarev et al., 2001.

Genus Cheravirus

The name has been derived from *Cherry rasp leaf virus* (CRLV), the type species.

This genus contains viruses that were earlier placed as tentative members of the genus *Nepovirus* in the family *Comoviridae.* Like nepoviruses, its members have a bipartite genome encapsidated in icosahedral particles and each genome segment is translated into a polyprotein. However, cheraviruses possess three CP species (20-25 kDa). Phylogenetic studies suggest them to be closer to the family *Sequiviridae*.

The purified particles sediment as three components. The T component has empty protein shells sedimenting at 56 *S*. The nucleoproteins, representing M and B components, sediment at 96 *S* and 128 *S* and have buoyant densities of 1.41 and 1.43 g/cm^3, respectively.

In this description, most features of cheraviruses pertain to those of *Apple latent spherical virus* (ALSV).

Virion Properties

Virions are isometric and 25 nm in diameter, with hexagonal outlines.

Genome Organization, Expression, and Replication

The genome organization and expression is presented in Figure A1.49. The ALSV has a positive-sense ssRNA genome with two RNA species, designated as RNA1, with 6,815 nt, and RNA2, with 3,394 nt. A poly (A) tail is present at the 3' end, while the 5' end has a VPg.

RNA1 contains two ORFs. ORF1 encodes a 23 kDa protein, while ORF2 codes for a 235 kDa protein, which has conserved motifs of proteinase (cysteine), helicase, and polymerase (from the N-terminus) as occur in

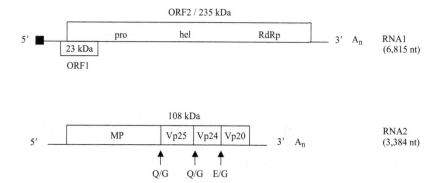

FIGURE A1.49. Organization and expression of the genome of a cheravirus *(Apple latent spherical virus)*. Open rectangles represent the open reading frames (ORF) with their translation products. Arrows indicate the putative cleavage sites. Hel, helicase domain; pro, proteinase domain; MP, movement protein; RdRp, RNA-dependent RNA polymerase; A$_n$, poly (A) tract; ■, virus protein genome-linked; for others, see text.

como-, nepo-, faba-, and sadwaviruses. Phylogenetic studies based on polymerase protein sequences suggest that ALSV is related to SDV, a sadwavirus, but distinct from members of the genera *Como-, Nepo-,* and *Fabavirus.* ALSV, however, differs from SDV in having three CP species and ORF1 (23 kDa).

RNA2 contains a single ORF which codes for a 108 kDa protein. By analogy with como- and nepoviruses, three components of CP species (designated as, Vp25, Vp24, Vp20 kDa) are likely to be encoded in the 3'-terminal region of RNA2. RNA2 contains the conserved MP motifs showing homology with those of nepovirus MP. The cleavage sites of the 108 kDa polyprotein are probably Q/G, between MP and Vp25 and Vp25 and Vp20, and E/G, between Vp20 and Vp24.

Each RNA species is translated into a polyprotein, which is sequentially processed into functional gene products.

UTRs of 101 nt and 90 nt are present at the 3'-ends of RNA1 and RNA2, respectively. They show about 81 percent homology, as reported for nepoviruses. At the 5' end, RNA1 has a UTR of 157 nt, while that of RNA2 is 311 nt long, showing 61 percent homology. The VPg is located between the helicase and proteinase domains.

Relations with Cells and Tissues

Parenchymatous cells of leaves from infected cucumber and *Chenopodium quinoa* plants show tubular structures containing rows of virus particles.

Host Range and Symptoms

CRLV causes "rasp" symptoms consisting of enations on the underside of cherry leaves and small enations, shortened internodes with stunted growth, in peach. CRLV infection of apple leads to deformed fruits (flat apple) with reduced quality and taste.

Transmission and Stability in Sap

Cheraviruses are transmissible through sap by mechanical inoculation and through seeds. The nematode, *Xiphinema americanum,* is a natural vector for CRLV. Vectors are not known for ALSV and AVBV.

Antigenic Properties and Relationships

Viruses are moderately immunogenic. ALSV and CRLV are serologically unrelated. Though cheraviruses resemble nepoviruses, they do not show serological relationships.

Relevant Literature

James and Upton, 2003, 2005; Li et al., 2000.

BIBLIOGRAPHY

Accotto GP, Riccioni L, Barba M, Boccardo G (1997). Comparison of some molecular properties of Ourmia melon and Epirus cherry viruses, two representatives of a proposed new virus group. *Journal of Plant Pathology* 78: 87-91.

Adams MJ, Antoniw JF, Fauquet CM (2005). Molecular criteria for genus and species discrimination within the family *Potyviridae*. *Archives of Virology* 150: 459-479.

Aguilar JM, Franco M, Marco CF, Berdiales B, Rodriguez-Cerezo E, Truniger V, Aranda MA (2003). Further variability within the genus *Crinivirus,* as revealed by determination of the complete RNA genome sequence of Cucurbit yellow stunting disorder virus. *Journal of General Virology* 84: 2555-2564.

An H, Melcher U, Doss P, Payton M, Guenzi AC, Verchot-Lubicz J (2003). Evidence that the 37 kDa protein of *Soil-borne wheat mosaic virus* is a virus movement protein. *Journal of General Virology* 84: 3153-3163.

Ayllon MA, López C, Navas-Castillo J, Gransy SM, Guerri J, Flores R, Moreno P (2001). Polymorphism of the 5'- terminal region of Citrus tristeza virus (CTV) RNA: Incidence of three sequence types in isolates of different origin and pathogenicity. *Archives of Virology* 146: 27-40.

Belin C, Schmitt C, Demangeat G, Komar V, Pinck L, Fuchs M (2001). Involvement of RNA2-encoded proteins in the specific transmission of *Grapevine fanleaf virus* by its nematode vector *Xiphinema index. Virology* 291: 161-171.

Boonham N, Henry CM, Wood KR (1995). The nucleotide sequence and proposed genome organization of Oat chlorotic stunt virus, a new soil-borne virus of cereals. *Journal of General Virology* 76: 2025-2034.

Boonham N, Henry CM, Wood KR (1998). The characterization of a subgenomic RNA and its in vitro translation products of Oat chlorotic stunt virus. *Virus Genes* 16: 141-145.

Bragg JN, Lawrence DM, Jackson AO (2004). The N-terminal 85 amino acids of the *Barley stripe mosaic virus* γb pathogenesis protein contain the three zinc-binding motifs. *Journal of Virology* 78: 7379-7391.

Brault V, Bergdoll M, Mutterer J, Prasad V, Pfeffer S, Erdinger M, Richards KE, Ziegler-Graff V (2003). Effects of point mutations in the major capsid protein of beet western yellows virus on capsid formation, virus accumulation, and aphid transmission. *Journal of Virology* 77: 3247-3256.

Cao Y, Cai Z, Ding Q, Li D, Han C, Yu J, Liu Y (2002). The complete nucleotide sequence of Beet black scorch virus (BBSV), a new member of the genus *Necrovirus*. *Archives of Virology* 147: 2431-2435.

Chen J, Chen J, Adams MJ (2001). Molecular characterization of a complex mixture of viruses in garlic with mosaic symptoms in China. *Archives of Virology* 146: 1841-1853.

Chen J, Shi Y-H, Adams MJ, Chen J-P (2005). The complete sequence of genomic RNA of an isolate of *Lily virus X* (genus *Potexvirus*). *Archives of Virology* 150: 825-832.

Chen YK, Derks AFL, Langeveld S, Goldbach R, Prins M (2001). High sequence conservation among cucumber mosaic virus isolates from lily. *Archives of Virology* 146: 1631-1636.

Choi I-R, Stenger DC, French R (2000). Multiple interactions among proteins encoded by the mite-transmitted wheat streak mosaic tritimovirus. *Virology* 267: 185-198.

Choi SA, Ryu KH (2003). The complete nucleotide sequence of the genome RNA of *Lily symptomless virus* and its comparison with that of other carlaviruses. *Archives of Virology* 148: 1943-1955.

Di Terlizzi B, Skrzeczkowski LJ, Mink GI, Scott SW, Zimmerman MT (2001). The RNA5 of *Prunus necrotic ringspot virus* is a biologically inactive copy of the 3'-UTR of the genomic RNA3. *Archives of Virology* 146: 825-833.

Dunoyer P, Pfeffer S, Fritsch C, Hemmer O, Voinnet O, Richards KE (2002). Identification, subcellular localization and some properties of a cysteine-rich suppressor of gene silencing encoded by *Peanut clump virus*. *Plant Journal* 29: 555-567.

Erhardt M, Stussi-Garaud C, Guilley H, Richards KE, Jonard G, Bouzoubaa S (1999). The first triple gene block protein of peanut clump virus localizes to the plasmodesmata during virus infection. *Virology* 264: 220-229.

Fazeli CF, Rezaian MA (2000). Nucleotide sequence and organization of ten open reading frames in the genome of Grapevine leafroll-associated virus 1 and identification of three subgenomic RNAs. *Journal of General Virology* 81: 605-615.

Fujisaki K, Hagihara F, Kaido M, Mise K, Okuno T (2003). Complete nucleotide sequence of spring beauty latent virus, a bromovirus infectious to *Arabidopsis thaliana*. *Archives of Virology* 148: 165-175.

Galiakparov N, Tanne E, Sela I, Gafny R (2003). Functional analysis of the grapevine virus A genome. *Virology* 306: 42-50.

Gentit P, Foissac X, Svanella-Dumus L, Peypelut M, Candresse T (2001). Characterization of two different apricot latent virus variants associated with peach asteroid spot and peach sooty ringspot disease. *Archives of Virology* 146: 1453-1464.

German-Retana S, Bergey B, Delbos RP, Candresse T, Dunez J (1977). Complete nucleotide sequence of the genome of a severe cherry isolate of apple chlorotic leaf spot trichovirus (ACLSV). *Archives of Virology* 142: 833-841.

Gorshkova EN, Erokhina TN, Stroganova TA, Yelina NE, Zamyatnin AA Jr, Kalinina NO, Schiemann J, Solovyev AG, Morozov SY (2003). Immunodetection and fluorescent microscopy of transgenically expressed hordeivirus TGBp3 movement protein reveals its association with endoplasmic reticulum elements in close proximity to plasmodesmata. *Journal of General Virology* 84: 985-994.

Grieco F, Dell'Orco M, Martelli GP (1996).The nucleotide sequence of RNA1 and RNA2 of olive latent virus 2 and its relationships in the family *Bromoviridae*. *Journal of General Virology* 77: 2637-2644.

Hammond RW, Ramirez P (2001). Molecular characterization of the genome of *Maize rayado fino virus,* the type member of the genus *Marafivirus. Virology* 282: 338-347.

Harsanyi A, Boddi B, Boka K, Almasi A, Gaborjanyi R (2002). Abnormal etioplast development in barley seedlings infected with BSMV by seed transmission. *Physiologia Plantarum* 114: 149-155.

Hartono S, Natsuaki T, Genda Y, Okuda S (2003). Nucleotide sequence and genome organization of Cucumber yellows virus, a member of the genus *Crinivirus. Journal of General Virology* 84: 1007-1012.

Hsu HT, Hsu YH, Bi IP, Lin NS, Chang BY (2004). Biological functions of the cytoplasmic TGBp1 inclusions of bamboo mosaic potexvirus. *Archives of Virology* 149: 1027-1035.

Iwanami T, Kondu Y, Kasarev AV (1999). Nucleotide sequences and taxonomy of satsuma dwarf virus. *Journal of General Virology* 80: 793-797.

Izadpanah K, Zhang YP, Daubert S, Masumi M, Rowhani A (2002). Sequence of the coat protein gene of Bermuda grass etched-line virus, and of the adjacent "marafibox." *Virus Genes* 24: 131-134.

Jaag HM, Kawchuk L, Rohde W, Fischer R, Emans N, Prufer D (2003). An unusual internal ribosomal entry site of inverted symmetry directs expression of a potato leafroll polerovirus replication-associated protein. *Proceedings of the National Academy of Sciences, USA* 100: 8939-8944.

Jacob T, Usha R (2001). 3'-terminal sequence analysis of the RNA genome of the Indian isolate of *Cardamom mosaic virus:* A new member of genus *Macluravirus* of *Potyviridae. Virus Genes* 23: 81-88.

James D, Jelkmann W, Upton C (2000). Nucleotide sequence and genome organization of cherry mottle leaf virus and its relationship to the members of the *Trichovirus* genus. *Archives of Virology* 145: 995-1007.

James D, Upton C (2000). Nucleotide sequence analysis of RNA-2 of flat apple isolate of *Cherry rasp leaf virus* with regions showing greater identity to animal picornaviruses than to related plant viruses. *Archives of Virology* 147: 1631-1641.

James D, Upton C (2005). Genome segment RNA-1 of a flat apple isolate of *Cherry rasp leaf virus:* Nucleotide sequence analysis and RT-PCR detection. *Archives of Virology* 150: 1469-1476.

Jones AT, McGavin WJ, Mayo MA, Angel-Diaz JE, Kärenlampi SO, Kokko H (2000). Comparisons of some properties of two variants of *Raspberry bushy dwarf virus* (RBDV) with those of three previously characterized RBDV isolates. *European Journal of Plant Pathology* 106: 623-632.

Kao CC, Sivakumaran K (2000). Brome mosaic virus, good for an RNA virologist's basic needs. *Molecular Plant Pathology* 1: 91-97.

Kasarev AV, Huan SS, Iwanami T (2001). Satsuma dwarf and related viruses belong to a new lineage of plant picorna-like viruses. *Virus Genes* 23: 45-52.

Kawakami S, Watanabe Y, Beachy RN (2004). Tobacco mosaic virus infection spreads cell to cell as intact replication complexes. *Proceedings of National Academy of Sciences USA* 101: 6291-6296.

Kim SM, Lee JM, Yim KO, Oh MH, Park JW, Kim KH (2003). Nucleotide sequences of two Korean isolates of *Cucumber green mottle mosaic virus*. *Molecular Cells* 16: 407-412.

Koenig R, Bergstrom GC, Gray SM, Loss S (2002). A New York isolate of *Soilborne wheat mosaic virus* differs considerably from the Nebraska type strain in the nucleotide sequences of various coding regions but not in the deduced amino acid sequences. *Archives of Virology* 147: 617-625.

Koev G, Liu S, Beckett R, Miller WA (2002). The 3'-terminal structure required for replication of Barley yellow dwarf virus RNA contains an embedded 3'-end. *Virology* 292: 114-126.

Koh LH, Cooper JI, Wong SM (2001). Complete sequences and phylogenetic analyses of a Singapore isolate of broad bean wilt fabavirus. *Archives of Virology* 146: 135-147.

Lee L, Telford EB, Batten JS, Scholthof KB, Rush CM (2001). Complete nucleotide sequence and genome organization of *Beet soilborne mosaic virus,* a proposed member of the genus *Benyvirus*. *Archives of Virology* 146: 2443-2453.

Li C, Yoshikawa N, Takahashi T, Ito T, Yoshika K, Koganezawa H (2000). Nucleotide sequence and genome organization of Apple latent spherical virus: A new virus classified into the family *Comoviridae*. *Journal of General Virology* 81: 541-547.

Lin MK, Chang BY, Liao JT, Lin NS, Hsu YH (2004). Arg-16 and Arg-21 in the N-terminal region of the triple-gene-block protein 1 of *Bamboo mosaic virus* are essential for virus movement. *Journal of General Virology* 85: 251-259.

Ling K-S, Zhu H-Y, Drong RF, Slightom JL, McFerson JR, Gonsalves D (1998). Nucleotide sequence of the 3'-terminal two-thirds of the grapevine leafroll-associated virus-3 genome reveals a typical monopartite closterovirus. *Journal of General Virology* 79: 1299-1307.

Liou RF, Yan HZ, Hong JL (2003). Molecular evidence that aphid-transmitted *Alpinia mosaic virus* is a tentative member of the genus *Macluravirus*. *Archives of Virology* 148: 1211-1218.

Lokesh GL, Gopinath K, Satheshkumar PS, Savithri HS (2001). Complete nucleotide sequence of Sesbania mosaic virus: A new species of the genus *Sobemovirus*. *Archives of Virology* 146: 209-223.

Lomonossoff GP, Shanks M (1999). Comoviruses (*Comoviridae*). In Granoff A, Webster RG (eds.), *Encyclopedia of virology,* Volume 1 (pp. 255-291). San Diego: Academic Press.

López-Moya JJ, García JA (1999). Potyviruses (*Potyviridae*). In Granoff A, Webster R (eds.), *Encyclopedia of virology,* Volume 2 (pp. 1369-1375). San Diego: Academic Press.

Magome H, Yoshikawa N, Takahashi T, Ito T, Miyakawa T (1997). Molecular variability of the genomes of capilloviruses from apple, Japanese pear, European pear and citrus trees. *Phytopathology* 87: 389-396.

Makinen K, Makelainen K, Arshava T, Tamm T, Mertis A, Truve E, Zavriev S, Saarma M (2000). Characterization of VPg and the polyprotein processing of *Cocksfoot mottle virus* (genus *Sobemovirus*). *Journal of General Virology* 81: 2783-2789.

Martelli GP, Grieco F (1997). *Oleavirus,* a new genus in the family *Bromoviridae. Archives of Virology* 142: 1933-1936.

Martelli GP, Jelkmann W (1998). *Foveavirus,* a new plant virus genus. *Archives of Virology* 143(6): 1245-1249.

Martelli GP, Russo M, Rubino L, Sabanadzovic S (1998). *Aureusvirus,* a novel genus in the family *Tombusviridae. Archives of Virology* 143(9): 1847-1851.

Martelli GP, Sabanadzovic S, Abou Ghanem-Sabanadzovic N, Saldarelli P (2002). *Maculavirus,* a new genus of plant viruses. *Archives of Virology* 147: 1847-1853.

Martelli GP, Sabanadzovics S, Abou Ghanem-Sabanadzovic N, Edwards MC, Dreher T (2002). The family *Tymoviridae. Archives of Virology* 147(9): 1837-1846.

Masuta C (2002). Recombination in plant RNA viruses. In Khan JA, Dijkstra J (eds.). *Plant viruses as molecular pathogens* (pp. 203-223). Binghamton, NY: The Haworth Press, Inc.

Meulewater F (1999). Necroviruses (*Tombusviridae*). In Granoff A, Webster RG (eds,), *Encyclopedia of virology,* Second edtion, Volume 2 (pp. 1003-1007). San Diego: Academic Press.

Minarfa A, Saldarelli P, Martelli GP (1997). Grapevine virus A: Nucleotide sequence, genome organization, and relationship in the *Trichovirus* genus. *Archives of Virology* 142: 417-423.

Mitchell EJ, Bond JM (2005). Variation in the coat protein sequence of *Turnip yellow mosaic virus* and comparison with previously published isolates. *Archives of Virology* 150: 2347-2355.

Mizumoto H, Hikichi Y, Okuno T (2002). The 3'-untranslated region of RNA1 as a primary determinant of temperature sensitivity of *Red clover necrotic mosaic virus* Canadian strain. *Virology* 293: 320-327.

Mukasa SB, Rubaihayo PR, Valkonen JP (2003). Sequence variability within the 3'-proximal part of the *Sweet potato mild mottle virus* genome. *Archives of Virology* 148: 487-496.

Nutter RC, Scheets K, Panganiban LC, Lommel SA (1989). The complete nucleotide sequence of the maize chlorotic mottle virus genome. *Nucleic Acids Research* 17: 3163-3177.

Okinaka Y, Mise K, Suzuki E, Okuno T, Furusawa I (2001). The C-terminus of Brome mosaic virus coat protein controls viral cell-to-cell and long-distance movement. *Journal of Virology* 75: 5385-5390.

Peremyslov VV, Andreev IA, Prokhnevsky AI, Duncan GH, Taliansky ME, Dolja VV (2004). Complex molecular architecture of beet yellows virus particles. *Proceedings of National Academy of Sciences USA* 101: 5030-5035.

Qiu P, Scholthof K-BG (2000). In vitro and in vivo generated defective RNAs of *Satellite panicum mosaic virus* define *cis*-acting RNA elements required for replication and movement. *Journal of Virology* 74: 2247-2254.

Qu F, Morris TJ (1999). Carmoviruses (*Tombusviridae*). In Granoff A, Webster RG (eds.), *Encyclopedia of virology,* Second edition, Volume 1 (pp. 243-247). San Diego: Academic Press.

Reavy B, Mayo MA, Turnbull-Russ AD, Murant AF (1993). *Parsnip yellow fleck* and *Rice tungro spherical viruses* resemble picornaviruses and represent two genera in a proposed new plant picornavirus family *(Sequiviridae). Archives of Virology* 131: 441-446.

Reddick BB, Habera LF, Law MD (1997). Nucleotide sequence and taxonomy of Maize chlorotic dwarf virus within the family *Sequiviridae. Journal of General Virology* 78: 1165-1174.

Rochon D'A (1999). Tombusviruses. In Granoff A, Webster RG (eds.), *Encyclopedia of virology,* Second edition, Volume 3 (pp. 1789-1798). San Diego: Academic Press.

Roossinck M (2001). *Cucumber mosaic virus,* a model for RNA virus evolution. *Molecular Plant Pathology* 2: 59-63.

Rubino L, Russo M (1997). Molecular analysis of the pothos latent virus genome. *Journal of General Virology* 78: 1219-1226.

Rush CM (2003). Ecology and epidemiology of *Benyviruses* and plasmodiophorid in vectors. *Annual Review of Phytopathology* 41: 567-592.

Rustici G, Accotto GP, Noris E, Masenga V, Luisoni E, Milne RG (2000). Indian citrus ringspot virus: A proposed new species with some affinities to potex-, carla-, fovea- and allexiviruses. *Archives of Virology* 145: 1895-1908.

Rustici G, Milne RG, Accotto GP (2002). Nucleotide sequence, genome organisation and phylogenetic analysis of Indian citrus ringspot virus. *Archives of Virology* 147: 2215-2224.

Sabanadzovic S, Abou Ghanem-Sabanadzovic N, Saldarelli P, Martelli GP (2001). Complete nucleotide sequence and genome organization of Grapevine fleck virus. *Journal of General Virology* 82: 2009-2015.

Saénz P, Qulot L, Quiot J-B, Candresse T, García JA (2001). Pathogenicity determinants in the complex virus population of a *Plum pox virus* isolate. *Molecular Plant-Microbe Interactions* 14: 278-287.

Saldarelli P, Minarfa A, Martelli GP (1996). The nucleotide sequence and genomic organization of grapevine virus B. *Journal of General Virology* 77: 2645-2652.

Sandgren M, Savenkov EI, Valkonen JPT (2001). The readthrough region of *Potato mop-top virus* (PMTV) coat protein encoding RNA, the second largest RNA of PMTV genome, undergoes structural changes in naturally infected and experimentally inoculated plants. *Archives of Virology* 146: 467-477.

Scheets K (2000). Maize chlorotic mottle machlomovirus expresses its coat protein from a 1.47-Kb subgenomic RNA and makes a 0.34-Kb subgenomic RNA. *Virology* 267: 90-101.

Shen P, Kaniewska M, Smith C, Beachy RN (1993). Nucleotide sequence and genome organization of Rice tungro spherical virus. *Virology* 193: 621-630.

Shirawski J, Voyatzakis A, Zacocomer B, Bernardi F, Haenni A-L (2000). Identification and functional analysis of the Turnip yellow mosaic tymovirus subgenomic promoter. *Journal of Virology* 74: 11073-11080.

Sit TL, Haikal PR, Callaway AS, Lommel SA (2001). A single amino acid mutation in the *Carnation ringspot virus* capsid protein allows virion formation but prevents systemic infection. *Journal of Virology* 75: 9538-9542.

Skaf JS, Schultz MH, Hirata H, de Zoeten GA (2000). Mutational evidence that the VPg is involved in the replication and not the movement of *Pea enation mosaic virus-1*. *Journal of General Virology* 81: 1103-1109.

Song SI, Song JT, Kim CH, Lee JS, Choi YD (1998). Molecular characterization of the garlic virus × genome. *Journal of General Virology* 79: 155-159.

Stenger DC, French R (2004). Complete nucleotide sequence of *Oat necrotic mottle virus:* A distinct *Tritimovirus* species (family *Potyviridae*) most closely related to *Wheat streak mosaic virus*. *Archives of Virology* 149: 633-640.

Stuart G, Moffett K, Bozarth RF (2004). A whole genome perspective on the phylogeny of the plant virus family *Tombusviridae*. *Archives of Virology* 149: 1595-1610.

Takeshita M, Suzuki M, Takanami Y (2001). Combination of amino acids in the 3a protein and the coat protein of Cucumber mosaic virus determines symptom expression and viral spread in bottle gourd. *Archives of Virology* 146: 697-711.

Taliansky ME, Robinson DJ (2003). Molecular biology of umbraviruses: Phantom warriors. *Journal of General Virology* 84: 1951-1960.

Taliansky ME, Robinson DJ, Murant AF (2000). Groundnut rosette disease virus complex: Biology and molecular biology. *Advances in Virus Research* 55: 357-400.

Thole V, Hull R (1998). Rice tungro spherical virus polyprotein processing: Identification of a virus-encoded protease and multifunctional analysis of putative cleavage sites. *Virology* 247: 106-114.

Turina M, Maruoka M, Morris J, Jackson AO, Scholthof K-BG (1998). Nucleotide sequence and infectivity of a full-length cDNA clone of Panicum mosaic virus. *Virology* 241: 141-155.

Turnbull-Ross AD, Mayo MA, Reavy B, Murant AF (1993). Sequence analysis of the *Parsnip yellow fleck virus* polyprotein: Evidence of affinities with picornaviruses. *Journal of General Virology* 74: 555-561.

Turnbull-Ross AD, Reavy B, Mayo MA, Murant AF (1992). The nucleotide sequence of Parsnip yellow fleck virus: A plant picorna-like virus. *Journal of General Virology* 73: 3203-3211.

Visser PB, Bol JF (1999). Nonstructural proteins of *Tobacco rattle virus* which have a role in nematode-transmission: Expression pattern and interaction with viral coat protein. *Journal of General Virology* 80: 3273-3280.

Wang J, Simon AE (1997). Analysis of the two-subgenomic RNA promoters for Turnip crinkle virus in vivo and in vitro. *Virology* 232: 174-186.

Wobus CE, Skaf JS, Schultz MH, de Zoeten GA (1998). Sequencing, genomic localization and initial characterization of the VPg of pea enation mosaic enamovirus. *Journal of General Virology* 79: 2023-2025.

Yoshikawa N, Sasaki E, Kato M, Takahashi T (1992). The nucleotide sequence of apple stem grooving capillovirus genome. *Virology* 191: 98-105.

Zheng HY, Chen J, Zhao MF, Lin L, Chen JP, Antoniw JF, Adams MJ (2003). Occurrence and sequences of *Lily mottle virus* and *Lily symptomless virus* in plants grown from imported bulbs in Zhejiang province, China. *Archives of Virology* 148: 2419-2428.

Zhou H, Wang H, Huang LF, Naylor F, Clifford P (2005). Heterogeneity in codon usage of sobemovirus genes. *Archives of Virology* 150: 1591-1605.

Ziegler A, Mayo MA, Murant AF (1993). Proposed classification of the bipartite-genomed raspberry bushy dwarf idaeovirus, with tripartite-genomed viruses in the family *Bromoviridae*. *Archives of Virology* 131: 483-488.

Description of Double-Stranded RNA Viruses

Jeanne Dijkstra
Jawaid A. Khan

FAMILY REOVIRIDAE

The name has been derived from respiratory enteric orphan, referring to diseases caused by these viruses in humans and animals.

The family *Reoviridae* comprises six genera infecting humans and animals and three genera, namely *Fijivirus, Phytoreovirus,* and *Oryzavirus,* of plant-infecting viruses that replicate both in plants and their arthropod vectors. The virions are icosahedral in shape and possess a double-stranded (ds) RNA genome with 10 to 12 segments, depending upon the genera. The viral RNA species are mostly monocistronic and present in equimol proportions. Each virion contains a single copy of the entire genome. The virion Mr is about 120×10^6 and has buoyant densities ranging from 1.36 to 1.39 g/cm^3 (CsCl). RNA constitutes 15-22 percent of particle weight. The plus strand of each duplex has a 5' terminal cap, while the minus strand has phosphorylated 5' termini. A poly (A) tail is absent at the 3' end. The distinction between the three genera of plant-infecting viruses is based on the number and size of genome segments, their electrophoretic profile, virion morphology, serology, and insect vectors. The genomic segments are numbered and designated S1-S12. Viroplasm or virus-inclusion bodies are present in the cytoplasm.

Genus Phytoreovirus

The first part of the name has been derived from the Greek word *phuton* (plant), indicating that they are plant reoviruses.

Type species: *Wound tumor virus* (WTV).

Virion Properties

The virions are icosahedral, apparently more angular than spherical, 65-70 nm in diameter, and double shelled; they have an outer layer of distinct capsomeres and an inner smooth core (50 nm in diameter); they probably lack spikes. WTV possesses three layers with an extra amorphous layer external to the outer shell. Phytoreoviruses have 12 ds RNA segments; the RNA constitutes 22 percent of particle weight.

Each virion possesses genomic RNA in supercoiled form.

The Mr of phytoreovirus is about 75×10^6 and the particles sediment at 510 S.

Genome Organization, Expression, and Replication

The dsRNA genome is about 25,000 base pairs long. All the RNA segments contain genus-specific, conserved sequences 5' GC (U/C) AUU . . . (U/C) GAU 3'. Adjacent to these sequences, RNA segment-specific inverted repeats of 6-14 bases are present. The *Rice dwarf virus* (RDV) genome has been well characterized. Ten of the viral genome segments are monocistronic, one (S11) has two ORFs, and S12 has three ORFs. Segments S1, 5, 7, and 12 are involved in RNA replication, segments S3 and 8 code for the structural protein, and S4, 6, 9, 10, 11, 12 encode nonstructural proteins.

Segment S1 encodes a polypeptide of about 155-164 kDa. It is located on the viral core particle and has sequence motifs conserved in RdRp. Segment S2 encodes a protein (P2) of 123-130 kDa. Present at the outer capsid, the P2 protein is essential for infection of the insect cells and influences transmission of the virus by the insect vector. The S3 genome segment encodes a major core capsid protein (CP) of 114 kDa in RDV. In the same virus, segments S4 and S6 encode nonstructural proteins of 79.8 and 57.4 kDa, respectively; the former possesses the nucleic-acid-binding zinc-finger motif. Messenger (m) RNA synthesized from purified RDV has a cap structure, m7GpppAm-, suggesting the presence of guanylyltransferase activity in the virion. An 89 kDa protein associated with the core and possessing mRNA guanylyltransferase activity is encoded by segment S5. Segment S7 encodes a core/nucleic-acid-binding protein (55-58 kDa). Segment S8 codes for the major outer shell protein (43-48 kDa), S9-12 codes for the nonstructural proteins of different sizes, segments and S9 (39 kDa) and S10 (35-39 kDa) are of unknown functions. Segment S11 (20-23 kDa) possesses nucleic-acid-binding properties and plays an important role in the virus replication and/or genome reassortment and another protein of unknown function. Segment S12 codes for three proteins of 37 kDa (nonstructural protein), 10.6 kDa (cytoplasmic phosphoprotein), and 9.6 kDa (function not known).

Associated with viroplasms, replication occurs in the cytoplasm of infected cells. Viroplasms consist of fibrillar and amorphous material. Dense particles of about 50 nm in diameter are embedded in the viroplasm, which is the site of virus synthesis.

Relations with Cells and Tissues

WTV and *Rice gall dwarf virus* are confined to the phloem tissues of the plant hosts. However, RDV is not limited to phloem and has been detected in cells adjacent to the vascular bundles. Infected and noninfected cells of either plant or insect hosts have a similar appearance, but RDV induces abnormalities in fat body cells and mycetocytes of the vector. The viruses are present in the cytoplasm with their particles arranged in scattered, clustered, or crystalline inclusions.

Immunodetection studies have demonstrated a greater accumulation of the virus-encoded proteins in rice leaves than in leafhoppers.

Host Range and Symptoms

Phytoreoviruses have narrow host ranges confined to the family Gramineae, but WTV can infect a large number of dicotyledonous hosts experimentally. WTV-infected plants initially show vein enlargements and characteristic tumors in roots and stems (few species). A tumor develops where plants are wounded or where lateral root initial breaks through the pericycle. RDV, on the other hand, does not induce such neoplasia; plants are stunted and develop chlorotic flecks. Symptoms induced by RGDV are almost similar to those found in fijivirus-infected plants, consisting of dark green coloring, leaf distortion, development of enations or galls on the veins of leaf and sheath, suppression of flowering, and stunting of plants.

Transmission and Stability in Sap

Natural transmission of phytoreoviruses is by cicadellid leafhoppers (Hemiptera, Cicadellidae; e.g., *Agallia, Agalliopsis,* and *Nephotettix*) in a circulative-propagative manner. The virus is transmitted transovarially and a viruliferous insect vector has a lifelong ability to transmit it.

Crude sap from virus-infected leaf tissues injected into vector hoppers retains infectivity at dilutions of 10^{-4} to 10^{-5} (RDV). But infectivity is lost when crude sap from infected plants is heated to about 45-50°C (RDV) or 50-60°C (WTV) for 10 min.

Relevant Literature

Anzola et al., 1989; Nakagawa et al., 2003; Suzuki et al., 1996.

Genus Fijivirus

The name has been derived from Fiji, the country where the type species was first isolated from infected sugarcane plants.
Type species : *Fiji disease virus* (FDV).

Virion Properties

The virions are icosahedral, 65-70 nm in diameter, and double-shelled, with a distinct type of projections. The outer shell with "A"-type spikes (about 11 nm in length and breadth, located at the 12 vertices on the icosahedron) and the inner shell with "B"-type spikes (about 8 nm long and 12 nm in diameter) comprise at least six structural proteins. Virus particles easily lose the outer layer, yielding spiked core particles 51-55 nm in diameter (with 12 "B"-spikes) that contain one copy of each genome segment and RNA-dependent RNA polymerase (RdRp).
Fijivirus species are classified into five groups on the basis of insect vectors, plant hosts, serological relationship, and nucleotide sequence homology.

Group 1: *Fiji disease virus*—Type species of the genus
Group 2: *Rice black streaked dwarf virus* (RBSDV), *Maize rough dwarf virus* (MRDV), *Mal de Rio Cuarto virus* (MRCV), *Pangola stunt virus* (PaSV)
Group 3: *Oat sterile dwarf virus* (OSDV)
Group 4: *Garlic dwarf virus* (GDV)
Group 5: *Nilaparvata lugens reovirus* (NLRV)

Genome Organization, Expression, and Replication

The dsRNA genome is 27,000 to 30,500 base pairs long.
Fijiviruses have 10 dsRNA segments with total Mr $18-20 \times 10^6$. All the RNA segments of RBSDV and sections S1 and S6-10 of MRDV possess the genus-specific conserved terminal sequences 5' AAGUUUUU . . . GUC 3',

which differ (both in length and sequence) from those of phytoreoviruses and oryzaviruses. Segment-specific inverted repeats are present immediately adjacent to them. These terminal-conserved sequences are suggested to be recognition signals for transcriptional and replication events.

Most of the dsRNA segments are monocistronic, with a few exceptions (where there are two open reading frames[ORF]). The functions of most of the proteins are unclear as segments carrying the same number in different viruses are often not homologous. For instance, the major outer structural protein is encoded on segment S10 of RBSDV and MRDV, but it is located on segment S8 of NLRV and OSDV. Although the coding functions are not fully defined, the largest genome segment S1 (NLRV) encodes a 165.9 kDa protein with characteristic RdRp sequence motifs; segment S2 (NLRV) encodes a protein of about 137 kDa, most likely the "B"-spike protein. Segment S3 (NLRV) encodes a major core protein of 139 kDa and segment S4, 5, and 6 encode 130, 106, and 95 kDa proteins, respectively, of unknown function. Segment S7 has nucleic-acid-binding properties in MRDV and NLRV, and in RBSDV it forms tubular structures. Segment S9 yields a nonstructural protein in RBSDV and NLRV, and in OSDV it encodes NTP-binding protein. Major outer-shell protein is encoded by segment S8 in NLRV, and by segment S10 in RBSDV and MRDV.

Not much is known about the replication of plant reoviruses. It is thought to be similar to that of animal reoviruses. Virus replication occurs in the cytoplasm of phloem-associated cells. Following infection, viroplasm appears in the cytoplasm. The virions are synthesized and assemble in viroplasm; later they migrate into the cytoplasm. During the infection, tubules (sometimes in the form of scrolls) of about 90 nm in diameter accumulate.

Relations with Cells and Tissues

Fijiviruses induce hyperplasia and hypertrophy of the phloem cells of infected plants, leading to enations and gall formation on the leaves. Inclusion bodies appear within neoplastic cells and viroplasm is found in the hypertrophied phloem cells. Two kinds of virus particles (i.e., 75-60 nm and 50-55 nm in diameter) are seen in cells of infected plants and insects. Small particles occur in the viroplasms, while the large particles are present in cytoplasm surrounding viroplasm.

Host Range and Symptoms

Fijiviruses have narrow host ranges restricted to the family Gramineae and Liliaceae. The virus produces characteristic symptoms consisting of vein swelling and gall formation on the lower side of leaves. Flowering is suppressed, plant becomes dwarfed, and formation of side shoots and induction of dark green coloring occurs.

Transmission and Stability in Sap

Fijiviruses can be transmitted vegetatively. Natural transmission is by delphacid planthoppers (Hemiptera, Delphacidae; e.g., *Laodelphax, Javesella, Delphacodes, Perkinsiella,* and *Unkanodes*) in a circulative-propagative manner. Transovarial transmission does not occur.

Relevant Literature

Distefano et al., 2003, 2005; Marzachi et al., 1995; Zhang et al, 2001.

Genus Oryzavirus

The name has been derived from the generic name of rice, *Oryza.*
Type species: *Rice ragged stunt virus* (RRSV).

Virion Properties

The virions are icoshedral and 57-65 nm in diameter with 12 "B"-type spikes, which are broader at the base. They are about 8-10 nm in length, 23-26 nm wide at the base, and 14-17 nm at the top. In purified preparations, particles resemble the B-spiked subviral particles of fijiviruses. However, double-shelled virions are not seen. It is likely that the outer shell is absent or it is labile and subsequently lost.

Genome Organization, Expression, and Replication

The dsRNA genome is 27,000 base pairs long.
Oryzaviruses have 10 dsRNA segments with total Mr of 18×10^6. All the RNA segments possess highly conserved nucleotide sequences at the 5' and 3' termini (5' GAUAAA ... GUGC 3'), respectively, and differ from those of phytoreoviruses or fijiviruses. Each RNA segment represents one ORF or at the most two. In this genus, RRSV has been studied well. Segment S1 encodes a core protein (B-spike protein) of 137 kDa, S2 codes for an inner core protein of 133 kDa, and S3 codes for a major core protein. Two proteins are encoded by segment S4, that is, 141 kDa by ORF1 with sequence motifs conserved in RdRp and 37 kDa protein by ORF2 with unknown

function. Segment S5 yields a protein of 91 kDa, which has guanyl transferase activity. A 66 kDa protein encoded by segment S6 is of unknown function. Segments S7 and S10 contribute toward the expression of nonstructural proteins of 68 kDa and 32 kDa, respectively. Section S8 encodes a protein of 67 kDa that is autocatalytically cleaved into 26 kDa and 46 kDa proteins. The 26 kDa protein is a self-cleavage protease and the 46 kDa is processed into the major CP (43 kDa). Segment S9 has a single ORF and encodes a 38 kDa spike protein involved in vector transmission.

Little is known about replication.

Relations with Cells and Tissues

An oryzavirus infects only the phloem tissues of the host. Virions are usually embedded in fibrillar viroplasms, virus particles are found in the cytoplasm of phloem-associated cells/tissues, such as parenchyma cells and sieve tubes. Further, enations developing on the leaves and leaf sheaths are due to hyperplasia and hypertrophy of parenchyma cells. Tubules containing viral antigens are also detected in infected plant/hopper tissues.

Host Range and Symptoms

Fijiviruses have narrow host ranges, restricted to the family Gramineae. Infected plants are stunted; leaves develop enations and get twisted and ragged. Young plants are darker green with excessive branching at the nodes, flowering is delayed, and emergence of the panicle is incomplete, with unfilled grains.

Transmission and Stability in Sap

Natural transmission of oryzaviruses is by *Nilaparvata lugens,* the rice brown planthopper, in a circulative-propagative manner. Though not transmitted through eggs, nymphs are found to be more efficient transmitters than adults. It replicates both in its host and vector.

Infectivity is lost when crude sap from infected plant, heated at approximately 60°C for 10 min, is injected into the vector. However, crude sap from virus-infected leaf tissues injected into vector hoppers retains infectivity at dilutions up to 10^{-5}.

Antigenic Properties and Relationships

Oryzaviruses are good but complex immunogens. RRSV and *Echinochloa ragged stunt virus* serologically cross-react with each other.

Relevant Literature

Hull, 2002; Upadhyaya et al., 1998.

FAMILY **PARTITIVIRIDAE**

The name has been derived from the Latin word *partitus,* meaning divided, and refers to the organization of the genome.

The family *Partitiviridae* comprises five genera, namely *Partitivirus, Chrysovirus, Alphacryptovirus, Betacryptovirus,* and *Varicosavirus.* The members of the first two genera infect fungi; those of the latter three genera, namely *Alphacryptovirus, Betacryptovirus,* and *Varicosavirus,* are plant-infecting viruses, often referred to as cryptic viruses or cryptoviruses, because they cause hardly any symptoms or are even latent. The virions are isometric in shape, 30-40 nm in diameter, and possess a dsRNA genome with two unrelated segments ranging in size between 1.4 and 3.0 kbp; a poly (A) tail is present at the 3' end. It is not known if different dsRNA molecules are encapsidated in one particle or in separate ones. Besides, smaller defective or satellite RNAs may also be present. The viral RNA segments are mostly monocistronic. The smaller RNA segment encodes the CP and the larger segment codes for the RNA polymerase. Two cistrons, CP and replication, successfully contribute toward the infection cycle of cryptoviruses, which occur frequently in their natural hosts. Most plant-infecting cryptic viruses cause latent infections and are present in low concentrations in their plant hosts. They are transmitted through seed and pollen with high frequency, but they are not transmitted mechanically, vegetatively, or by grafting, and no natural vectors are known. Because of their less-practical importance, cryptoviruses have not received much attention.

Genus **Alphacryptovirus**

Type species: *White clover cryptic virus 1* (WCCV 1).

Virion Properties

The virions are isometric, 30 nm in diameter and contain bipartite, linear, dsRNA. The capsids are made up of a single peptide species of Mr 55×10^3. The particles sediment at about 118 S and have a buoyant density of about 1.37 g/cm^3 (CsCl).

Genome Organization, Expression, and Replication

The genome consists of two segments of dsRNA, 1.7 and 2.0 Kbp in size; they are monocistronic. The smaller segment encodes the CP and the larger one codes for a putative RdRp. In *Beet cryptic virus 3* (BCV 3), one strand was found to contain a single ORF (1,431 nucleotides), which encodes a putative polypeptide of 54.9 kDa. This polypeptide showed sequence motifs present in the RdRp of other RNA viruses.

Mostly, no sequence homology exists at the generic or intergeneric level. However, *Alfalfa cryptic virus 1* and *Hop trefoil cryptic virus 1* (HTCV 1) reveal some homology. The 5'-terminus sequences in RNA1 and RNA2 of BCV 3 have sequences suggesting involvement in the replication and packaging processes.

Virus accumulates in the cytoplasm, and in vitro studies have shown that transcription and replication takes place by a semiconservative mechanism.

Relations with Cells and Tissues

Presence of virions in all plant parts, in the cytoplasm or in nucleoli has been demonstrated. They may be associated with cytoplasmic inclusion bodies. Owing to their low concentration, virus particles are not seen in infected cells.

Host Range and Symptoms

An alphacryptovirus does not produce apparent symptoms. Due to its lack of transmissibility (except through seeds), the experimental host range is not known.

Transmission

An alphacryptovirus can be transmitted mechanically or by grafting. They are transmitted by ovule and pollen to the seed embryo. Cell-to-cell spread does not occur, except at the cell division. No vector is known. Since virus assay species are lacking, biophysical properties such as thermal inactivation and longevity in vitro could not be established.

Antigenic Properties and Relationships

Alphacryptoviruses are good immunogens; some are serologically related within the genus. They do not show relationships with betacryptoviruses or mycoviruses (partiti- and chrysoviruses). The genomic dsRNA can react with an anti-dsRNA antibody.

Relevant Literature

Boccardo and Candresse, 2005a, b; Milne and Marzachi, 1999; Xie et al., 1993.

Genus **Betacryptovirus**

Type species: *White clover cryptic virus 2* (WCCV 2).

Virion Properties

The virions are isometric, 38 nm in diameter, and bigger than those of alphacryptoviruses, with prominent subunits.

The particles sediment as one component and have a buoyant density of 1.37g/cm^3 (CsCl).

Genome Organization and Replication

No molecular data are available on the betacryptovirus genome. It consists of two segments of dsRNA, 2.1 and 2.25 kbp in size, which are monocistronic. Virus accumulates in the cytoplasm and in vitro studies have shown that transcription and replication takes place by a semiconservative mechanism.

Relation with Cells and Tissues

The virion may be found in the cytoplasm of cells in all plant parts. Inclusion bodies are not seen.

Host Range and Symptoms

A betacryptovirus does not cause apparent symptoms.

Transmission and Stability in Sap

The virus can be transmitted by mechanical inoculation or by grafting. In nature they are transmitted by ovule and pollen. Cell-to-cell spread does not occur except at the cell division. No vector is known. Since virus assay species are lacking, their biophysical properties, such as thermal inactivation and longevity in vitro, could not be established.

Antigenic Properties and Relationships

Betacryptoviruses are good immunogens. Some of them are serologically related within the genus, as *Hop trefoil cryptic virus 2* (HTCV2) shows relationship with WCCV2 and *Red clover cryptic virus 2;* the latter two are related to each other. No relationships with alphacryptoviruses or mycoviruses (partiti- and chrysoviruses) have been established.

Relevant Literature

Brunt et al., 1996; Milne and Marzachi, 1999.

UNASSIGNED GENUS

Genus Endornavirus

The first part of the name has been derived from the Greek word *endon* (within).
Type species: *Vicia faba endornavirus.*

Virion Properties

Members of the genus do not produce virus particles.

Genome Organization, Expression, and Replication

The length of the dsRNA genome ranges from approximately 14,000 to 18,000 nucleotides (nt). Each genome encodes a polypeptide with amino acid sequences typical of helicase (hel) and RNA-dependent RNA polymerase (RdRp) (Figure A2.1).
Viral RNA replicates in cytoplasmic vesicles which are bounded by a membrane. The vesicles contain genomic dsRNA and RdRp, thus forming replication complexes comparable to those of positive-sense single-stranded RNA viruses.

Relations with Cells and Tissues

Endornavirus RNA is present in cells of all tissues.

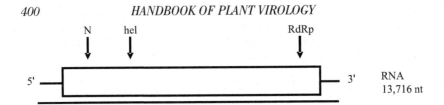

FIGURE A2.1. Organization and expression of the genome of an endornavirus *(Oryza sativa endornavirus)*. Rectangle represents the open reading frame with the positions of the break in the coding strand (N), helicase domain (hel), and RNA-dependent RNA polymerase domain (RdRp) indicated by arrows.

Host Range and Symptoms

Vicia faba, Phaseolus vulgaris, Oryza rufipogon (wild rice), and culti-vars of *Oryza sativa* have been found to be infected.

Most host plants are infected symptomlessly except *V. faba* in which *Vicia faba endornavirus* causes cytoplasmic male sterility.

Transmission and Stability in Sap

Endornaviruses cannot be transmitted mechanically, and they have no known vectors. Transmission is through seed and pollen.

Antigenic Properties and Relationships

Monoclonal antibodies raised against purified cytoplasmic vesicles from *V. faba* plants infected with *Vicia faba endornavirus* could detect the dis-ease (male sterility) in an early stage. RdRp and helicase sequences of endornaviruses point into the direction of a relationship with viruses of the "alpha-like" supergroup.

Relevant Literature

Horiuchi et al., 2001; Pfeiffer, 2002.

BIBLIOGRAPHY

Anzola JV, Dall DJ, Xu Z, Nuss DL (1989). Complete nucleotide sequence of wound tumor virus genomic segments encoding nonstructural polypeptides. *Vi-rology* 171: 222-228.

Boccardo G, Candresse T (2005a). Complete sequence of the RNA1 of an isolate of *White clover cryptic virus 1*, type species of the genus *Alphacryptovirus*. *Archives of Virology* 150: 399-402.

Boccardo G, Candresse T (2005b). Complete sequence of the RNA2 of an isolate of *White clover cryptic virus 1*, type species of the genus *Alphacryptovirus*. *Archives of Virology* 150: 403-405.

Brunt AA, Crabtree K, Dalwitz MJ, Gibbs AJ, Watson L (1996). *Viruses of plants: Description and lists from the VIDE database*. United Kingdom: CAB International.

Distéfano AJ, Conci LR, Hidalgo MM, Guzmán FA, Hop HE, del Vas M (2003). Sequence and phylogenetic analysis of genome segments S1, S2, S3 and S6 of *Mal de Rio Cuarto virus,* a newly accepted Fijivirus species. *Virus Research* 92: 113-121.

Distéfano AJ, Hopp HE, del Vas M (2005). Sequence analysis of genome segments S5 and S10 of Mal de Rio Cuarto virus *(Fijivirus, Reoviridae). Archives of Virology* 150(6): 1241-1248.

Horiuchi H, Udagawa T, Koga R, Moriyama H, Fukuhara T (2001). RNA-dependent RNA polymerase activity associated with endogenous double-stranded RNA in rice. *Plant Molecular Biology* 31: 713-719.

Hull R (2002). *Matthews' plant virology.* San Diego, San Francisco, New York, Boston, London, Sydney, Tokyo: Academic Press.

Marzachi C, Boccardo G, Milne R, Isogai M, Uyeda I (1995). Genome structure and variability of fijiviruses. *Seminars in Virology* 6: 103-108.

Milne G, Marzachi C (1999). Cryptoviruses *(Partitiviridae).* In Granoff A, Webster RG (eds.), *Encyclopedia of virology,* Volume 2, Second edition (pp. 312-315). San Diego: Academic Press.

Nakagawa A, Miyazaki N, Toka J, Naitow H, Ogawa A, Fujimoto Z, Mizuno H, Higashi T, Watanabe Y, Omura T, Cheng RH, Tsukihara T (2003). The atomic structure of *Rice dwarf virus* reveals the self-assembly mechanism of component proteins. *Structure* 11: 1227-1238.

Pfeiffer P (2002). Large dsRNA genetic elements in plants and the novel dsRNA associated with the "447" cytoplasmic male sterility in *Vicia faba.* In Tavantzis SM (ed.), *dsRNA genetic elements: Concepts and applications in agriculture, forestry, and medicine* (pp. 259-274). Boca Raton, FL: CRC Press.

Suzuki N, Sugawara M, Nuss DL, Matsura Y (1996). Polycistronic (tri- or bicistronic) phytoreoviral segments translatable in both plants and insect cells. *Journal of Virology* 70: 8155-8159.

Upadhyaya NM, Ramm K, Gellatly JA, Li Z, Kositratana W, Waterhouse PM (1998). Rice ragged stunt oryzavirus genome segment S4 could encode an RNA-dependent RNA polymerase and second protein of unknown function. *Archives of Virology* 143: 1815-1822.

Xie WS, Antoniw JF, White RF (1993). Nucleotide sequence of beet cryptic virus 3 ds RNA2 which encodes a putative RNA-dependent RNA polymerase. *Journal of General Virology* 74: 1467-1470.

Zhang HM, Chen JP, Adams MJ (2001). Molecular characterization of segments 1 to 6 of Rice black-streaked dwarf virus from China provides the complete genome. *Archives of Virology* 146: 2331-2339.

Appendix 3

Description of Negative-Sense, Single-Stranded RNA Viruses

Jeanne Dijkstra
Jawaid A. Khan

FAMILY RHABDOVIRIDAE

The name has been derived from the Greek word *rhabdos* (rod) and refers to the morphology of the virions. However, in essence, the virus particles are usually bacilliform (sometimes bullet-shaped) rather than rod-shaped.

The members of this family infect animals (both invertebrates and vertebrates) and plants. The family comprises four genera of animal-infecting viruses and two genera, namely *Cytorhabdovirus* and *Nucleorhabdovirus,* of plant-infecting viruses. Viruses of these two genera, however, also replicate in their vector insects.

The distinction between the two genera of plant-infecting viruses is based on the site of their replication and maturation, namely in the cytoplasm (genus *Cytorhabdovirus*) or in the nucleus *(Nucleorhabdovirus).*

Genus Cytorhabdovirus

Type species: *Lettuce necrotic yellows virus* (LNYV).

Virion Properties

The virions are bacilliform (seldom bullet-shaped), 225-380 nm long, and 60-95 nm in diameter. The nucleocapsid, visible in cross sections as an electron-dense core of 35-60 nm in diameter, is surrounded by a lipid, host-derived membrane (envelope) with surface projections (spikes, peplomers). In negatively stained preparations for the electron microscope the helically wound nucleocapsid gives the virions the characteristic cross-striations with a periodicity of 4-5 nm.

The nucleocapsid consists of a negative-sense, single-stranded (ss) RNA, the nucleocapsid (N) protein, and the large (L) and the phosphorylated (P) proteins, both with polymerase activity. Besides the N, L, and P proteins, the virions contain the matrix (M) protein and the glycosylated (G) protein. The M protein is associated with the inside of the envelope. It binds to the N protein and is supposed to regulate transcription of the genome by playing a role in the coiling of the nucleocapsid. The M protein also interacts with the glycosylated (G) protein, which forms the spikes.

The virion Mr is 500-$1,100 \times 10^6$. The particles sediment at 800-1050 S and have buoyant densities of 1.17-1.19 g/cm^3.

Genome Organization, Expression, and Replication

The RNA genome consisting of 11,000-14,000 nucleotides (nt), encodes six proteins, namely the five structural proteins N, P, M, G, and L, and additional nonstructural proteins, such as, for instance, sc4, probably a movement protein identified in a nucleorhabdovirus (Figure A3.1).

The genome is transcribed into a number of subgenomic (sg) messenger (m) RNAs that are sequential along the genome. Transcription starts from the 3' terminus in the following order:-leader-N-P-sc4-M-G-L-trailer.

The mRNAs thus synthesized are polyadenylated and acquire caps at the outset of their synthesis.

Infection Process

Infection starts with attachment of the spikes to the cell surface, where the virus particles penetrate into the cell by endocytosis. The viral membrane is

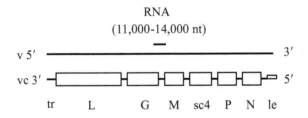

FIGURE A3.1. Organization and expression of the genome of a plant-infecting rhabdovirus. Rectangles represent the subgenomic messenger RNAs with their gene products. G, glycosylated protein; L, large protein with polymerase activity; le, leader RNA; M, matrix protein; N, nucleocapsid protein; P, phosphorylated protein with polymerase activity; sc4, a probably nonstructural protein translated from an additional gene present in nucleorhabdoviruses; v, viral strand with negative (–) polarity; vc, viral complementary strand.

removed and the nucleocapsids are released into the cytoplasm. After production and translation of the mRNAs the viral RNA will be replicated, first full-length positive strands followed by full-length negative strands in viroplasms (i.e., characteristic structures in the cytoplasm consisting of strands of granular material). The viroplasm is the site in cytoplasm where replication takes place. The newly formed nucleocapsids mature into virions by budding into vesicles of the endoplasmic reticulum, often at the edges of viroplasms. Most likely, virus particles are assembled into virions by simultaneous coiling of the nucleocapsids into corelike structures and envelopment.

Relations with Cells and Tissues

Cores and virions are found in cells of all leaf tissues. In cells of plants infected with LNYV, the outer nuclear membrane develops "blisters" containing vesicles.

Host Range and Symptoms

Cytorhabdoviruses have narrow host ranges. Many species are restricted to the family Gramineae.

The viruses do not produce characteristic symptoms (only some chlorosis and yellowing) on any of their sensitive host plants.

Transmission and Stability in Sap

Most of the virus species are not sap transmissible, LNYV being one of the exceptions. The viruses are not very stable in sap from infected plants. Their infectivity in sap is rapidly lost at a temperature of approximately 52°C.

Natural transmission of cytorhabdoviruses is by leafhoppers (Homoptera, Cicadellidae), planthoppers (Homoptera, Delphacidae), or aphids in a circulative-propagative manner. Both adults and larvae can transmit. Transovarial transmission has been reported for *Barley yellow striate mosaic virus*.

Antigenic Properties and Relationships

The virions and their structural proteins are moderately immunogenic. Some serological relationships have been established between two cytorhabdovirus species. These viruses also showed a weak serological reaction with a nucleorhabdovirus.

Relevant Literature

Jackson et al., 2005; Tanno et al., 2000; Wetzel et al, 1994.

Genus Nucleorhabdovirus

Type species: *Potato yellow dwarf virus* (PYDV).
For virion properties and genome organization, expression, and replication, see previous section on genus *Cytorhabdovirus*.

Infection Process

After penetration of the virions into the cell, the viral membrane is removed and the nucleocapsids are released into the cytoplasm. Thereafter, the nucleocapsids enter the nucleus, where transcription into mRNAs will take place. The mRNAs thus formed are then transported to the cytoplasm. Subsequently, the translation products N, P, and L proteins are entering the nucleus for replication of the genome. The site of replication is characterized by granular structures, viroplasms, located near the periphery of the nucleus. Newly formed nucleocapsids simultaneously coil and bud in the presence of M and G proteins at the inner nuclear membrane. The virions thus formed are released into the perinuclear space (i.e., the space between the inner and outer nuclear membrane) from which they spread into the cytoplasm.

Relations with Cells and Tissues

Virions and nucleocapsids are found both in the perinuclear space and in the cytoplasm. An infected nucleus shows depletion of chromatin and a uniform granular nucleoplasm.

Host Range and Symptoms

Most nucleorhabdoviruses have narrow host ranges and many are restricted to members of one plant family only; PYDV is one of the exceptions in having a wide host range.
Symptoms range from yellowing, vein yellowing, mosaic, and mottling to crinkling of the leaves, necrosis, and severe stunting.

Transmission and Stability in Sap

Many nucleorhabdoviruses are not readily transmitted mechanically. The viruses are not very stable in sap from infected plants, and infectivity is lost within a few hours at a temperature of 20°C.

Natural transmission is by leafhoppers or aphids in a circulative-propagative manner.

Antigenic Properties and Relationships

Nucleorhabdoviruses are poorly immunogenic. No serological relationships have been reported between members of the genus.

Relevant Literature

Scholthof et al., 1994; Van Beek et al., 1985.

FAMILY BUNYAVIRIDAE

The family has been named after the genus *Bunyavirus,* whose name had been derived from Bunyamwera, a place in Uganda where the first isolate of the type species, *Bunyamwera virus,* originated.

The family comprises five genera. Four of them contain only animal viruses, whereas the viruses in the fifth (genus *Tospovirus*) infect plants, although they also replicate in their vector insect (thrips).

Genus Tospovirus

The siglum has been derived from the name of the type species, *Tomato spotted wilt virus* (TSWV).

Virion Properties

The virions are spherical (80-110 nm in diameter) and possess a lipid membrane with spikes (peplomers) consisting of two types of glycoproteins, G1 and G2, on its surface.

The membrane envelops three pseudocircular nucleocapsids and each of them consists of an ssRNA genome segment, which is encapsidated by an N protein. The latter is closely associated with the viral polymerase

(L protein). During repeated mechanical transfers of TSWV nucleocapsids are often generated that remain nonenveloped.

The virion Mr is $300\text{-}400 \times 10^6$. The particles sediment at 350-500 *S* and have buoyant densities of 1.16-1.18 g/cm³ in sucrose and 1.20-1.21 g/cm³ in CsCl.

Genome Organization, Expression, and Replication

The ssRNA genome consists of three segments, large (L), medium (M) and small (S), of 8,897, 4,821, and 2,916 nt, respectively. The segments have complementary 3' and 5' termini thus forming stable panhandle structures.

The L-RNA is of negative polarity and codes for the viral RNA-dependent RNA polymerase (RdRp) of 320 kDa (Figure A3.2).

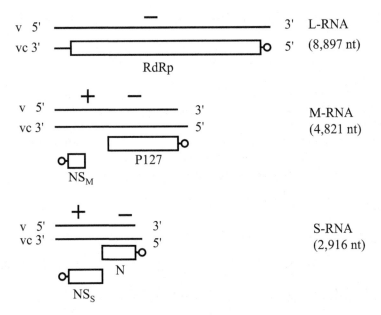

FIGURE A3.2. Organization and expression of the genome of a tospovirus *(Tomato spotted wilt virus)*. Rectangles represent the messenger RNAs with their gene products. N, nucleocapsid protein; NS_M and NS_S, nonstructural proteins translated from M-RNA and S-RNA, respectively; P127, precursor to the G1 and G2 proteins; RdRp, RNA-dependent RNA polymerase; v, viral strand with negative (−) or ambisense (+ −) polarity; vc, viral complementary strand; O, cap.

Both the M- and S-RNAs have an ambisense coding strategy. The viral (v) strand of the M-RNA encodes a nonstructural protein (NS_M) of 34 kDa involved in the cell-to-cell movement of the nucleocapsids, whereas the viral complementary (vc) strand codes for a protein of 127 kDa, from which the G1 and G2 proteins are processed (Figure A3.2).

The S-RNA encodes a nonstructural protein (NS_S) of 52 kDa in the v-RNA and the N protein of 29 kDa in the vc-RNA (Figure A3.2). The function of the NS_S is still unknown.

A deletion in the M-RNA strand is responsible for a defective form of the virus consisting of nonenveloped particles.

At low concentrations of N protein, the polymerase transcribes the genomic RNAs with the help of caps snatched from host mRNAs. The mRNAs thus formed are translated into various proteins. Thereafter, the viral genome is replicated.

Infection Process

This process takes place in the cytoplasm. After the virus has entered the cell, the virions shed their membrane and infectious nucleocapsids are released. After replication of the viral genome, the G1 and G2 proteins formed earlier are transported to the site where budding will take place. The newly formed viral RNAs assemble with the N protein, and the L protein binds to the nucleocapsids thus formed. Thereafter, the nucleocapsids associate with the NS_M protein and they are transported from cell to cell via plasmodesmata through NS_M-induced tubular structures.

Nucleocapsids may also form new virions by association with the G1 and G2 proteins, followed by budding into the endoplasmic reticulum or Golgi apparatus.

Relations with Cells and Tissues

The virions are clustered in large aggregates in the cisternae of the endoplasmic reticulum. In some cases, accumulations of NS_S protein can be observed in the cytoplasm as paracrystalline structures. Defective forms of the virus lacking the membrane usually appear as electron-dense amorphous structures in ultrathin sections of infected leaf cells.

Host Range and Symptoms

More than 1,000 plant species belonging to more than 90 families are known to be susceptible to TSWV. Most of the other tospoviruses, however, have a much narrower host range.

Tospovirus-infected, sensitive hosts show a wide variety of symptoms ranging from mild chlorosis and reddish-brown discoloration to leaf distortion, necrosis, and cessation of growth.

Transmission and Stability in Sap

Tospoviruses can be transmitted mechanically, but they are very unstable. At approximately 20°C their infectivity in crude sap from infected plants is lost within a couple of hours, and after exposure for 10 min at approximately 45°C the viruses are completely inactivated.

Natural transmission is by thrips species in the genera *Frankliniella* and *Thrips* in a circulative-propagative manner. Only larvae can acquire the viruses, but larvae and adults can transmit them. Only virions can be acquired by the vector, not the nucleocapsids. The viruses replicate in the midgut of larvae, but the rate of replication and the ability of the virus to move from the midgut to muscle cells and salivary glands depend on the vector.

Antigenic Properties and Relationships

Antisera can be raised against purified suspensions of virus, nucleocapsids or viral proteins. Three different serogroups of tospoviruses are distinguished by using the N protein as a parameter. Within serogroup II two serotypes are recognized based on different reactions with monoclonal antibodies. A third serogroup comprising only *Impatiens necrotic spot virus* does not show any cross-reactions with viruses in serogroups I and II.

No serological relationships have been found between TSWV and the animal-infecting bunyaviruses.

Relevant Literature

Nagata et al., 2002.

UNASSIGNED GENERA

Genus Ophiovirus

The name has been derived from the Greek word *ophis* (snake) and refers to the snaky structure of the viruses.

Type species: *Citrus psorosis virus* (CPsV).

Virion Properties

The morphology of ophioviruses resembles that of tenuiviruses. The nucleocapsids are pleiomorphic and appear in the electron microscope as filamentous particles, approximately 3 nm in diameter, forming circles of different contour lengths. The circles can collapse to form pseudolinear, double-stranded (ds) structures, approximately 10 nm in diameter. Virions of *Mirafiori lettuce big-vein virus* (MLBVV) contain nearly equimolar amounts of RNA molecules of both polarities.

The virions have a buoyant density of 1.22 g/cm^3 (in Cs$_2$SO$_4$).
The CP varies in size from 43 to 50 kDa.

Genome Organization, Expression, and Replication

The genome consists of three ssRNA segments of 7,500-9,000, 1,600-1,800, and 1,500 nt, but that of MLBVV consists of four RNA molecules of approximately 7,800, 1,700, 1,500, and 1,400 nt (Figure A3.3). Unlike

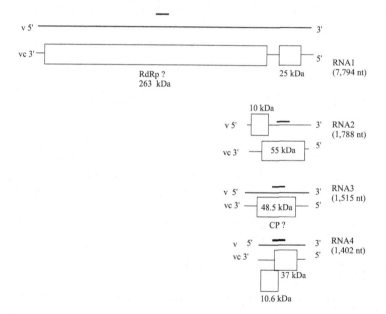

FIGURE A3.3. Organization and expression of the genome of an ophiovirus *(Mirafiori lettuce big-vein virus)*. Rectangles represent the open reading frames with their translation products. CP?, putative capsid protein; RdRp?, putative RNA-dependent RNA polymerase; v, viral strand; vc, complementary strand.

tospoviruses and tenuiviruses, multipartite genomes are largely negative-stranded. The genome of MLBVV contains seven ORFs. RNA segment 1 contains two ORFs on the vcRNA sequence. The putative ORF1 product is a polypeptide of 24,996 Da. The large ORF2 encodes a protein of 262,635 Da.

On the vcRNA segment 2 one ORF is present with a putative product of 54,586 Da. An additional minor ORF, found at the vcRNA, has a theoretical coding capacity of 9,960 Da.

The RNA segment 3 comprises one single, large ORF at the vc strand coding for a putative product of 48,544 Da.

The vc RNA4 contains two overlapping ORFs in different reading frames, encoding putative proteins of 37,261 and 10,618 Da.

The large product of RNA1 and the putative protein of RNA3 most likely represent the RdRp and CP, respectively.

Host Range and Symptoms

Ophioviruses have a restricted host range. In contrast with tenuiviruses they do not infect members of the family Gramineae. Natural hosts of the three ophiovirus species are citrus, *Ranunculus,* and tulip.

Symptoms range from local lesions to mottling in the leaves, but CPsV causes serious symptoms in the bark and wood of citrus trees, consisting of formation of scales on the outer bark and discoloration of tissue and gum-like deposits, in both the bark and the wood. Eventually the affected tree dies. The recently identified MLBVV causes the typical dilated veins in lettuce formerly attributed to a varicosavirus.

Transmission and Stability in Sap

The virus can be transmitted mechanically, but it is not very stable. It is inactivated in crude sap from infected plants heated at a temperature of 50°C for 10 min.

No natural vectors are known, but MLBVV has been found to be transmitted by the fungus *Olpidium brassicae.*

Antigenic Properties and Relationships

The N protein is weakly immunogenic and no definite serological relationships have been established.

Relevant Literature

Lot et al., 2002; Van der Wilk et al., 2002.

Genus Tenuivirus

The name has been derived from the Latin word *tenuis* (slender, fine) and refers to the morphology of the virions.

Type species: *Rice stripe virus* (RSV).

Tenuiviruses have some characteristics in common with members of the family *Bunyaviridae*, in particular with the animal-infecting viruses in the genus *Phlebovirus*.

Virion Properties

The virions are characterized by their long, threadlike, spiral-shaped structures of various lengths that are often arranged in circles. They consist of nucleocapsids, 3-10 nm in diameter. Their lengths depend on the size of the RNA segments they contain.

The viruses have a buoyant density of 1.282-1.288 g/cm^3 (in CsCl).

Genome Organization, Expression, and Replication

The genome organization and expression is presented in Figure A3.4. The ssRNA genome consists of 4-6 segments (RNAs 1-6). The largest segment (RNA1) has negative polarity, the others ambisense coding strategy (Figure A3.4).

The RNAs 1-4 of RSV contain 8,970, 3,514, 2,504, and 2,157 nt, respectively. The 3' and 5' terminal sequences of each RNA strand show complementarity. The ambisense RNAs are transcribed into mRNAs with the help of caps. The negative-sense RNA1 codes for the RdRp of 337 kDa. RNA2 encodes two proteins of 23 kDa and 94 kDa in an ambisense arrangement. Their functions are not known. Ambisense RNA3 encodes the N protein of 35 kDa in its vc strand and a protein of 24 kDa of unknown function in its v strand. Ambisense RNA4 codes for a nonstructural protein (NSP) of 20 kDa in its v strand and another nonstructural protein of 32 kDa in its vc strand.

For some tenuiviruses cap-snatching strategy in the production of subgenomic mRNAs has been demonstrated.

Relations with Cells and Tissues

Both granular and crystalline inclusions are found in cells of sensitive plants infected with RSV. The shape of the inclusions varies from rings to rods. NSP is abundantly present in the inclusions.

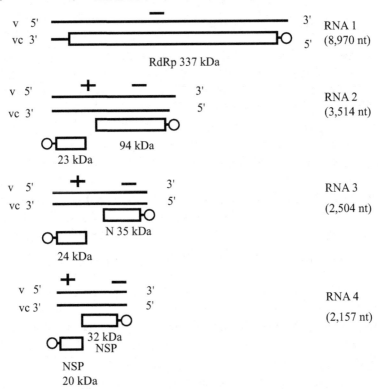

FIGURE A3.4. Organization and expression of the genome of a tenuivirus *(Rice stripe virus)*. Rectangles represent the messenger RNAs with their translation products. N, nucleocapsid protein; NSP, nonstructural protein; RdRp, RNA-dependent RNA polymerase; v, viral strand; vc, viral complementary strand; O, cap.

Ultrathin sections of infected plant cells often show granular regions in the cytoplasm. The presence of N protein has been established serologically in phloem tissue and mesophyll of RSV-infected wheat leaves.

N protein and other translation products except NSP have also been detected in viruliferous planthoppers infected with *Maize stripe virus* (MSpV).

Host Range and Symptoms

Tenuiviruses infect only members of the family Gramineae.

Chlorotic stripes and spots on the leaves are the main type of symptoms, but yellowing, stunting, and wilting may also occur.

Transmission and Stability in Sap

Most tenuiviruses are not mechanically transmissible. Transmission can be achieved artificially by injection of the vector with sap from diseased plants. Virus-containing sap is inactivated at a temperature of approximately 50°C.

In nature tenuiviruses are readily transmitted by delphacid planthoppers (Homoptera, Delphacidae) belonging to, among others, the genera *Laodelphax* and *Peregrinus* in a circulative-propagative manner.

The viruses can be acquired by the vector in a short time (about 15 sec), but an acquisition period of one day is required for optimal transmission. A latent period ranges from 4 to 30 days. Thereafter, the vector can transmit the virus, and it remains infective for the rest of its life. Most organs of the vector insect have been found infected. In general, nymphs are more efficient vectors than adults. Most tenuiviruses are transmitted transovarially.

Antigenic Properties and Relationships

Tenuiviruses are good immunogens. The species are serologically interrelated. A serological relationship exists between the N protein and NSP of some species.

Relevant Literature

Chomchan et al, 2002, 2003; Falk and Tsai, 1998.

Genus Varicosavirus

The name has been derived from the Latin word *varix* (dilated vein) and refers to the symptoms observed in lettuce plants affected by big-vein disease.

Type species: *Lettuce big-vein associated virus* (LBVaV).

Virion Properties

The virions are rod-shaped with modal lengths of 320-360 nm, diameter of about 18 nm, axial canal obvious, nonenveloped. Virus particles are highly unstable and present in low concentration in infected plants and form aggregates. The capsids are made up of single-protein species with a Mr of 48,000. Positive- and negative-sense RNAs are encapsidated separately.

The virus particles have a buoyant density of 1.27 g/cm^3 (Cs$_2$SO$_4$).

Genome Organization and Replication

The genome consists of two dsRNA species that are monocistronic. The smaller segment codes for the CP and the larger one for a putative RdRp. Information available on genome organization of LBVaV shows that it contains a two-segmented, negative-strand, probably ssRNA genome. Purified RNA preparations reveal two RNA species, namely RNA1 (7.3 kb) and RNA2 (6.6 kb). RNA1 contains one large ORF encoding an L protein (2,040 amino acids, predicted Mr 232,092) with sequence homology to those of L polymerases of non-segmented negative-strand RNA viruses within plant rhabdoviruses. Though LBVaV is segmented and nonenveloped, most of the conserved motifs found in the L polymerases of plant rhabdovirus are present. Further, LBVaV contains transcription termination/polyadenylation signals—like poly (U) tracts in the L protein gene resembling those of rhabdoviruses and paramyxovirus.

The gene coding for the CP is located on negative-sense, ssRNA2. An amino acid sequence analysis shows a high degree of homology to the nucleoprotein of rhabdoviruses.

Virus accumulates in the cytoplasm and in vitro studies have shown that transcription and replication take place by a semiconservative mechanism.

Relations with Cells and Tissues

The virions may be found in leaves, roots, mesophyll, and phloem parenchyma; inclusions in the form of crystals are also seen in cytoplasm.

Host Range and Symptoms

Few natural hosts of LBVaV are reported; experimental host range includes several cultivars of lettuce. LBVaV, formerly called lettuce big-vein virus, was believed to be the causal agent of the big-vein disease of lettuce. Recent data, however, has shown that an ophiovirus, isolated from big-vein–diseased lettuce plants, induces the characteristic symptoms. This ophiovirus has been named *Mirafiori lettuce big-vein virus.* In several lettuce cultivars, LBVaV can cause necrotic rings and spots.

Transmission and Infectivity in Sap

The virus species can be transmitted by mechanical inoculation, grafting, or the soil-inhabiting fungus *Olpidium brassicae* and survives in resting spores, which can persist in soil for a period of 20 years or so and retain the ability to transmit disease for over 15 years. Thermal inactivation of virus particles is approximately 50°C; longevity in vitro is for one day.

Antigenic Properties and Relationships

A serological relationship has been established between LBVaV and Tobacco stunt virus, a tentative species in the genus *Varicosavirus*.

Relevant Literature

Sasaya et al., 2001a,b.

BIBLIOGRAPHY

Chomchan P, Li SF, Shirako Y (2003). *Rice grassy stunt tenuivirus* nonstructural protein p5 interacts with itself to form oligomeric complexes in vitro and in vivo. *Journal of Virology* 77: 769-775.

Chomchan P, Miranda GJ, Shirako Y (2002). Detection of rice grassy stunt tenuivirus nonstructural proteins p2, p5 and p6 from infected rice plants and from viruliferous brown planthoppers. *Archives of Virology* 147: 2291-2300.

Falk BW, Tsai JH (1998). Biology and molecular biology of viruses in the genus *Tenuivirus*. *Annual Review of Phytopathology* 36: 139-163.

Jackson AO, Kietzgen RG, Goodin MM, Bragg JN, Deng M (2005). Biology of plant rhabdoviruses. *Annual Review of Phytopathology* 43: 623-660.

Lot H, Campbell RN, Souche S, Milne RG, Roggero P (2002). Transmission by *Olpidium brassicae* of *Mirafiori lettuce virus* and *Lettuce big-vein virus* and their roles in lettuce big-vein etiology. *Phytopathology* 92: 288-293.

Nagata T, Inoue-Nagata AK, Van Lent J, Goldbach R, Peters D (2002). Factors determining vector competence and specificity for transmission of TSWV. *Journal of General Virology* 83: 663-671.

Sasaya T, Ishikawa K, Koganezawa H (2001a). Nucleotide sequence of the coat protein gene of *Lettuce big-vein virus*. *Journal of General Virology* 82: 1509-1515.

Sasaya T, Ishikawa K, Koganezawa H (2001b). The nucleotide sequence of RNA1 of *Lettuce big-vein virus,* Genus *Varicosavirus,* reveals its reaction to nonsegmented negative-strand RNA viruses. *Virology* 297: 289-297.

Scholthof K-BG, Hillman BI, Modrell B, Heaton LA, Jackson AO (1994). Characterization and detection of sc4—A 6th gene encoded by Sonchus yellow net virus. *Virology* 204: 279-288.

Tanno F, Nakatsu A, Toriyama S, Kojima M (2000). Complete nucleotide sequence of Northern cereal mosaic virus and its genome organization. *Archives of Virology* 145: 1373-1384.

Van Beek NAM, Lohuis D, Dijkstra J, Peters D (1985). Morphogenesis of sonchus yellow net virus in cowpea protoplasts. *Journal of Ultrastructure Research* 90: 294-303.

Van der Wilk F, Dullemans AM, Verbeek M, van den Heuvel JFJM (2002). Nucleotide sequence and genome organization of an ophiovirus associated with lettuce big-vein disease. *Journal of General Virology* 83: 2869-2877.

Wetzel T, Dietzgen RG, Dale JL (1994). Genomic organization of *Lettuce necrotic yellows rhabdovirus*. *Virology* 200: 401-412.

Appendix 4

Description of Single-Stranded DNA Viruses

Jeanne Dijkstra
Jawaid A. Khan

FAMILY GEMINIVIRIDAE

The family has been named after the characteristic twinned (geminate, derived from the Latin word *gemini,* meaning twins) virus particles. The virions measure 18-20 by 30-35 nm and consist of two quasi-icosahedral capsids with T = 1 symmetry. Each geminate particle contains one single-stranded (ss) DNA molecule that replicates in the nuclei of infected cells. The genomes may be monopartite or bipartite.

The family comprises four genera, namely *Mastrevirus, Curtovirus, Topocuvirus,* and *Begomovirus,* which differ from one another in their genome organization and their vectors.

Genus Mastrevirus

The siglum has been derived from the name of the type species, *Maize streak virus* (MSV).

Virion Properties

The geminate particles of approximately 20 × 30 nm have a sedimentation coefficient of about 75 S and a buoyant density of 1.35 g/cm^3 (in CsCl).

Genome Organization, Expression, and Replication

Mastreviruses possess a monopartite genome of approximately 2600-2800 nucleotides (nt). The genome contains four open reading frames (ORFs), two each on the viral (v)-sense and viral complementary (vc)-sense strand. The former two ORFs (*CP* and *V2*) encode the capsid protein (CP) and movement protein (MP). The ORFs on the vc-sense strand (*Rep* and *Rep A*) code for replication initiation proteins (Rep) involved in virus replication and transcription. For more details, see Chapter 7 in this book.

Relations with Cells and Tissues

Virus particles often occur in aggregates in the nucleus, but also in the cytoplasm and vacuole. In the vacuole, the geminate particles sometimes form geometrical crystalline arrays.

Host Range and Symptoms

Mastreviruses have narrow host ranges. Out of the 12 species in the genus, only two infect dicotyledonous plants, whereas the others are restricted to the family Gramineae. Symptoms consist of streaking, mosaic, and stunting in gramineous hosts and severe dwarfing and necrosis in dicotyledonous hosts.

Transmission and Stability in Sap

Mastreviruses are not mechanically transmissible in the conventional way, but MSV has been transmitted by vascular puncture of maize seeds.

The viruses are inactivated in sap heated at a temperature of about 50-60°C.

Natural transmission is by leafhoppers (Homoptera, Cicadellidae) such as *Cicadulina* spp. and *Orosius* spp., in a circulative-nonpropagative manner.

Antigenic Properties and Relationships

The viruses are moderately immunogenic. *Tobacco yellow dwarf virus* (TYDV) is distantly serologically related to *Beet curly top virus*, a curtovirus. Monocot-infecting and dicot-infecting mastreviruses are not serologically related. However, grass-infecting viruses can be grouped according to the continent from which they originate.

Relevant Literature

Bosque-Perez, 2000; Dinant et al, 2004.

Genus **Curtovirus**

The siglum has been derived from the name of the type species, *Beet curly top virus* (BCTV). The genus most likely arose from a recombination event between a mastrevirus and a begomovirus.

Virion Properties

The geminate particles of about 18-22 × 30 nm have a sedimentation co-efficient of approximately 76 *S*. The virus concentration is low in infected plants.

Genome Organization, Expression, and Replication

Curtoviruses possess a monopartite genome of approximately 2,900-3,000 nt. The genome contains six to seven ORFs: three on the v-sense strand and three or four on the viral complementary (c)-sense strand. The three ORFs on the v-sense strand *(V1, V2, V3)* encode the CP *(V1)*, and two proteins for virus movement (MP) *(V2, V3)*. The four ORFs on the c-sense strand *(C1, C2, C3, C4)* are involved in virus replication and transcription, and they encode Rep *(C1)*, TrAP (transactivator protein) *(C2)*, REn (replication enhancer protein *(C3)* (not present in *Horseradish curly top virus*), and a host activation protein *(C4)*. For more details, see Chapter 7 in this book.

Relations with Cells and Tissues

Virus particles have been demonstrated electron microscopically in the nuclei of phloem parenchyma cells, but not in the cytoplasm of these cells, and also not in sieve elements or companion cells.

Host Range and Symptoms

The host range of BCTV is very wide, with more than 300 species in 44 dicot families, especially the Chenopodiaceae, Compositae, Cruciferae, Leguminosae, and Solanaceae. Curtoviruses induce a wide variety of symptoms, consisting of vein clearing, yellowing, vein swelling, malformation, curling of leaves, phloem necrosis, and growth of axillary buds.

Transmission and Stability in Sap

Curtoviruses are not mechanically transmissible in the conventional way, but they have been transmitted by pin-pricking and injection.

BCTV is inactivated in sap heated at a temperature of approximately 80°C for 10 min, but its infectivity is retained for eight days in sap kept at 20°C.

Natural transmission is by leafhoppers (Homoptera, Cicadellidae), such as *Circulifer* spp., in a circulative-nonpropagative manner.

Antigenic Properties and Relationships

The viruses are strongly immunogenic.

BCTV and *Tomato pseudo-curly top virus,* a topocuvirus, are relatively closely related serologically. Distant relationships exist between curtoviruses, mastreviruses, and begomoviruses.

Relevant Literature

Park et al., 2002; Sunter et al., 2001.

Genus **Topocuvirus**

The siglum has been derived from the name of the type species, *Tomato pseudo-curly top virus* (TPCTV), the only species in the genus.

This genus has been split off from the genus *Curtovirus.* Its genome resembles that of mastreviruses and begomoviruses. The organization of the c-sense genes is similar to that of the monopartite begomoviruses.

The host plant range of TPCTV is rather narrow.

Natural transmission is by tree hoppers (Homoptera, Membracidae) in a circulative-nonpropagative manner.

Relevant Literature

Briddon and Markham, 2001.

Genus **Begomovirus**

The siglum has been derived from the name of the type species, *Bean golden mosaic virus-Puerto Rico* (BGMV-PR).

Virion Properties

The geminate particles of approximately 18×30 nm have an Mr of about 2.6×10^6 and both are required for infection.

Genome Organization, Expression, and Replication

All "New World" begomoviruses possess a bipartite genome, but the genomes of some of those from the other continents contain only one ssDNA component.

A satellite circular DNA has been found associated with a monopartite begomovirus (Tomato leaf curl virus satellite DNA) and encapsidated by the CP of this virus *(Tomato leaf curl virus)*. Recently satellite component designated as DNA β has been found associated with leaf curl disease of cotton and other begomoviruses. DNA β components are a group of symptom-modulating, ss satellite molecules. This DNA species depends upon DNA-A for its replication and is encapsidated in the CP of the associated begomovirus.

The bipartite genome consists of two DNA circles, designated as components A and B. DNA-A encodes four proteins, namely the CP on the v-sense strand and the Rep, TrAP, and REn on the vc-sense strand. DNA B component encodes virus MP and nuclear shuttle protein (NSP). For more details, see Chapter 7 in this book.

Relations with Cells and Tissues

Virus particles are present in the nuclei of phloem and adjacent parenchyma cells, either packed in paracrystalline array or in loose aggregates. Virus particles have also been observed in vacuoles of the sieve elements. Nucleoli of infected nuclei form electron-dense rings.

Host Range and Symptoms

In general, begomoviruses have a narrow host range and they are restricted to dicotyledonous species, mostly in the families Leguminosae, Malvaceae, and Solanaceae. They cause bright-yellow mosaic, chlorotic and necrotic spots, vein chlorosis, and epinasty (downcurling of the leaves).

Transmission and Stability in Sap

Some species are mechanically transmissible, for example, BGMV and *Mungbean yellow mosaic virus*.

The viruses are inactivated in sap heated at temperatures ranging from 40 to 50°C or from 50 to 55°C. BGMV retains its infectivity for approximately three days in sap stored at 23°C, but other begomoviruses lose their infectivity between one and two days at 20°C.

Natural transmission is by the whitefly, *Bemisia tabaci* or *B. argentifolii* (the silver leaf biotype) (Aleyrodidae) in a circulative-nonpropagative manner.

Antigenic Properties and Relationships

Begomoviruses are moderately immunogenic. All species are relatively closely related serologically.

Relevant Literature

Brown, 2002; Brown et al., 2001; Monsalve-Fonnegra et al., 2002.

FAMILY NANOVIRIDAE

The name has been derived from the generic name *Nanovirus*.

Virus particles are 17 to 22 nm in diameter and have icosahedral symmetry. Each particle contains one circular ssDNA molecule. The genomes are multipartite, consisting of 6 to 11 DNA molecules, each of approximately 1,000 nt. However, it is not known how many of these DNA molecules are required for a fully functional genome. The viral CP has an Mr of approximately 20×10^3.

Each individual DNA molecule encodes proteins involved in replication (Rep), or in other functions (non-Rep), such as the CP, MP, cell-cycle link protein (Clink), and NSP.

The replication of the viruses is very similar to that of the geminiviruses.

The family comprises two genera, namely *Nanovirus* and *Babuvirus*.

Genus Nanovirus

The name has been derived from the Greek word *nanos*(dwarf) and pertains to both the small size of the virions and the symptoms (dwarfing, stunting) the viruses induce in their hosts.

The genus comprises three species, namely the type species *Subterranean clover stunt virus* (SCSV), *Faba bean necrotic yellows virus* (FBNYV), and *Milk vetch dwarf virus* (MDV).

Virion Properties

The icosahedral particles have a buoyant density of 1.24-1.30 g/cm^3 (in Cs_2SO_4).

Genome Organization, Expression, and Replication

FBNYV, MDV, and SCSV possess 11, 10, and 7 different DNA components, respectively. For more details, see also description of the family and Chapter 7 in this book.

Relations with Cells and Tissues

Phloem cells show abnormally shaped nuclei.

Host Range and Symptoms

Each of the three species has a narrow host range, restricted to a number of leguminous species. Symptoms consist of stunting, leafroll, severe chlorosis, interveinal necrosis, thick and brittle leaves, and premature death.

Transmission and Stability in Sap

Nanoviruses are not mechanically transmissible. Natural transmission is by aphids (e.g., *Aphis craccivora*) in a circulative-nonpropagative manner. A helper factor is required for its transmission by aphids.

Antigenic Properties and Relationships

The viruses are moderately immunogenic. The three species are serologically related. Analysis of non-Rep proteins of the three species has shown that they are closely related.

Genus **Babuvirus**

The siglum has been derived from the name of the type species, *Banana bunchy top virus* (BBTV), the only species in the genus.

Virion Properties

The icosahedral particles have a sedimentation coefficient of 46 *S*.

Genome Organization, Expression, and Replication

BBTV possesses six or nine (depending on the isolate) different DNA components. The genome organization and expression is presented in Figure A4.1. Each DNA component of the BBTV genome contains a conserved stem-loop structure, major common region (CR-M), and stem-loop common region (CR-SL). A potential TATA box at the 3' end of the stem-loop and at least one major gene in the virion sense associated with the polyadenylation signals are also present. The stem-loop structure contains highly conserved nonanucleotides similar to those of all geminiviruses. The stem-loop structure is incorporated in the CR-SL and shows about 60 percent identity in all the BBTV DNA components. The CR-M displays similarities to the promoter of *Wheat dwarf virus,* a mastrevirus.

DNA-1 contains one large ORF. It contains the minimal replicative unit of BBTV and encodes the "master" viral replicase. No ORF has been identified in DNA-2. DNA-3 encodes a 20.11 kDa CP. DNA-4 encodes a 13.74 kDa polyprotein with similarity to the MP of *Maize streak virus,* a mastrevirus. DNA-5 encodes an 18.97 kDa protein possessing retinoblastoma protein (Rb)-binding activity and facilitates in DNA replication. DNA-6 codes for a 17.4 kDa protein, an NSP that is involved in virus movement. DNA-4 and -6 are very similar to those of the BC1 (MP) and BV1 (NSP) of begomoviruses.

It appears that babuvirus follows the same replication strategy as geminiviruses. They encode Rep proteins and possess nicking and ligating properties, and their genome contains stem-loop-like structure with conserved nonanucleotide sequence.

Relations with Cells and Tissues

Phloem cells show abnormally shaped nuclei.

Host Range and Symptoms

BBTV has a very narrow host range, restricted to the genus *Musa.*

In the early stages of infection, symptoms consist of irregular, dark green streaks on the leaves, later followed by congestion of the leaves at the apex of the pseudostem, resulting in a "rosetting" (bunchy top). Symptomatic plants rarely bear fruits.

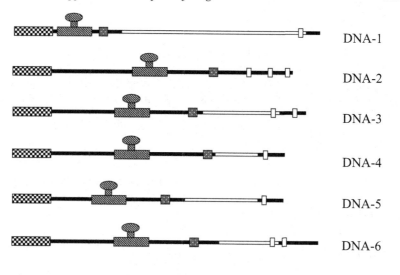

A.

B.

☐ Polyadenylation signal   ▨ CR-M   ▨ CR-SL

🍄 Stem-loop   ▨ TATA box   ☐ ORF

FIGURE A4.1. Organization and expression of the genome of a babuvirus *(Banana bunchy top virus)*. A, linear representation of each component; B, general organization of the genome.

Transmission and Stability in Sap

BBTV is not mechanically transmissible.

Natural transmission is by the aphid *Pentalonia nigronervosa* in a circulative-nonpropagative manner. Plant propagating material acts as a source of the long-distance virus dissemination.

Antigenic Properties and Relationships

BBTV is not serologically related to nanoviruses. Analysis of the Rep and non-Rep proteins has shown is only a distant relationship between these proteins and those of the nanoviruses.

Relevant Literature

Franz et al., 1999; Katul et al,. 1998; Wanitchakorn et al., 2000.

BIBLIOGRAPHY

Bosque-Perez NA (2000). Eight decades of *Maize streak virus* research. *Virus Research* 71: 107-121.

Briddon RW, Markham PG (2001). Complementation of bipartite begomovirus movement functions by topocuviruses and curtoviruses. *Archives of Virology* 146: 1811-1819.

Brown JK (2002). The molecular epidemiology of begomoviruses. In Khan JA, Dijkstra J (eds.), *Plant viruses as molecular pathogens* (pp. 279-315). Binghamton, NY: The Haworth Press, Inc.

Brown JK, Torres-Jerez I, Idris AM, Bank GK, Wyatt SD (2001). The core region of the coat protein gene is highly useful for establishing the provisional identification and classification of begomoviruses. *Archives of Virology* 146: 1581-1598.

Dinant S, Ripoll C, Pieper M, David C (2004). Phloem specific expression driven by wheat dwarf geminivirus V-sense promoter in transgenic dicotyledonous species. *Physiologia Plantarum* 121: 108-116.

Franz AWE, Van der Wilk F, Verbeek M, Dullemans AM, Van den Heuvel JFJM (1999). Faba bean necrotic yellows virus (genus *Nanovirus*) requires a helper factor for its aphid transmission. *Virology* 262: 210-219.

Katul L, Timchenko T, Gronenborn B, Vetten HJ (1998). Ten distinct circular ssDNA components, four of which encode putative replication associated proteins, are associated with the faba bean necrotic yellows virus genome. *Journal of General Virology* 79: 3101-3109.

Monsalve-Fonnegra ZI, Argüello-Astorga GR, Rivera-Bustamante RF (2002). Geminivirus replication and gene expression. In Khan JA, Dijkstra J (eds.), *Plant viruses as molecular pathogens* (pp. 257-277). Binghamton, NY: The Haworth Press, Inc.

Park SH, Hur J, Park J, Lee S, Lee TK, Chang M, Davi KR, Kim J, Lee S (2002). Identification of a tolerant locus on *Arabidopsis thaliana* to hypervirulent *Beet curly top virus* CFH strain. *Molecular Cells* 13: 252-258.

Sunter G, Sunter JL, Bisaro DM (2001). Plants expressing Tomato Golden Mosaic Virus AL2 or Beet Curly Top Virus L2 transgenes show enhanced susceptibility to infection by DNA and RNA viruses. *Virology* 285: 59-70.

Wanitchakorn R, Hafner GJ, Harding RM, Dale JL (2000). Functional analysis of proteins encoded by banana bunchy top virus DNA-4 to -6. *Journal of General Virology* 81: 299-306.

Appendix 5

Description of Reverse-Transcribing Viruses

Jeanne Dijkstra
Jawaid A. Khan

FAMILY CAULIMOVIRIDAE

The name has been derived from the generic name *Caulimovirus*. The family comprises six genera, namely *Caulimovirus, Soymovirus, Cavemovirus, Petuvirus, Badnavirus,* and *Tungrovirus.*

Virus particles are either isometric (in members of the first four genera) or bacilliform (in members of the last two genera). They possess a single molecule of circular double-stranded (ds) DNA of 7,200 to 8,100 nucleotides (nt). Depending on the genus, the genomes contain two to seven ORFs. The members of all genera have genomes that encode the capsid protein (CP), an aspartate protease (PR), a reverse transcriptase (RT), and a ribonuclease H. The members of the family have sedimentation coefficients ranging from 200 S to 220 S. The six genera are mainly distinguished on the basis of their genome organization. Replication is by reverse transcription. For more details about genome organization, expression, and replication, see Chapter 7 in this book.

Genus Caulimovirus

The siglum has been derived from the name of the type species, *Cauliflower mosaic virus* (CaMV).

Virion Properties

The stable virions are isometric, about 50 nm in diameter. The CP is composed of 420 subunits (T = 7 symmetry). The virions have a buoyant density of 1.37 g/cm^3 (in CsCl) and an Mr of approximately 22 × 10^6.

Genome Organization, Expression, and Replication

The genome of caulimoviruses contains six open reading frames (ORFs). See also description of the family.

Relations with Cells and Tissues

Virions produce conspicuous cytoplasmic electron-dense inclusion bodies consisting of an ORF 6-encoded protein in which virus particles are embedded. These inclusions are especially prominent in epidermal tissue stripped from leaves of infected plants. Translucent inclusions consisting of ORF 2-encoded protein can also be seen in the cytoplasm. Sometimes virus particles are observed in the nucleus. The protein encoded by ORF1 may be involved in cell-to-cell transport, whereas the product of ORF2 is present in viroplasm.

Host Range and Symptoms

The host range of caulimoviruses is restricted to dicotyledons. The viruses are inducing mostly mosaic and mottling, but some cause spots and rings *(Blueberry red ringspot virus)* or necrotic flecks and line pattern *(Carnation etched ring virus)*.

Transmission and Stability in Sap

The viruses are readily transmissible mechanically. The virions are inactivated in sap heated at temperatures ranging from 75 to 85°C, and at 20°C their infectivity is retained from five to seven days.

Natural transmission is by aphids in a nonpersistent manner. The product encoded by ORF2 acts as a helper protein in vector transmission (aphid-transmission factor; ATF).

Antigenic Properties and Relationships

The viruses are moderately immunogenic. Cross-reactions occur among many species in the genus.

Relevant Literature

Pooggin et al., 2002.

Genus Soymovirus

The siglum has been derived from the name of the type species, *Soybean chlorotic mottle virus* (SbCMV), one of the two species in this genus.

Virion Properties

The virions have a buoyant density of 1.37 g/cm^3 (in CsCl) and an Mr of about 5.45×10^6. For other properties, see description of the family.

Genome Organization, Expression, and Replication

The genome of soymoviruses contains seven ORFs. See also description of the family.

Relations with Cells and Tissues

Inclusions are similar to those induced by caulimoviruses.

Host Range and Symptoms

SbCMV has a very narrow host range, restricted to a few leguminous species. It induces chlorotic mottling and mosaic on the leaves and some stunting.

Transmission and Stability in Sap

The viruses are mechanically transmissible. No vector is known.

Antigenic Properties and Relationships

SbCMV is moderately immunogenic. It is not serologically related to caulimoviruses.

Relevant Literature

Glasheen et al., 2002; Stavolone et al., 2003.

Genus **Cavemovirus**

The siglum has been derived from the name of the type species, *Cassava vein mosaic virus* (CsVMV), the only species in the genus.

Virion Properties

The virus has a buoyant density of 1.37 g/cm^3. For other properties, see description of the family.

Genome Organization, Expression, and Replication

The genome of CsVMV is 8,158 base pair (bp) long. It contains five ORFs encoding for 164, 9, 77, 46, and 6 kDa proteins. The putative 164 kDa protein includes the capsid protein (CP) and the movement proteins (MP). It shows similarities to the zinc fingerlike-RNA binding domains (CP) and the intercellular transport domains (MP) of pararetroviruses. The 77 kDa protein contains motifs such as PR, RT, and RNAseH of pararetroviruses. The 46 kDa protein is a putative transactivator factor, while 6 kDa is a putative transcription regulator.

Relations with Cells and Tissues

Inclusions are present in the cytoplasm of infected cells.

Host Range and Symptoms

Symptoms include a chlorosis that follows the veins and coalesces to form a mosaic pattern. Leaf distortion and epinasty of young leaves often occurs.

Transmission and Stability in Sap

Virus is transmitted by mechanical inoculation and grafting.

Relevant Literature

Calvert et al., 1995; de Kochko et al., 1998.

Genus **Petuvirus**

The siglum has been derived from the first part of the name of the type species, *Petunia vein clearing virus* (PVCV), the only species in the genus.

Virion Properties

The virus has a buoyant density of 1.37 g/cm^3. For other properties, see description of the family.

Genome Organization, Expression, and Replication

The genome of PVCV is 7,205 bp long. It contains two large ORFs in the transcribed strand. In the intergenic region a transfer (t) RNA-primer binding site is present. The ORFI encodes a large protein of 126 kDa. Its N-terminal part shows similarity to the MP of caulimoviruses, whereas two distinctive sequence elements of the integrase function of retroviruses and retrotransposons are present at the C termini. The ORFII encodes a 125 kDa protein that contains domains for an RNA-binding element (common to the *gag* gene of retroelements) followed by consensus sequence for an acid protease, RT, and RNAseH.

Relations with Cells and Tissues

Infected cells show proliferation of endoplasmic reticulum.

Host Range and Symptoms

Symptoms include vein clearing and leaf malformation.

Transmission and Stability in Sap

The virus is transmitted by mechanical inoculation and grafting.

Relevant Literature

Richert-Poggeler and Shepherd, 1997; Richert-Poggeler et al., 2003.

Genus **Badnavirus**

The siglum has been derived from "bacilliform DNA viruses."
Type species: *Commelina yellow mottle virus* (ComYMV).

Virion Properties

Virions possess bacilliform particles with rounded ends. The particles have a modal length of 130 nm, but their lengths range from 60 to 900 nm. The virions have a buoyant density of 1.31 g/cm³ (in Cs_2SO_4) and a sedimentation coefficient of 218 S.

Genome Organization, Expression, and Replication

The badnavirus genome contains three ORFs. See also description of the family.

Relations with Cells and Tissues

No inclusion bodies have been observed, but virus particles are found in the cytoplasm in groups or singly.

Host Range and Symptoms

The viruses have narrow host ranges. They infect both dicotyledonous and monocotyledonous species. Badnaviruses may induce a variety of symptoms, ranging from mottle, streak, and mosaic to swellings in stems or roots, necrosis, and defoliation (in cacao affected by *Cacao swollen shoot virus* (CSSV).

Transmission and Stability in Sap

CSSV is mechanically transmissible. It is inactivated in sap heated at temperatures ranging from 55 to 60°C for 10 min.
Natural transmission of badnaviruses is by mealybugs (*Planococcus* spp.) (CSSV), aphids, or leafhoppers in a nonpersistent manner.

Antigenic Properties and Relationships

The viruses are moderately immunogenic.

Relevant Literature

Harper et al., 2004; Matsuda et al., 2002.

Genus **Tungrovirus**

The name has been derived from that of the type species, *Rice tungro bacilliform virus* (RTBV), the only species in the genus.

Virion Properties

Virions possess bacilliform particles with dimensions comparable to those of the badnaviruses. The virions have a buoyant density of 1.31 g/cm^3 (in CsSO$_4$) and the sedimentation coefficient is 175 *S*.

Genome Organization, Expression, and Replication

The genome of RTBV contains four ORFs. See also description of the family.

Relations with Cells and Tissues

No inclusion bodies have been observed, but virus particles are found in the cytoplasm.

Host Range and Symptoms

The host range of the virus complex RTBV and *Rice tungro spherical virus* (RTSV), a waikavirus, is very narrow and restricted to members of the Gramineae.

Symptoms (caused by RTBV) consist of yellow-orange discoloration of the lower leaves of rice plants, followed by stunting.

Transmission and Stability in Sap

The virus is not mechanically transmissible. RTBV/RTSV complex is inactivated in sap heated at 63°C for 10 min. Infectivity is retained in sap stored for 24 h at room temperature.

Natural transmission is by leafhoppers *(Nephotettix virescens)* in a nonpersistent manner. Vector transmission of RTBV is only possible in presence of RTSV.

Antigenic Properties and Relationships

RTBV is moderately immunogenic.

Relevant Literature

Frischmuth, 2002; Pooggin et al., 2002.

FAMILY PSEUDOVIRIDAE

The family has been named after its genus *Pseudovirus*. The name refers to the nature of its members, which replicate in a (retro)viruslike way but do not show infectivity ("false" viruses). They are often also referred to as LTR (long terminal repeats)-retrotransposons (mobile genetic elements) of the *Ty1-copia* family, which become integrated into the genomes of many eukaryotic organisms. The retrotransposons produce isometric, viruslike particles containing positive-sense, ssRNA. They have now acquired the status of viruses, because of the essential role they play in the replication cycle of the retrotransposons. The family comprises three genera, namely *Pseudovirus, Hemivirus,* and *Sirevirus.* As the members of genus *Hemivirus* have been found only in some algae, yeast, and the insect *Drosophila melanogaster,* no description of this genus will be included here.

Genus Pseudovirus

Type species: *Saccharomyces cerevisiae Ty1 virus* (SceTy1V).

Virion Properties

The virions have ovoid to spheroid particles with a mean radius of 20-30 nm. The particles have an Mr of 14×10^6 and a sedimentation coefficient of 200-300 *S* (wild-type).

Genome Organization, Expression, and Replication

The major virion RNA species is capped and polyadenylated and consists of an LTR-to-LTR transcript of 5,900 nt. The type species also packages host-derived primer tRNA.

The RNA transcript possesses two overlapping ORFs, *gag* and *pol. Pol* is expressed as a +1 frameshift (FS) (Figure A5.1).

Besides ssRNA, preparations of virus particles also contain various DNA forms, both intermediates and the integrated dsDNA "provirus." The proteins encoded by the *gag* gene are the CP and another putative small protein. The *pol* gene codes for PR, integrase (IN), and reverse transcriptase/RNase H (RT/RH) (Figure A5.1). These primary translation products are processed by PR.

The first step in the replication process is transcription of SceTy1V to generate the previously mentioned full-length RNA. In the cytoplasm, RNA is encapsidated in unprocessed *gag* and *gag-pol* proteins, where PR

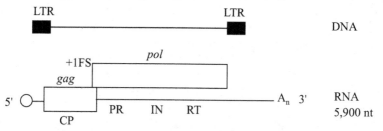

FIGURE A5.1. Organization and expression of the genome of a pseudovirus *(Saccharomyces cerevisae Ty1virus)*. Open rectangles represent the open reading frames, *gag* and *pol,* with their translation products (CP, PR, IN, RT). CP, capsid protein; +1 FS, +1 ribosomal frameshift; IN, integrase; LTR, long terminal repeats; PR, aspartate proteinase; RT, reverse transcriptase/RNaseH; O, cap; A_n, poly (A) tract.

processes them into mature capsids. Thereafter, the reverse transcription process starts by extension of the tRNA primer, which binds to the (–)-strand primer-binding site in the RNA. A (–)-strand ssDNA is formed. After priming of the plus strand, a (+)-strand ssDNA is produced, which corresponds to the retroviral intermediate. Finally, a dsDNA is formed, but indications suggest that stretches of RNA rather than DNA are also in the (+) strand. The dsDNA is then imported into the cell nucleus. The IN inserts the DNA into the host genome.

Relations with Cells and Tissues

No cytopathic effects have been observed. Virus particles are present in the cytoplasm.

Host Range and Symptoms

Members of this genus are found in the genomes of plants, yeasts, and insects.

Transmission and Stability

The viruses lose their activity at 65°C. There is only vertical virus transmission.

Relevant Literature

Boeke and Stoye, 1997; Voytas et al., 1992.

Genus Sirevirus

Type species: *Glycine max SIRE 1 virus.*

FAMILY METAVIRIDAE

The family comprises two genera, namely *Metavirus* (after which the family has been named) and *Errantivirus*. As the members of genus *Errantivirus* have been found only in invertebrates, no description of this genus will be given here.

The spherical particles are very heterogeneous, especially the intracellular ones, which are in different stages of maturation. They are, therefore, referred to as viruslike particles (VLPs). Some species also develop more homogeneous extracellular particles, and they are referred to as virions. Both VLPs and virions have a diameter of about 50 nm and possess polyadenylated, positive-sense RNA. The virons are not enveloped in the genus *Metavirus*.

Genus Metavirus

The name has been derived from the Greek word *metathesis* (transposition).

Type species: *Saccharomyces cerevisiae Ty3 virus* (SceTy3V).

Virion Properties

The virions are spherical but irregular and have a diameter of about 50 nm.

Genome Organization, Expression, and Replication

The major virion RNA species consists of 5,200 nt. It contains two overlapping ORFs, *gag 3* and *pol 3*, comparable to *gag* and *pol* of the pseudoviruses. However, the order of functional domains differs from the latter. *Pol 3* is expressed as a +1 FS (see Figure A5.2).

The protein encoded by the *gag 3* gene is the CP of 26 kDa. The *pol 3* gene codes for PR of 15 kDa, IN of 58 and 61 kDa, and an RT-IN fusion protein of approximately 115 kDa. Some VLPs contain a nucleocapsid (NC) protein of 15 kDa.

The integrated form (dsDNA) of SceTy3V is 5,400 nt long and possesses an internal domain flanked by two LTRs.

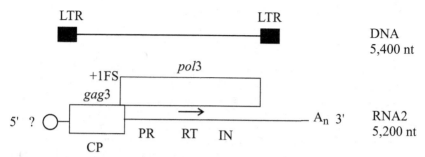

FIGURE A5.2. Organization and expression of the genome of a metavirus *(Saccharomyces cerevisae Ty3virus)*. Open rectangles represent the open reading frames, *gag3* and *pol3* with their translation products (CP, PR, RT, IN). CP, capsid protein; +1FS, +1 ribosomal frameshift; IN, integrase; LTR, long terminal repeats; PR, aspartate proteinase; RT, reverse transcriptase/RNaseH; ? O, putative cap; A_n, poly (A) tract; →, translational readthrough.

SceTy3V is transcribed into a genomic RNA of 5,200 nt. In the case of *Lilium henryi del1 virus* (LheDel1V), the genomic RNA possesses a single long ORF from which CP and *pol* proteins are expressed.

Relations with Cells and Tissues

The irregular spherical VLPs are present in the cytoplasm as clusters or single particles.

Host Range and Symptoms

Members of the genus are found in the genomes of yeasts, fungi, insects and a plant.

Transmission and Stability

Only vertical transmission of the viruses occurs.

Relevant Literature

Boeke and Stoye, 1997.

BIBLIOGRAPHY

Boeke JD, Stoye JP (1997). Retrotransposons, endogenous retroviruses, and the evolution of retroelements. In Varmus H, Hughes S, Coffin J (eds.), *Retroviruses* (pp. 343-435). New York: Cold Spring Harbor Laboratory.

Calvert LA, Ospina MD, Shepherd RJ (1995). Characterization of cassava vein mosaic virus: A distinct pararetrovirus. *Journal of General Virology* 76: 1271-1278.

de Kochko A, Verdaguer B, Taylor N, Carcamo R, Beachy RN, Fauquet C (1998). Cassava vein mosaic virus (CsVMV), type species for a new genus of plant double stranded DNA viruses? *Archives of Virology* 143: 945-962.

Frischmuth T (2002). Recombination in plant DNA viruses. In Khan JA, Dijkstra J (eds.), *Plant viruses as molecular pathogens* (pp. 339-363). Binghamton, NY: The Haworth Press, Inc.

Glasheen BM, Polashock JJ, Lawrence DM, Gillett JM, Ramsdell DC, Vorsa N, Hillman BI (2002). Cloning, sequencing, and promoter identification of *Blueberry red ringspot virus,* a member of the family *Caulimoviridae* with similarities to the "Soybean chlorotic mottle-like" genus. *Archives of Virology* 147: 2169-2186.

Harper G, Hart D, Moult S, Hull R (2004). Banana streak virus is very diverse in Uganda. *Virus Research* 100: 51-56.

Matsuda Y, Liang G, Zhu Y, Ma F, Nelson RS, Ding B (2002). The Commelina yellow mottle virus promoter drives companion-cell-specific gene expression in multiple organs of transgenic tobacco. *Protoplasma* 220: 51-58.

Pooggin MM, Ryabova LA, Hohn T (2002). Translational strategies in members of the *Caulimoviridae.* In Khan JA, Dijkstra J (eds.), *Plant viruses as molecular pathogens* (pp. 317-337). Binghamton, NY: The Haworth Press, Inc.

Richert-Poggeler KR, Noreen F, Shwarzacher T, Harper G, Hohn T (2003). Induction of infectious *Petunia vein clearing* (pararetro) *virus* from endogenous provirus in petunia. *EMBO Journal* 15(22): 4836-4845.

Richert-Poggeler KR, Shepherd RJ (1997). *Petunia vein-clearing virus:* A plant pararetrovirus with the core sequences for an integrase function. *Virology* 236: 137-146.

Stavalone L, Ragozzino A, Hohn T (2003). Characterization of *Cestrum yellow leaf curling virus:* A new member of the family *Caulimoviridae. Journal of General Virology* 84: 3459-3464.

Voytas DF, Cummings MP, Konieczny A, Ausubel FM, Rodermal SR (1992). Copia-like retrotransposons are ubiquitous among plants. *Proceedings of the National Academy of Sciences USA* 89: 7124-7128.

Index

Page numbers followed by the letter "f" indicate figures; those followed by "pl" indicate a plate; and those followed by the letter "t" indicate a table.

Order a copy of this book with this form or online at:
http://www.haworthpress.com/store/product.asp?sku=5530

HANDBOOK OF PLANT VIROLOGY

_____in hardbound at $89.95 (ISBN-13: 978-1-56022-978-0; ISBN-10: 1-56022-978-0)

_____in softbound at $69.95 (ISBN-13: 978-1-56022-979-7; ISBN-10: 1-560222-979-9)

Or order online and use special offer code HEC25 in the shopping cart.

COST OF BOOKS_____

POSTAGE & HANDLING_____
(US: $4.00 for first book & $1.50
for each additional book)
(Outside US: $5.00 for first book
& $2.00 for each additional book)

SUBTOTAL_____

IN CANADA: ADD 7% GST_____

STATE TAX_____
(NJ, NY, OH, MN, CA, IL, IN, PA, & SD
residents, add appropriate local sales tax)

FINAL TOTAL_____
(If paying in Canadian funds,
convert using the current
exchange rate, UNESCO
coupons welcome)

☐ **BILL ME LATER:** (Bill-me option is good on US/Canada/Mexico orders only; not good to jobbers, wholesalers, or subscription agencies.)

☐ Check here if billing address is different from shipping address and attach purchase order and billing address information.

Signature_____

☐ **PAYMENT ENCLOSED: $**_____

☐ **PLEASE CHARGE TO MY CREDIT CARD.**

☐ Visa ☐ MasterCard ☐ AmEx ☐ Discover
☐ Diner's Club ☐ Eurocard ☐ JCB

Account # _____

Exp. Date_____

Signature_____

Prices in US dollars and subject to change without notice.

NAME_____

INSTITUTION_____

ADDRESS_____

CITY_____

STATE/ZIP_____

COUNTRY_____ COUNTY (NY residents only)_____

TEL_____ FAX_____

E-MAIL_____

May we use your e-mail address for confirmations and other types of information? ☐ Yes ☐ No
We appreciate receiving your e-mail address and fax number. Haworth would like to e-mail or fax special discount offers to you, as a preferred customer. **We will never share, rent, or exchange your e-mail address or fax number.** We regard such actions as an invasion of your privacy.

Order From Your Local Bookstore or Directly From

The Haworth Press, Inc.

10 Alice Street, Binghamton, New York 13904-1580 • USA
TELEPHONE: 1-800-HAWORTH (1-800-429-6784) / Outside US/Canada: (607) 722-5857
FAX: 1-800-895-0582 / Outside US/Canada: (607) 771-0012
E-mail to: orders@haworthpress.com

For orders outside US and Canada, you may wish to order through your local
sales representative, distributor, or bookseller.
For information, see http://haworthpress.com/distributors

(Discounts are available for individual orders in US and Canada only, not booksellers/distributors.)

PLEASE PHOTOCOPY THIS FORM FOR YOUR PERSONAL USE.

http://www.HaworthPress.com   BOF06

Printed in the United States
by Baker & Taylor Publisher Services